PHARMACOCHEMISTRY LIBRARY - VOLUME 30

THE HISTAMINE H_3 RECEPTOR

A target for New Drugs

PHARMACOCHEMISTRY LIBRARY, edited by H. Timmerman

Other titles in this series

Volume 18 Trends in Receptor Research. Proceedings of the 8th Noordwijkerhout-Camerino Symposium, Camerino, Italy, 8-12 September, 1991
edited by P. Angeli, U. Gulini and W. Quaglia

Volume 19 Small Peptides. Chemistry, Biology and Clinical Studies
edited by A.S. Dutta

Volume 20 Trends in Drug Research. Proceedings of the 9th Noordwijkerhout-Camerino Symposium, Noordwijkerhout (The Netherlands), 23-27 May, 1993
edited by V. Claassen

Volume 21 Medicinal Chemistry of the Renin-Angiotensin System
edited by P.B.M.W.M. Timmermans and R.R. Wexler

Volume 22 The Chemistry and Pharmacology of Taxol® and its Derivatives
edited by V. Farina

Volume 23 Qsar and Drug Design: New Developments and Applications
edited by T. Fujita

Volume 24 Perspectives in Receptor Research
edited by D. Giardinà, A. Piergentili and M. Pigini.

Volume 25 Approaches to Design and Synthesis of Antiparasitic Drugs
edited by Nitya Anand

Volume 26 Stable Isotopes in Pharmaceutical Research
edited by Thomas R. Browne

Volume 27 Serotonin Receptors and their Ligands
edited by B.Olivier et al.

Volume 28 Proceedings XIVth International Symposium on Medicinal Chemistry
edited by F. Awouters

Volume 29 Trends in Drug Research II. Proceedings of the 11th Noordwijkerhout-Camerino Symposium, Noordwijkerhout (The Netherlands), 11-15 May,1997
edited by H. van der Goot

PHARMACOCHEMISTRY LIBRARY
Editor: H. Timmerman

Volume 30

THE HISTAMINE H_3 RECEPTOR
A Target for New Drugs

Edited by:

ROB LEURS and HENK TIMMERMAN

Department of Pharmacochemistry, Free University Amsterdam, The Netherlands

1998
ELSEVIER
Amsterdam – Lausanne – New York – Oxford – Shannon – Singapore - Tokyo

ELSEVIER SCIENCE B.V.
Sara Burgerhartstraat 25
P.O. Box 211, 1000 AE Amsterdam, The Netherlands

First edition 1998

Library of Congress Cataloging in Publication Data
A catalog record from the Library of Congress has been applied for.

ISBN: 0-444-82936-9

∞ The paper used in this publication meets the requirements of ANSI/NISO Z39.48-1992 (Permanence of Paper).

Printed and bound by CPI Antony Rowe, Eastbourne
Transferred to digital printing 2006

To the memory of our friend and colleague, Giulio Bertaccini

PREFACE

It took about fifteen years after the role of histamine in allergic diseases had been established before the first clinically useful antihistamine was available in the late thirties. When the H_2 receptor had been defined, it took less time until the H_2 antagonist cimetidine was ready for clinical use.

So, when in the early eighties the H_3 receptor was identified, many thought that soon an H_3 ligand, an agonist or an antagonist, would become available as a therapeutic agent. Such has not happened, however. One might wonder why. One factor is without doubt the fact that many investigators do consider histamine mainly, if not only, as a mediator present in e.g. mast cells, being released during allergic events. Histamine is, as has become very clear, an important neurotransmitter, though. Its role in the nervous system, especially in the central part of it, is rather extensive.

The H_3 receptor is mainly found as a presynaptic one, both on histaminergic neurons (the auto-type) and on other neuronal systems (the hetero-type). Both the H_3 agonist and the H_3 antagonist cause important pharmacological effects. Several ligands have become available by now, induding radiolabelled analogues.

In this book the current state of affairs with regard to the medicinal chemistry and pharmacology of the H_3 receptor and the several ligands available are presented by a number of experts in the field. The book presents an extended review of what has happened since the first H_3 paper appeared. We hope that this work will lead to an increase of the interest, of both academia and industry, for the H_3 receptor, especially as a target for drug development.

During the preparation of the book we were shocked by the sudden death of Professor Giulio Bertaccini. Prof. Bertaccini had just finished, with his associates, his important contribution to the present work. We have dedicated this book to the memory of Giulio, an excellent scientist, an extremely good friend and a superb human being.

H. Timmerman
R. Leurs

CONTENTS

R. Leurs and H. Timmerman (Editors)
The Histamine H_3 Receptor

Subclassification of histamine receptors, H_3-receptor subtypes ? Localization of H_3 receptors in the brain

J-M. Arrang[a], S. Morisset[a], C. Pillot[b] and J-C. Schwartz[a]

[a]Unité de Neurobiologie et Pharmacologie Moléculaire (U.109), Centre Paul Broca de l'INSERM, 2ter rue d'Alésia, 75014 Paris, France

[b]Laboratoire de Physiologie, Faculté des Sciences Pharmaceutiques et Biologiques, 4 Avenue de l'Observatoire, 75006 Paris, France

Histamine is released in the brain from neurons projecting in a diffuse manner to widely divergent cerebral areas and arising from the tuberomammillary nucleus of the posterior hypothalamus. It affects target cells via activation of three receptor subtypes termed H_1, H_2 and H_3 [1-3]. Understanding the roles of a central neurotransmitter requires the identification of the neuronal populations expressing its various receptor subtypes. Detailed mappings of H_1, H_2 and H_3 receptors have been established in rodent brain [4]. Recently, the cloning of H_1 and H_2-receptor subtypes has allowed to further study via *in situ* hybridization the phenotype of neurons expressing these receptors. The H_3 receptor was evidenced as an autoreceptor. Its cloning is still awaited and should clarify its molecular pharmacology, the putative existence of H_3 subtypes, as well as the phenotype of neurons expressing its gene transcripts.

1. SUBCLASSIFICATION OF HA RECEPTORS

The H_1, H_2 and H_3-histamine receptor subtypes have been characterized by means of functional assays and design of selective agonists and antagonists [1,2]. The recent cloning of cDNAs encoding H_1 and H_2 receptors has provided further information about the molecular pharmacology of these two receptors. Although the H_3 receptor has not yet been isolated, all three receptors seem to belong to the superfamily of receptors coupled to G proteins. Their main properties are summarized in Table 1.

1.1. The histamine H_1 receptor

The H_1 receptor was initially defined in functional assays (e.g. smooth muscle contraction) and the design of potent antagonists, the so-called « antihistamines » (e.g. mepyramine), most of which display prominent sedative properties.

Biochemical and localisation studies of the H_1 receptor were made feasible with the design of reversible and irreversible radiolabelled probes such as [^{3}H]mepyramine, [^{125}I]iodobolpyramine and [^{125}I]iodoazidophenpyramine [4-6].

Initial biochemical studies indicate that agonist binding was regulated by guanyl nucleotides, implying that the receptor belongs to the superfamily of receptors coupled to G proteins. In addition, various intracellular responses were found to be associated with H_1-receptor stimulation : inositol phosphate release, increase in Ca^{2+} fluxes, cyclic AMP or cyclic GMP accumulation in whole cells and arachidonic acid release [1].

The deduced amino acid sequence of a bovine H_1 receptor was recently disclosed after expression cloning of a corresponding cDNA [7]. Starting from the bovine sequence, the H_1 receptor DNA was also cloned in the guinea pig [8,9], a species in which the pharmacology of the receptor is well established.

When stably expressed in transfected fibroblasts, the guinea pig H_1 receptor was found to trigger a large variety of intracellular signals involving or not coupling to pertussis toxin-sensitive G proteins (Gi or Go) – namely Ca^{2+} transients, inositol phosphates or arachidonate release [10]. H_1 receptor stimulation potentiates cAMP accumulation induced by forskolin in the same transfected fibroblasts, a response which resembles the H_1-potentiation of histamine H_2- or adenosine A_2-receptor-induced accumulation of cAMP in brain slices. All these responses mediated by a single H_1 receptor were known to occur in distinct cell lines or brain slices but could have been due to stimulation of isoreceptors. The H_1 receptor mediates various excitatory responses in brain [11]. The widespread distribution of H_1 receptors has been analyzed in the guinea pig brain by autoradiography [12] and *in situ* hybridization [8,9].

1.2. The histamine H_2 receptor

Molecular properties of the H_2 receptor have remained largely unknown for a long time. For instance reversible labelling of the H_2 receptor was achieved only recently using [^{3}H]tiotidine or, more reliably, [^{125}I]iodoaminopotentidine [13].

By screening cDNA or genomic libraries with homologous probes, the gene encoding the H_2 receptor was first identified in dogs [14] and, subsequently, in other species including human [15, 16].

Using transfected cell lines not only the well established positive linkage of the H_2 receptor with adenylyl cyclase [2] was confirmed but also the unexpected inhibition of arachidonate release [17] and stimulation of Ca^{2+} transients [18]. Hence H_2 receptor stimulation can trigger intracellular signals either opposite or similar to those evoked by H_1 receptor stimulation. Interestingly, transfected cell lines displayed spontaneous agonist-independent H_2-receptor activity, with some H_2 antagonists displaying inverse agonism in those cell lines [19].

Helmut Haas and colleagues showed that, in hippocampal pyramidal neurons, H_2-receptor stimulation potentiates excitatory signals by decreasing a Ca^{2+}-activated K^+ conductance, presumably via cAMP production [11]. H_2-receptor activation depolarizes thalamic relay neurons slightly, increasing markedly apparent membrane conductance, a response due to enhancement of the hyperpolarization-activated cation current I_h [20].

A detailed mapping of the histamine H_2 receptor and its gene transcripts has been performed in guinea pig. H_2 receptors are widely but heterogeneously distributed, in some cases in a way suggesting their potential co-expression with H_1 receptors within the same neuronal populations [21].

Table 1
Properties of three histamine receptor subtypes

	H_1	H_2	H_3
Coding sequence	491 a.a. (bovine) 488 a.a. (guinea pig) 486 a.a. (rat) 487 a.a. (human)	358 a.a. (rat) 359 a.a. (dog, human, guinea pig)	?
Chromosome Localization	Chromosome 3	Chromosome 5	?
Highest brain Densities	Thalamus Cerebellum Hippocampus	Striatum Cerebral cortex Amygdala	Striatum Frontal cortex Subst. Nigra
Autoreceptor	No	No	Yes
Affinity for Histamine	Micromolar	Micromolar	Nanomolar
Characteristic Agonist	2-(3-trifluoromethyl) histamine	Impromidine	(R)α-methylhistamine
Characteristic Antagonist	Mepyramine	Cimetidine	Thioperamide
Radioligands	$[^3H]$Mepyramine $[^{125}I]$Iodobolpyramine	$[^3H]$Tiotidine $[^{125}I]$Iodoamino-potentidine	$[^3H](R)\alpha$-methylhistamine $[^{125}I]$Iodophenpropit $[^{125}I]$Iodoproxyfan
Second Messengers	Inositol phosphates (+) Arachidonic acid (+) cAMP (potentiation)	cAMP (+) Arachidonic acid (-) Ca^{2+} (+)	Inositol phosphates (-)

The histamine H_3 receptor

The H_3 receptor was initially detected as an autoreceptor controlling histamine synthesis and release in brain [22]. Thereafter it was shown to inhibit presynaptically the release of other monoamines in brain and peripheral tissues as well as of neuropeptides from unmyelinated C-fibers [23].

Since attempts to isolate the H_3-receptor gene have not yet been successful, the molecular structure of the H_3 receptor remains to be established. Reversible labeling of this receptor was first achieved using the highly selective agonist [^{3}H](R)α-methylhistamine [24], then [^{3}H]Nα-methylhistamine, a less selective agonist, was also proposed [5] as well as, more recently, [^{125}I]iodophenpropit [25] and [^{125}I]iodoproxyfan [26], two antagonists.

The agonist binding on H_3-receptor sites is regulated by guanyl nucleotides [26-29], strongly suggesting that the H_3 receptor, like the other histamine receptors, belongs to the superfamily of receptors coupled to G proteins. Several studies indicate the involvement of Gi/Go proteins in the coupling of the H_3 receptor to its effector system. For example, pertussis toxin abolished the H_3-receptor mediated stimulation of GTPγS binding [30]. However, H_3 receptors in vascular smooth muscle increase voltage-dependent Ca^{2+} currents via a pertussis-insensitive G protein [31], suggesting that other G-protein subfamilies might be involved in H_3-receptor mediated signal transduction. Constitutive H_3 receptors in a gastric cell line appear to be negatively coupled to phospholipase C [32].

2. THE HISTAMINE H_3 AUTORECEPTOR

The H_3 receptor was first characterized as an autoreceptor regulating histamine release from histaminergic neurons [24, 33]. This feedback control mechanism is now well documented [3, 22, 34]. The existence of a selective neuronal histamine transporter remaining to be established, the autoreceptor-regulated modulation of [^{3}H]histamine release and formation was evidenced by taking advantage of a methodology developed earlier [35] and consisting in labeling the endogenous pool of histamine using the ^{3}H-precursor.

Exogenous histamine decreases the release of [^{3}H]histamine induced by potassium [33, 36, 37], veratridine [33, 36] or field electrical stimulation [38-41]. The inhibition was stronger with stimulations of low intensity. The autoinhibitory effect of histamine displayed a high pharmacological specificity and the analysis of the response led to the pharmacological definition of H_3 receptors.

That H_3 receptors are directly located on histaminergic nerve terminals was shown by the persistence of the autoinhibitory effect of histamine in the presence of tetrodotoxin, in kainate-injected animals and in synaptosomes [36]. The autoregulation was found in various brain regions known to contain histamine nerve-endings, suggesting that all terminals are endowed with H_3 autoreceptors. The extent of the autoinhibition can be regulated in a complex manner by extracellular calcium, suggesting that H_3 receptors modulate histamine release via a control of intraneuronal Ca^{2+} fluxes [36]. The various steps involved between H_3-receptor activation and final changes in calcium influx within histamine neurons still remains to be clarified. H_3 receptors also mediate the autoinhibition of histamine release in human brain with a pharmacology apparently similar to that of corresponding receptors in rodents [42].

Verdière et al. [35] had demonstrated that depolarization of brain slices was also accompanied by an enhancement in [^{3}H]histamine synthesis. Again, exogenous histamine was found to reduce by up to 60-70% this stimulation, with an EC_{50} of 0.3 μM, similar to that observed at H_3 receptors regulating histamine release [24, 43]. The pharmacological analysis of this response confirmed the involvement of H_3 receptors. A regulation of histamine synthesis was also observed in the posterior hypothalamus [43] and tuberomammillary neurons themselves are sensitive to histamine and to an H_3-receptor agonist which inhibits their firing, through a hyperpolarization accompanied by an increased input resistance [11], strongly suggesting the existence of autoreceptors at the level of histaminergic perikarya or dendrites. The chain of events leading to the presynaptic regulation of histamine formation is not yet clarified and may be related to the control of release, possibly via intracellular calcium. A direct feedback inhibition of histidine decarboxylase by histamine has been excluded [34], but an increase in histidine decarboxylase activity has been reported in mice pretreated with H_3-receptor antagonists [44, 45].

3. H_3-RECEPTOR SUBTYPES ?

3.1. Heterogeneity of H_3-receptor binding sites

Several studies have reported that some H_3-receptor antagonists could discriminate between two specific binding sites for radiolabelled H_3-agonists. Biphasic competition curves have been observed for burimamide and thioperamide [27, 28, 46-48]. However, the affinity of thioperamide has been reported to be 10-fold higher in phosphate buffer as compared to Tris buffer [49] and the heterogeneity of thioperamide binding is lost in the presence of sodium ions [50]. These data are unlikely to reflect binding to distinct subtypes of the receptor, but rather reflect conformational states of the same receptor dependent on the ionic composition of the medium or the radiolabelled agonist, and for each of which some antagonists display a differential affinity.

In contrast to what could be expected, binding studies with radiolabelled antagonists did not bring further evidence for H_3-receptor heterogeneity. For example, burimamide biphasically displaced [^{125}I]iodophenpropit but monophasically displaced [^{3}H]GR 168320 from rat cortex membranes [29, 48] and yielded a steep competition curve in rat striatal membranes using [^{125}I]iodoproxyfan as the radioligand [26].

3.2. Functional studies

In contrast to thioperamide, phenylbutanoylhistamine, an H_3-receptor antagonist with moderate potency [51], failed to antagonize the inhibitory effect of (R)α-methylhistamine on acetylcholine release from slices of rat entorhinal cortex [52]. In addition, this compound also failed to antagonize the (R)α-methylhistamine-induced inhibition of non-adrenergic non-cholinergic contractions of the guinea pig ileum [53].

It was suggested from these data that this compound might discriminate H_3-receptor subtypes. However, it was subsequently shown that additional non-histaminergic properties of phenylbutanoylhistamine greatly underestimated its moderate potency at H_3 receptors and do not allow any conclusion about H_3-receptor subtypes [54].

Two functional H_3-receptor models, i.e. the inhibition of noradrenaline release in the mouse brain and of acetylcholine release in the guinea pig small intestine, have been studied thoroughly in order to compare their pharmacological profiles. A large number of H_3-receptor antagonists belonging to various chemical classes did not reveal pharmacological differences between the two models [55, 56]. In a series of histamine homologues, impentamine displayed partial agonism in the mouse brain cortex whereas it acted as a true competitive antagonist at the guinea pig jejunum [57]. A similar observation was made with iodoproxyfan. The latter compound and its deiodo analogue act as partial agonists at the functional H_3 receptor in the mouse brain cortex, when only iodoproxyfan behaves as a partial agonist in the guinea pig ileum [56]. At present, whether these apparent differences are related to differences in the efficiency of receptor coupling, or to species variants or receptors remain unclear.

The binding and functional studies mentioned above do not provide any clear evidence for the putative existence of H_3-receptor subtypes. Whereas it was probably the last receptor to be characterized through « classical » pharmacological methods, cloning of the H_3-receptor gene is not yet achieved. Only this cloning is expected to provide us a direct conclusion about this important issue.

4. DISTRIBUTION OF CENTRAL H_3-RECEPTOR BINDING SITES

Consistent data have been reported with respect to the autoradiographic localization of H_3-receptor binding sites using various radioligands, i.e. [^{3}H](R)α-methylhistamine [24, 58, 59], [^{3}H]Nα-methylhistamine [60, 61], [^{125}I]iodophenpropit [29] and [^{125}I]iodoproxyfan [26].

Among these various H_3 receptor ligands, only [^{3}H](R)α-methylhistamine was used for a very detailed mapping of rat brain [62] and the main data summarized in Table 2.

H_3 receptors are widely and heterogeneously distributed in almost all regions of the rat central nervous system [4]. They are expressed in most areas receiving histaminergic projections, for instance, cerebral cortex, hippocampus and substantia nigra pars reticula [63-66]. In some regions, their distribution is consistent with that of histaminergic nerve terminals and H_3 receptors might largely represent autoreceptors. For instance, in the hippocampal formation, the higher number of [^{3}H](R)α-methylhistamine binding sites found in the subicular complex and in the molecular layer of the dentate gyrus, is consistent with the distribution of histaminergic terminals. The high number of H_3 receptors in the bed nucleus of stria terminalis, in the lateral and central amygdala and in the medial part of the thalamus might also reflect the high density of HA terminals [62-66].

Table 2
Compared densities of histamine H_3 receptors and histaminergic innervation in various areas of rat brain

Brain structures	H_3 receptors	HA-innervation
Cortex		
Isocortex		
Layers I-III	1-2+	2-3+
IV-V	2-4+	2-3+
VI	2-3+	2-3+
Allo- and periallo-cortices	1-3+	1-3+
Hippocampal formation		
Ammon's horn	1+	1+
Dentate gyrus	1-2+	1+
Subicular complex	2-3+	1-3+
Amygdaloid complex		
Medial amygdaloid group	2-3+	3+
Basal amygdaloid group	2-3+	2-3+
Bed nucleus of the stria terminalis	3+	3+
Basal formation		
Accumbens nucleus	3-4+	2-3+
Caudate putamen	2-4+	1-2+
Globus pallidus	3+	1-2+
Olfactory tubercle	3-4+	2-3+
Lateral septal nucleus	2+	2+
Thalamus		
Anterior nuclei	2+	1-2+
Median and lateral nuclei	1-3+	2-3+
Ventral and posterior nuclei	1+	1+
Hypothalamus		
Anterior and lateral groups	1-3+	3-4+
Intermediate group	2-3+	3-4+
Posterior and mammillary groups	2-3+	3-4+
Mesencephalon		
Colliculi	1-2+	2+
Substantia nigra, pars compacta	1+	2-3+
pars reticulata	3-4+	2-3+
Central grey	2-3+	1-3+
Rhombencephalon		
Cerebellum	1-2+	0-1+
Locus coeruleus	2+	2+
Raphe nuclei	1-2+	1+
Pontine nuclei	0-1+	0-1+
Vestibular nuclei	1-2+	2+

Densities of H_3 receptors in rat brain were obtained from Pollard and Bouthenet [4], Pollard et al. [62], and histaminergic innervation from Steinbusch and Mulder [64], Inagaki et al. [65] and Panula et al. [66].

However as initially noted [24], large differences between the density of H_3 receptors and that of histamine axons in various brain regions, indicate that the majority of H_3 receptors are not autoreceptors.

For instance, H_3 receptor density is relatively low in the hypothalamus which contains a high density of histaminergic axons and perikarya. However, the detection of some H_3 receptors at the level of the tuberomammillary nuclei possibly indicates the existence of autoreceptors at the level of histamine perikarya or dendrites [62-66].

In the cerebral cortex and basal ganglia, in which the highest densities were found, H_3-receptor binding sites display again a distribution pattern different from that of histaminergic neurons.

In the cerebral cortex, a high density of H_3 receptors is observed in the deep layers of various cortical areas in which a moderately dense network of histamine-immunoreactive fibres was found [62-66] and the decreasing rostrocaudal gradient of H_3 receptors is opposite to that of L-histidine decarboxylase activity, a marker of histaminergic axons [34]. Some of these binding sites are likely to represent H_3 heteroreceptors presynaptically localized on noradrenergic and serotonergic nerve terminals [67, 68]. However, several lines of evidence indicate that the binding in the cerebral cortex is predominantly localized on intrinsic neurons, rather than on afferents : lesions of the medial forebrain bundle increased, and cortical quinolinic and kainic acid lesions reduced cortical H_3 binding [60, 62]. Some of these postsynaptic sites might represent H_3 receptors modulating acetylcholine release [54]. The increase in histamine H_3-receptor ligand binding in the visually deprived cortex and superior colliculus may also represent an up-regulation of postsynaptic H_3 receptors [69].

The high density of H_3 receptors in the striatal complex also contrasts with the low density of histaminergic fibres. They also most likely represent postsynaptic H_3 receptors as indicated by their increase in the striatum after interruption of histaminergic (and possibly other aminergic) inputs following lesions in the lateral hypothalamic area [62]. The selective depletion of neuronal histamine with α-fluoromethylhistidine as well as the lesion of nigrostriatal dopaminergic neurons with 6-hydroxydopamine, also increased H_3 binding in the striatum [59, 70, 71]. In addition, intrastriatal administration of kainic or quinolinic acids markedly decreased striatal H_3 binding [60, 62, 72]. Therefore, these observations suggest that a major part of the striatal H_3 receptors are localized on intrinsic neurons and/or striatal efferents. In most of the studies mentioned above, the same lesions not only induced modifications of the H_3 binding in the striatum but also in the substantia nigra, strongly suggesting a localization of H_3 receptors on GABAergic striatonigral nerve endings. In agreement with the dopamine D_1-receptor gene expression of these neurons [73, 74], the increased H_3 binding observed in the striatum and substantia nigra following destruction of the ascending dopaminergic neurons was reversed by a selective D_1-receptor agonist [71]. Moreover, H_3 receptor activation was recently shown to inhibit dopamine D_1-receptor dependent [^{3}H]GABA release in superfused slices of rat substantia nigra pars reticulata [75].

The olfactory system is another area where the density of H_3 receptors is not correlated with histaminergic nerve fibres. H_3 receptors are more abundant in the olfactory nuclei than in the main olfactory formation, a distribution opposite to that of histamine immunoreactivity. H_3 receptors are less abundant in the mesencephalon and lower brainstem but they are observed in nuclei receiving histaminergic inputs, such as vestibular nuclei, locus coeruleus or dorsal raphe [62, 66]. In these latter two nuclei, they could be either autoreceptors or, more likely, they could be expressed on noradrenergic or serotoninergic perikarya and/or dendrites as they apparently do at the level of their axons. H_3 receptors are found in low or very low density in cerebellum, spinal cord and pituitary gland.

Binding sites of $[^3H]N^{\alpha}$-methylhistamine have been also shown on autoradiograms from mouse and guinea pig forebrain with a distribution qualitatively similar to the rat [76]. In the brain of human and non-human primates the distribution of $[^3H](R)$-α-methylhistamine binding sites resembles that of the rat, H_3 receptors being particularly abundant in the caudate putamen and globus pallidus [77]. However, their high density in the external layers of cerebral cortex, which receive a dense histaminergic innervation [78], differs from the rat and suggests that in human the autoreceptors might be important in such functions like sleep and wakefulness for which histamine is involved [34].

In conclusion, the distribution of H_3 receptors and the effects of lesions are consistent with functional studies showing that they are inhibitory receptors not only on histaminergic nerve terminals, but also on various aminergic and non-aminergic cerebral neurons and therefore they may be involved in a large variety of functions.

REFERENCES

1. S.J. Hill, Pharmacol. Rev., 42 (1990) 45.
2. J.C. Schwartz and H.L. Haas (eds.),The Histamine Receptor, Wiley Liss, New York, 1992.
3. J.C. Schwartz, J.M. Arrang, M. Garbarg and E.Traiffort, in Psychopharmacology: The Fourth Generation of Progress (eds. F.E. Bloom and D.J. Kupfer), Raven Press, New-York,1995, pp. 397-405.
4. H. Pollard and M.L. Bouthenet, in The Histamine Receptor (eds. J.C.Schwartz and H.L. Haas) Wiley Liss, New-York, 1992, pp.179-192.
5. M. Garbarg, E. Traiffort, M. Ruat, J.M. Arrang and J.C. Schwartz, in The Histamine receptor (eds. J.C. Schwartz and H.L. Haas) Wiley Liss, New-York, 1992, pp. 73-95.
6. M. Ruat, E. Traiffort and J.C. Schwartz, in The Histamine Receptor (eds. J.C. Schwartz and H.L. Haas) Wiley Liss, New-York, 1992, pp.97-107.
7. M. Yamashita, H. Fukui, K. Sugawa, Y. Horio, S. Ito, H. Mizuguchi and H. Wada, Proc. Natl. Acad. Sci. USA, 88 (1991) 11515.
8. Y Horio, Y Mori, I Higuchi, K. Fujimoto, S. Ito and H. Fukui, J. Biochem., 114 (1993) 408.
9. E. Traiffort, R. Leurs, J.M. Arrang, J.Tardivel-Lacombe, J. Diaz, J.C. Schwartz and M. Ruat, J.Neurochem., 62 (1994) 507.
10. R. Leurs, E.Traiffort, J.M. Arrang, J. Tardivel-Lacombe, M.Ruat and J.C. Schwartz, J.Neurochem., 62 (1994) 519.
11. H.L. Haas, in The Histamine Receptor (eds. J.C. Schwartz and H.L. Haas) Wiley Liss, New-York, 1992, pp. 161-177.

12. M.L Bouthenet, M. Ruat, N. Salès, M. Garbarg and J.C. Schwartz, Neuroscience, 26 (1988) 553.
13. M. Ruat, E.Traiffort, M.L Bouthenet, J.C Schwartz, J.Hirschfeld, A. Buschauer and W. Schunack, Proc. Natl. Acad. Sci. USA, 87 (1990) 1658.
14. I. Gantz, M.Schaffer, J. Delvalle, C. Logsdon, V. Campbell, M.Uhler and T. Yamada, Proc. Natl. Acad. Sci. USA, 88 (1991) 429.
15. I. Gantz, G. Munzert, T. Tashiro, M. Schaffer, L. Wang, J. Delvalle and T. Yamada, Biochem. Biophys. Res. Commun.,178 (1991) 1386.
16. E. Traiffort, M.L. Vizuete, J. Tardivel-Lacombe, E. Souil, J.C. Schwartz and M. Ruat. Biochem. Biophys. Res. Commun., 211 (1995) 570.
17. E. Traiffort, M. Ruat, J.M. Arrang, R. Leurs, D. Piomelli and J.C.Schwartz, Proc. Natl. Acad. Sci. USA, 89 (1992) 2649.
18. J. Delvalle, L. Wang, I. Gantz and T. Yamada, Am. J. Physiol., 263 (1992) G967.
19. M.J. Smit, R. Leurs, A.E. Alewijnse, J. Blauw, G.P. Van Nieuw Amerongen, Y. Van De Vrede, E. Roovers and H. Timmerman. Proc. Natl. Acad. Sci. USA, (1996) 6802.
20. D.A McCormick and A. Williamson, J. Neurosci., 11 (1991) 3188.
21. M.L. Vizuete, E. Traiffort, M.L. Bouthenet, M.Ruat, E. Souil, J. Tardivel-Lacombe and J.C. Schwartz, Neuroscience, 80 (1997) 321.
22. J.M. Arrang, M. Garbarg and J.C. Schwartz, in The Histamine Receptor (eds. J.C Schwartz and H.L. Haas) Wiley Liss, New-York, 1992, pp.145-159.
23. E. Schlicker, B. Malinowska, M. Kathmann, and M. Göthert, Fund. Clin. Pharmacol. 8 (1994) 128.
24. J.M. Arrang, M. Garbarg, J.C. Lancelot, H. Pollard, M. Robba, W. Schunack and J.C. Schwartz, Nature, 327 (1987) 117.
25. F.P. Jansen, B. Rademaker, A. Bast and H. Timmermann, Eur. J. Pharmacol., 217 (1992) 203.
26. X. Ligneau, M. Garbarg, M.L.Vizuete, J. Diaz, K. Purand, H. Stark, W. Schunack and J.C. Schwartz, J. Pharmacol. Exp. Ther., 271 (1994) 452.
27. R.E. West, A. Zweig, N.Y.Shih, M.I. Siegel, R.W. Egan and M.A. Clark, Mol. Pharmacol., 38 (1990) 610.
28. J.M. Arrang, J. Roy, J.L. Morgat, W. Schunack and J.C. Schwartz, Eur. J. Pharmacol. Mol. Pharmacol. Sect.,188 (1990) 219.
29. F.P. Jansen, T.S. Wu, H.P.Voss, H.W.M. Steinbusch, R.C. Vollinga, B. Rademaker, A. Bast and H. Timmerman, Br. J. Pharmacol.,113 (1994) 355.
30. E.A. Clark and S.J. Hill, Eur. J. Pharmacol., 296 (1996) 22
31. M. Oike, K. Kitamura, H and H. Kuriyama, J. Physiol. 448 (1992) 133.
32. Y. Cherifi, C. Pigeon, M. Le Romancier, A. Bado, F. Reyl-Desmars and M.J.M. Lewin, J. Biol. Chem. 267 (1992) 25315.
33. J.M. Arrang, M. Garbarg and J.C. Schwartz, Nature, 302 (1983) 832.
34. J.C. Schwartz, J.M. Arrang, M. Garbarg H. Pollard and M. Ruat, Physiol. Rev. 71 (1991) 1.
35. M. Verdière, C. Rose and J.C. Schwartz, Eur. J. Pharmacol. 34 (1975) 157.
36. J.M. Arrang, M. Garbarg and J.C. Schwartz, Neuroscience, 15 (1985) 553.
37. S.J. Hill and R.M. Straw, Br. J. Pharmacol., 95 (1988) 1213.

38. J.F. Van der Werf, A. Bast, G.J. Bijloo, A Van der Vliet and H. Timmerman, Eur. J. Pharmacol. 138 (1987) 199.
39. J.F. Van der Werf, G.J. Bijloo, A. Van der Vliet, A. Bast and H. Timmerman, Agents Actions, 20 (1987) 239.
40. A. Van der Vliet, J.F. Van der Werf, A. Bast and H. Timmerman, J. Pharm. Pharmacol., 40 (1988) 577.
41. A. Van der Vliet, A. Bast and H. Timmerman, Agents Actions, 30 (1990) 206.
42. J.M. Arrang, B. Devaux, J.P. Chodkiewicz and J.C. Schwartz, J. Neurochem., 51 (1988) 105.
43. J.M. Arrang, M. Garbarg and J.C. Schwartz, Neuroscience, 23 (1987) 149.
44. N. Sakai, A. Sakurai, E. Sakurai, K. Yanai, K. Maeyama and T. Watanabe, Biochem. Biophys. Res. Commun., 185 (1992) 121.
45. K.I. Meguro, K.Yanai, N.Sakai, E. Sakurai, K. Maeyama, H. Sasaki and T. Watanabe, Pharmacol. Biochem. Behav., 50 (1995) 321.
46. M. Kathmann, E. Schlicker, M.Detzner and H. Timmerman, Naunyn-Schmiedeberg's Arch. Pharmacol.,348 (1993) 498.
47. P. Cumming and A. Gjedde, Brain Res.,641 (1994) 203.
48. J.D. Brown, C.T. O'Shaughnessy, G.J. Kilpatrick, D.I.C. Scopes, P. Beswick, J.W.Clitherow and J.C. Barnes, Eur.J. Pharmacol., 311 (1996) 305.
49. R.E. West, A. Zweig, R.T. Granzow, M.I. Siegel and R.W. Egan. J. Neurochem., 55 (1990) 1612.
50. E.A. Clark and S.J. Hill, Br. J. Pharmacol.,114 (1995) 357.
51. R. Lipp, H. Stark and W. Schunack , in The Histamine Receptor (eds. J.C. Schwartz and H.L. Haas) Wiley Liss, New-York, 1992, pp. 57-72.
52. J. Clapham and G.J. Kilpatrick, Br. J. Pharmacol., 107 (1992) 919.
53. S.J. Taylor and G.J. Kilpatrick, Br. J. pharmacol., 105 (1992) 667.
54. J.M.Arrang, G. Drutel and J.C. Schwartz, Br. J. Pharmacol.,114 (1995) 1518.
55. E. Schlicker, M. Kathmann, S. Reidemeister, H. Stark and W. Schunack, Br. J. Pharmacol.,112 (1994) 1043.
56. E. Schlicker, M. Kathmann, H. Bitschnau, I. Marr, S. Reidemeister, H. Stark and W. Schunack,Naunyn-Schmiedeberg's Arch. Pharrmacol.,353 (1996) 482.
57. R. Leurs, M. Kathmann, R.C. Vollinga, W.M. P. B. Menge, E. Schlicker and H. Timmerman, J. Pharmacol. Exp. Ther., 276 (1996) 1009.
58. K. Yanai, J.H. Ryu, T. Watanabe, R. Iwata and T. Ido, Neuroreport, 3 (1992) 961.
59. J.H. Ryu, K. Yanai, E. Sakurai, C-Y. Kim and T. Watanabe, Dev. Brain Res., 87 (1995) 101.
60. P. Cumming, C. Shaw and S.R. Vincent,Synapse, 8 (1991) 144.
61. P. Cumming, A. Gjedde and S.R. Vincent, Brain Res., 641 (1994) 198.
62. H. Pollard, J. Moreau, J.M. Arrang and J.C. Schwartz, Neuroscience, 52 (1993) 1
63. T. Watanabe, Y. Taguchi, S. Shiosaka, J. Tanaka, H. Kubota, Y. Terano, M. Tohyama and H. Wada, Brain Res., 295 (1984) 13.
64. H.W.M. Steinbusch and A.H. Mulder, in Handbook of Chemical Neuroanatomy. Classical Transmitters and Transmitter Receptors in the CNS. Vol. 3, Part II (eds. A. Björklund, T. Hökfelt and M.J. Kuhar) Elsevier, Amsterdam, 1984, pp.126-140.

65. N. Inagaki, A. Yamatodani, M. Ando-Yamamoto, M. Tohyama, T. Watanabe and H. Wada, J. Comp. Neurol., 273 (1988) 283.
66. P. Panula, U. Pirvola, S. Auvinen and M.S. Airaksinen, Neuroscience, 28 (1989) 585.
67. E. Schlicker, K. Fink, M. Hinterthaner and M. Göthert, Naunyn-Schmiedeberg's Arch. Pharmacol., 340 (1989) 633.
68. K. Fink, E. Schlicker, A. Neise and M. Göthert, Naunyn-Schmiedeberg's Arch. Pharmacol., 342 (1990) 513.
69. Y. Nakagawa, K. Yanai, J.H. Ryu, M. Kiyosawa, M. Tamai and T. Watanabe, Brain Res., 643 (1994) 74.
70. J.H. Ryu, K. Yanai and T. Watanabe, Neurosci. Lett., 178 (1994) 19.
71. J.H. Ryu, K. Yanai, X-L. Zhao and T.Watanabe, Br. J. Pharmacol.,118 (1996) 585.
72. J.H. Ryu, K. Yanai, R. Iwata, T. ido and T. Watanabe, Neuroreport, 5 (1994) 621.
73. C.R. Gerfen, T.M. Engber, LC Mahan, Z. Susel, T.N.Chase, F.J.Monsma and D.R. Sibley.Science, 250 (1990) 1429.
74. C. Lemoine, E. Normand, and B. Bloch, Proc. Natl. Acad. Sci. USA, 88 (1991) 4205.
75. M. Garcia, B. Floran, J.A. Arias-Montano, J.M. Young and J. Aceves, Neuroscience, 80 (1997) 241.
76. P. Cumming, C. Laliberte and A. Gjedde, Brain Res,664 (1994) 276.
77. M.I. Martinez-Mir, H. Pollard, J. Moreau, J.M. Arrang, M. Ruat, E. Traiffort, J.C. Schwartz and J.M. Palacios, Brain Res., 526 (1990) 322.
78. P. Panula, M. Airaksinen and U.Pirvola. Neuroscience, 34 (1990) 127.

R. Leurs and H. Timmerman (Editors)
The Histamine H_3 Receptor

Modulation of in vitro neurotransmission in the CNS and in the retina via H_3 heteroreceptors

E. Schlicker and M. Kathmann

*Institut für Pharmakologie und Toxikologie der Universität Bonn, Reuterstrasse 2b, D-53113 Bonn, Germany

The term "H_3 receptor" has been coined by Arrang et al.[1] H_3 receptors are located on paracrine cells and on neurones; activation of H_3 receptors usually causes inhibition of the release of the respective mediator or neurotransmitter. The receptor characterized by Arrang et al.[1] is an example of an autoreceptor, i.e. of a receptor via which the transmitter released from a given neurone influences its own release. H_3 receptor-mediated inhibition of the release of transmitters other than histamine has also been described; such receptors are known as heteroreceptors. The present review will focus on H_3 heteroreceptors in the central nervous system (CNS); in separate chapters of this book, H_3 autoreceptors, H_3 heteroreceptors in the neuroendocrine system as well as H_3 receptor-mediated modulation of transmitter release in vivo will be considered. A separate article will also deal with H_3 heteroreceptors in peripheral tissues although an example of an H_3 receptor in the retina will be covered in our chapter, due to the close relationship between CNS and retina[2].

In the first part of our chapter the occurrence of H_3 heteroreceptors in the CNS and in the retina will be described. Then the location of the H_3 heteroreceptors will be discussed. (The term "heteroreceptor" will be used in a relatively broad sense in this article, i.e. regardless of whether the presynaptic location on the nerve endings themselves has been proven or not.) Next, interactions between H_3 heteroreceptors and other types of presynaptic receptors will be considered. Finally, some general remarks with respect to H_3 heteroreceptors as targets for new drugs will be given.

---

*The authors' work was supported by grants from the Deutsche Forschungsgemeinschaft (Bonn, Germany) (grants Schl 266/1-1, 1-2, 1-3 and 1-4) and from the Herbert-Reeck-Stiftung (Bonn, Germany).

1. OCCURRENCE OF H_3 HETERORECEPTORS IN THE CNS AND IN THE RETINA

A map of the CNS, in which the H_3 heteroreceptors so far identified are depicted, is given in Fig. 1. H_3 receptor-mediated inhibition of the release of the monoamines dopamine[3] (DA), serotonin[4-6] (5-hydroxytryptamine, 5-HT) and noradrenaline[6-12] (NA), of γ-aminobutyric acid[13] (GABA) and of acetylcholine[14,15] (ACh) has been studied in superfused brain preparations. Briefly, slices or synaptosomes are preincubated with the radioactively labelled neurotransmitter itself (in the case of the monoamines or GABA) or its precursor (in the case of acetylcholine), which are taken up into the respective nerve endings via the respective neuronal transporters. Quasi-physiological release of the respective neurotransmitter can be elicited by electrical stimulation or by increasing the K^+ concentration. The release of glutamate (Glu) has not been measured directly;

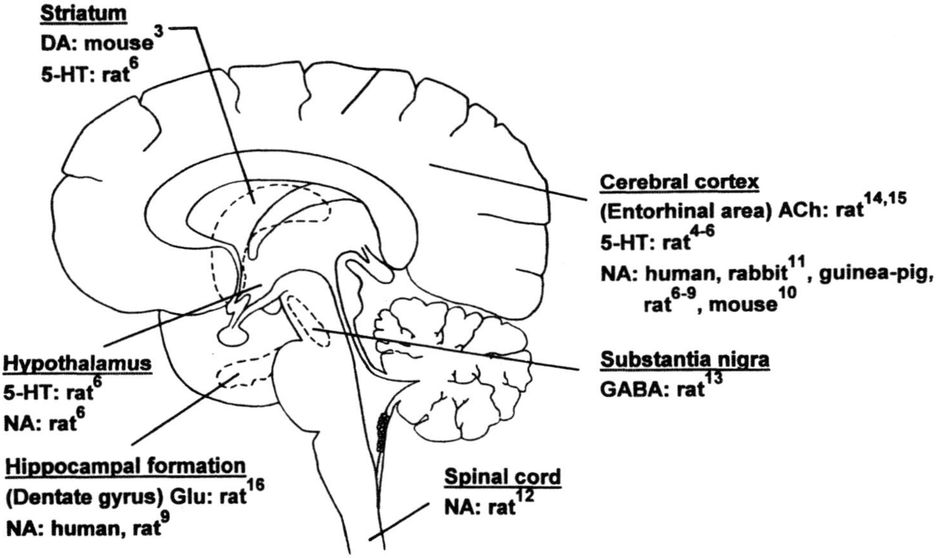

Fig. 1. Occurrence of H_3 receptors inhibiting release of acetylcholine, of amino acid and monoamine neurotransmitters in the mammalian CNS in vitro. The schematic drawing represents a midsagittal section of the human brain; three areas with a more lateral position are shown by broken line (substantia nigra and part of the hippocampus and of the striatum). For each of the six regions of the CNS (subregions given in brackets), in which H_3 heteroreceptors have been identified, the neurotransmitter(s) and the species are indicated. The superscripts refer to the numbers of the papers as listed under References. Own unpublished data suggest that an H_3 receptor-mediated inhibition of noradrenaline release also occurs in the human cerebral cortex and hippocampus and in the guinea-pig cerebral cortex. Note that a *presynaptic* location has not been verified for each of the H_3 heteroreceptors or has been even excluded (for details, see Table 1). *Abbreviations*: ACh, acetylcholine; DA, dopamine; GABA, γ-aminobutyric acid; Glu, glutamate; 5-HT, 5-hydroxytryptamine, serotonin; NA, noradrenaline

rather the field excitatory postsynaptic potentials in hippocampal slices or the excitatory postsynaptic currents in whole-cell patches evoked by Glu, which was released in response to electrical stimulation, were recorded[16]. An H_3 heteroreceptor causing inhibition of DA release in the guinea-pig retina was recently identified in superfusion experiments (Fig. 2; unpublished results). The experimental model may be of interest inasmuch as the maximum inhibitory effect mediated via H_3 receptors is very marked (about 80%) and exceeds that mediated via H_3 autoreceptors in rat brain cortex slices[1] or via H_3 heteroreceptors on noradrenergic neurones in mouse brain cortex slices[10] (about 60% in both instances).

The H_3 heteroreceptor causing inhibition of NA release has been found in four regions of the CNS and in five species including humans. The H_3 receptor-mediated inhibition of NA release in superfused mouse brain cortex slices, which is more pronounced than that in cerebral cortex slices from the other four species, has been examined in a series of studies[10,17-20], which show the criteria which should be fulfilled if a receptor is classified as an H_3 receptor. (Basically the same, although usually less, experiments have been carried out for the identification of other H_3 heteroreceptors, including that in the retina; Fig. 2). First, the effect of histamine should occur in the high nanomolar/low micromolar concentration range and the maximum effect should be obtained at 1-10 μmol/l. The effect of histamine should be mimicked by the selective H_3 receptor agonists R-α-methylhistamine[21] and imetit[22-24] at about 10- and 50-fold higher potency, respectively. The S-enantiomer of α-methylhistamine should be about 100-fold less potent than the R-enantiomer. Second, the effect of histamine (or a selective H_3 receptor agonist) should be antagonized by H_3 receptor antagonists like thioperamide[21] (pA_2 of about 8.5), iodoproxyfan[25] (pA_2 of 8.5-9) or clobenpropit[24] (pA_2 of 9.5-10). (3) The effect of histamine should not be mimicked by H_1 and H_2 receptor agonists and not be blocked by H_1 and H_2 receptor antagonists. This criterion is very relevant for studies in tissues in which an H_3 and, in addition, an H_1 or H_2 receptor is present. Guinea-pig cerebral cortex slices represent such an example. In this preparation, noradrenaline release is inhibited by histamine in a manner sensitive to clobenpropit, which, in addition, unmasks a facilitatory H_2 receptor (unpublished results).

Fourth, the potencies of H_3 receptor ligands should be correlated with their affinities at H_3 binding sites and with their potencies in other H_3 receptor models. In two cooperation projects[18,19] with Professor W. Schunack and coworkers (Freie Universität Berlin), we have determined the potencies of thioperamide, iodoproxyfan and 21 newly synthetized H_3 receptor antagonists in mouse brain cortex slices; the range of apparent pA_2 values was 7.02-9.20. The potencies in this H_3 receptor model were significantly correlated with their affinities for H_3 binding sites labelled by ^{3}H-N$^\alpha$-methylhistamine in rat brain cortex membranes and with their potencies at the H_3 heteroreceptor causing inhibition of the electrically induced contraction in the guinea-pig ileum (the contraction is elicited by the release of ACh) (Fig. 3). The H_3 heteroreceptors in the mouse brain cortex and in the guinea-pig ileum appear to be pharmacologically identical.

In a cooperation project[20] with Professor H. Timmerman, Dr. R. Leurs and coworkers (Vrije Universiteit Amsterdam), we compared the effects of the higher homologues of histamine (i.e. replacement of the ethylene side chain by n-propylene, n-butylene, n-pentylene etc.) in mouse brain cortex slices and in guinea-pig jejunum.

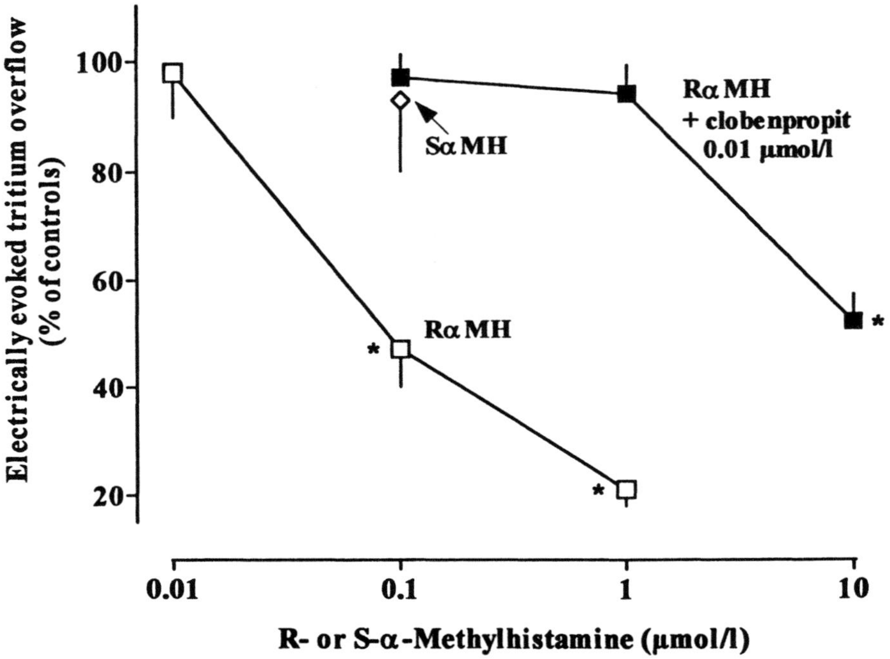

Fig. 2. Effects of H_3 receptor ligands on the electrically evoked tritium overflow from superfused guinea-pig retinal discs preincubated with ^{3}H-noradrenaline. The guinea-pig retina is an exception among mammalian retinae in that it is avascular and therefore also devoid of sympathetic (noradrenergic) vascular neurones. Since, in addition, intrinsic noradrenergic neurones are virtually lacking, the intrinsic *dopaminergic* neurones can not only be labelled with ^{3}H-dopamine but also with ^{3}H-noradrenaline (which is also a substrate for the neuronal dopamine transporter). Hence, the electrically evoked tritium overflow represents quasi-physiological dopamine release in this model[44]. The figure shows that the inhibitory effect of the H_3 receptor agonist R-α-methylhistamine (RαMH) (1) is not mimicked by its S-enantiomer (SαMH) 0.1 µmol/l and (2) is shifted to the right by the H_3 receptor antagonist clobenpropit, yielding an apparent pA_2 value of 10.05. *Experimental details*: Following preincubation with ^{3}H-noradrenaline 0.025 µmol/l for 60 min, the retinal discs were superfused with medium routinely containing vanoxeamine (GBR 12909) 0.5 µmol/l, an inhibitor of the neuronal dopamine transporter, and (-)-sulpiride 3.2 µmol/l, a D_2 receptor antagonist. The latter two drugs were used to increase the electrically evoked tritium overflow. Two 2-min periods of electrical stimulation (1 Hz, 100 mA, 2 ms) were administered after 60 and 100 min of superfusion and the ratio of the overflow evoked by S_2 over that evoked by S_1 was determined; the S_2/S_1 values in controls (not shown) were near unity. RαMH or SαMH was added to the medium before and during S_2 whereas clobenpropit was present throughout superfusion. Tritium overflow evoked by S_1 was 8.47±0.96% of tissue tritium and was not affected by clobenpropit 0.01 µmol/l. Means±SEM of 5-10 experiments. $^*P<0.001$ (compared to the corresponding RαMH-free control; Student's t-test)

(The H_3 receptor in the latter preparation causes inhibition of the electrically induced contraction which is due to the release of ACh.) Interesting enough, the propylene, butylene and pentylene homologues of histamine proved to be partial agonists with an intrinsic activity of 0.3-0.6 in mouse brain cortex slices but behaved as pure antagonists in the guinea-pig jejunum. The hexylene and octylene homologues were pure antagonists in both models. These findings suggest that the H_3 heteroreceptor in the mouse brain cortex exhibits a more efficient receptor coupling than the H_3 heteroreceptor in the guinea-pig jejunum without excluding that the differences may be

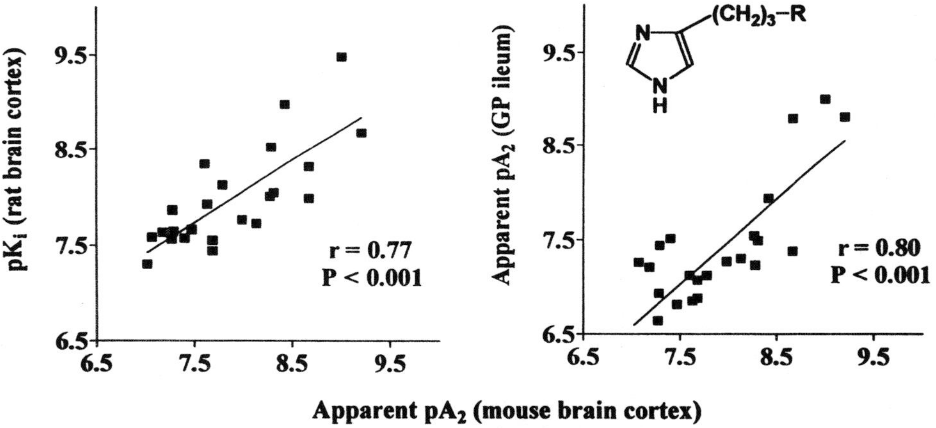

Fig. 3. Correlation of the potencies of thioperamide, iodoproxyfan and 21 recently synthetized H_3 receptor antagonists at the H_3 heteroreceptor in mouse brain cortex slices with their affinities for H_3 binding sites in rat brain cortex membranes (left panel) and their potencies at the H_3 heteroreceptor in guinea-pig ileum strips (right panel). In mouse brain cortex slices preincubated with ^{3}H-noradrenaline the potencies of the test compounds to antagonize the inhibitory effect of histamine on the electrically evoked tritium overflow were determined. The affinities of the compounds for H_3 binding sites were studied in rat brain cortex membranes labelled with ^{3}H-N$^\alpha$-methylhistamine. In guinea-pig ileum strips, the potencies of the compounds to antagonize the inhibitory effect of R-α-methylhistamine on the electrically induced contraction (elicited by the release of acetylcholine) were examined. The 23 antagonists resemble histamine in that they comprise an imidazole ring substituted in position 4(5). However, the ethylene side chain of histamine is replaced by a longer chain (propylene chain in 19 of the 23 compounds; see general formula in the right panel) and the amino functionality is replaced by another functional group which, in turn, is connected directly or via a short alkyl chain to an aromatic or aliphatic ring system. The compounds are amides, carbamates, esters, ethers, guanidines, guanidine esters and thioamides as well as urea and thiourea derivatives. The apparent pA_2 and pK_i values are taken from ref. 18 and 19; for thioperamide the pK_i at the high affinity site is used. The correlation coefficients (r) and their significance levels (P) are given in the two panels. The equations of the regression lines are: $y = 0.65x + 2.85$ (left panel) and $y = 0.90x + 0.30$ (right panel).

explained by the occurrence of species variants of the H_3 receptor or even of H_3 receptor subtypes. There are several hints to H_3 receptor subtypes in the literature[14,26-28]; however, there is so far no general agreement with respect to the occurrence of H_3 receptor heterogeneity.

2. ARE THE RELEASE-INHIBITING H_3 HETERORECEPTORS LOCATED PRESYNAPTICALLY ON THE RESPECTIVE NEURONES?

Standard superfusion and electrophysiological techniques do not allow to decide whether an H_3 heteroreceptor involved in the inhibition of release of a given transmitter is actually located presynaptically (i.e. on the axon terminals) of the respective neurone. Recent (unpublished) data from our laboratory may serve to illustrate this point. In guinea-pig cerebral cortex slices, noradrenaline release is inhibited by histamine via H_3 receptors and facilitated via H_2 receptors. Using an appropriate technique (see next paragraph) we found that only the H_3 but not the H_2 receptor is located presynaptically.

For studies in which transmitter release is measured directly, two experimental approaches are commonly used to determine the location of the receptor[29]. Thus, the release may be studied in isolated nerve endings (synaptosomes) obtained by homogenization and differential centrifugation of CNS tissue. Alternatively, brain slices may be superfused with K^+-rich Ca^{2+}-free medium containing tetrodotoxin (to inhibit impulse flow along the axons) and transmitter release is evoked by introduction of Ca^{2+} ions into the medium. If modulation of transmitter release still occurs under one of the two experimental approaches, a presynaptic location of the receptor under study can be postulated (although, in the case of the latter method, the location of the receptor on structures immediately adjacent to the nerve ending cannot be excluded with absolute certainty[29]). Using these two approaches, we found that the H_3 heteroreceptors causing inhibition of DA release in the mouse striatum, of NA release in the guinea-pig, rat and mouse cortex and of 5-HT release in the rat brain cortex are located presynaptically on their respective neurones (Table 1). On the other hand, the H_3 receptor-mediated inhibition of ACh release previously found in slices[14,15] did not occur in synaptosomes[15] suggesting that this H_3 receptor is not located presynaptically on the cholinergic neurones themselves.

In the study by Brown and Reymann[16] Glu release and its H_3 receptor-mediated inhibition were not determined directly but rather the electrophysiological signal caused by endogenously released Glu was examined. The basal electrophysiological approach leaves open the question as to whether the receptor is located (i) presynaptically on the medial perforant path, (ii) on a more proximal site of the axon of the perforant path or (iii) postsynaptically on the target cells of the perforant path. Additional experiments were carried out by the authors to identify the location of the H_3 heteroreceptor in the rat dentate gyrus. The third possibility could be excluded by two experimental approaches. Thus, the effect of histamine at a postsynaptic site is unlikely since its effect was still observed when the K^+ currents (via which histamine exerts its *post*synaptic effects) were blocked. Moreover, when the effect of endogenously released Glu on the postsynaptic neurone was mimicked by pressure ejection of a Glu receptor agonist, the effect of this latter compound was not affected by histamine. The second possibility could also be excluded since the amplitude of the fibre volley of the

Table 1. Location of H_3 heteroreceptors inhibiting the release of monoamines, acetylcholine and glutamate in the brain. To prove or disprove the presynaptic location of H_3 receptors, transmitter release was studied in isolated nerve endings (synaptosomes) or in brain slices superfused with K^+-rich Ca^{2+}-free medium containing tetrodotoxin (TTX) (in the latter case, transmitter release was evoked by introduction of Ca^{2+} ions into the medium). The experimental approaches used in the electrophysiological study to show the presynaptic location of H_3 receptors on glutamatergic neurones are described in the text.

Transmitter	Species	Brain region	Presynaptic location?	Ref.
Dopamine	Mouse	Striatum	Yes: Ca^{2+}, TTX	3
Noradrenaline	Guinea-pig	Cerebral cortex	Yes: Ca^{2+}, TTX	Unpublished
	Rat	Cerebral cortex	Yes: Ca^{2+}, TTX	8
	Mouse	Cerebral cortex	Yes: Ca^{2+}, TTX	30
Serotonin	Rat	Cerebral cortex	Yes: Synaptosomes; Ca^{2+}, TTX	31
Acetylcholine	Rat	Entorhinal area	No: Synaptosomes	15
Glutamate	Rat	Dentate gyrus	Yes: See text	16

perforant path was not affected by histamine. Two findings are in favour of a presynaptic location of the H_3 receptor. Thus, the two components of the excitatory postsynaptic potential in response to endogenously released glutamate (mediated via AMPA- [α-amino-3-hydroxy-5-methyl-4-isoxazolepropionic acid] and NMDA- [N-methyl-D-aspartic acid] receptors) were affected by histamine in an identical manner. Furthermore, if paired pulses were administered, histamine attenuated the depression of the second compared to the first pulse. Both phenomena are typical for presynaptic receptors on glutamatergic neurones.

3. INTERACTIONS OF PRESYNAPTIC H_3 HETERORECEPTORS WITH OTHER TYPES OF PRESYNAPTIC RECEPTORS

Axon terminals are usually endowed with several types of presynaptic receptors which do not act independently. It has been described, in particular, for *noradrenergic* neurones that activation of a given presynaptic receptor blunts the effect mediated via another type of presynaptic receptor[32-34] activated subsequently. H_3 heteroreceptors on noradrenergic neurones also participate in such receptor interactions. When, e.g., the presynaptic α_2-adrenoceptor (i.e. the autoreceptor on the noradrenergic neurone) in the cerebral cortex of the rabbit[11], rat[35] and mouse[35] was activated first and the H_3 receptor was activated subsequently, the H_3 receptor-mediated effect on NA release was attenuated. The H_3 receptor-mediated effect on NA release in the mouse brain cortex was also blunted by activation of the presynaptic EP_3 receptor (i.e. the receptor for the prostaglandins of the E series)[30]. The reverse receptor interactions were also shown in the mouse brain cortex[30,35]; thus, when the H_3 receptor was activated first and the α_2 (or EP_3) receptor was activated subsequently, the α_2 (or EP_3) receptor-mediated inhibition of NA release was slightly attenuated. The interactions between the three types of presynaptic receptors on the noradrenergic neurones of the mouse brain cortex are shown in Fig. 4. In the rat brain cortex another receptor interaction was found[36]; thus, activation of the H_3 receptor blunted the NMDA receptor-mediated facilitation of NA release. For the *dopaminergic* neurone in the mouse striatum a receptor interaction between the H_3 heteroreceptor and the dopamine autoreceptor was shown. The inhibitory effect of the former receptor on DA release was blunted by simultaneous activation of the latter receptor[3]. With respect to the *GABAergic* neurones in the rat substantia nigra, an interaction between the (presynaptically located?) H_3 and the facilitatory dopamine D_1 receptor was described[13]. Thus, only that component of GABA release was inhibited by H_3 receptors which was elicited by activation of D_1 receptors. Possible interactions between presynaptic receptors were also studied for the *glutamatergic* neurone in the rat dentate gyrus. However, the H_3 receptor-mediated effect was not affected (or even slightly increased) if the Glu autoreceptor was activated simultaneously[16].

Some of the problems inherent to the identification of such interactions will be discussed in more detail for the interaction between the H_3 heteroreceptor and the α_2 autoreceptor on the noradrenergic neurone in the mouse brain cortex[35]. The (H_3 receptor-mediated) effect of histamine on the electrically induced noradrenaline release was attenuated by the α_2-adrenoceptor agonist talipexole (B-HT 920) and increased by the α_2-adrenoceptor antagonist rauwolscine. Does the effect of rauwolscine only occur if the α_2-autoreceptors are activated simultaneously by endogenously released NA or is blockade of the α_2-autoreceptor sufficient, irrespective of whether the receptor is subject to an endogenous tone or not? To clarify this problem, the stimulation protocol used to evoke NA release was modified. Usually NA release is evoked by a 2-min period of electrical stimulation under which an endogenous tone of NA is building up at the α_2-adrenoceptor. When however few electrical pulses are administered at a very high frequency (3 pulses at 100 Hz) such an endogenous tone will not develop. Rauwolscine failed to increase the effect of histamine under the latter stimulation protocol suggesting that the effect of the drug only occurs if the presynaptic α_2-adrenoceptor is simultaneously activated by endogenous NA.

Another point is that talipexole and rauwolscine do not only decrease and increase, respectively, the effect of histamine but of course, in addition, decrease and increase NA release per se. One might argue that the alteration in the amount of NA release (rather than the activation or blockade of the α_2-adrenoceptor) is the reason why the effect of histamine is decreased and increased, respectively. For clarification of this problem, experiments were done in which the current strength or the duration of the electrical pulses was changed in order to adjust the amount of NA release to the level obtained in brain slices not exposed to the α_2-adrenoceptor agonist or antagonist.

Under this modified condition, the same results were obtained suggesting that the activation/blockade of the α_2-adrenoceptor and not the level of NA release is responsible for the observed alteration of the effect of histamine by talipexole or rauwolscine.

We do not know whether the interactions between the presynaptic receptors occur on the level of the receptors or on a site beyond the receptor level, e.g. on the level of G proteins, ion channels or other second messengers. Such receptor interactions may explain drug interactions in vivo. They are also of importance for planning in vitro experiments. H_3 receptor-mediated effects in superfused slice preparations are frequently small; the inhibitory effect on NA and DA release can be increased by simultaneous blockade of the respective autoreceptor. There are even examples of H_3 heteroreceptors which could only be identified if the respective autoreceptor was blocked[3,12,36].

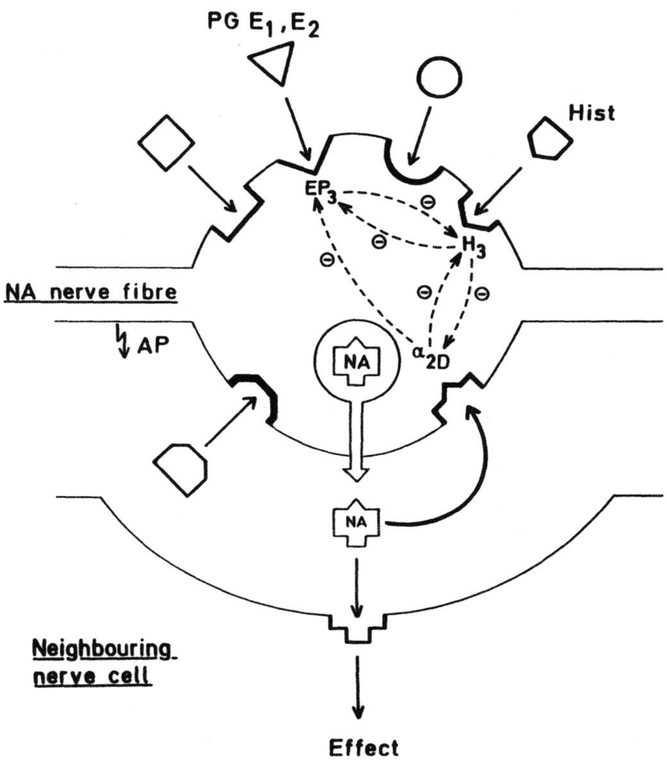

Fig. 4. Schematic drawing of an axon terminal of the noradrenergic neurone in the mouse brain cortex. The axon terminal is endowed with several types of presynaptic receptors including the autoreceptor for noradrenaline itself (which is an α_{2D}-adrenoceptor in this species[45]) and a variety of heteroreceptors (only the H_3 and EP_3 heteroreceptors are identified). Each of the three receptors causes inhibition of noradrenaline release. With respect to the interactions between the three receptors, e.g., "$EP_3 \xrightarrow{\ominus} H_3$" means that, if the EP_3 receptor is activated first and the H_3 receptor is activated subsequently, the effect mediated via the H_3 receptor is attenuated. The interactions between the three receptors have been shown in the following studies: H_3 and α_2, ref. 35; H_3 and EP_3, ref. 30; α_2 and EP_3, ref. 46. It is so far not known whether activation of the EP_3 receptor also affects the effect mediated via the α_2-autoreceptor activated subsequently. *Abbreviations*: AP, action potential; Hist, histamine; NA, noradrenaline; PG E_1, E_2, prostaglandins E_1 and E_2.

4. H_3 HETERORECEPTORS - TARGETS FOR NEW DRUGS?

H_3 receptor ligands have been proposed as drugs for several indications in the CNS[28,37,38], e.g. as cognition/arousal enhancing, antiepileptic and anti-appetite drugs and as drugs for the treatment of schizophrenia and narcolepsy. Both release-modulating H_3 auto-[39] and heteroreceptors (own unpublished results) have been identified in human brain preparations. We will focus on H_3 heteroreceptors.

If one considers H_3 receptor *agonists* as potential drugs, one has to realize that H_3 heteroreceptor-mediated effects are frequently small or moderate. In addition, one has to take into consideration that the release of at least six transmitters is affected and that the effects occur in at least six regions of the CNS (Fig. 1). On the other hand, it is so far unknown whether all the examples of H_3 heteroreceptors identified in animals also occur in humans. Moreover, one must be cautious when extrapolating from in vitro data to the situation in vivo. To give an example. Clonidine, which affects presynaptic α_2-adrenoceptors on several types of neurones and in a variety of CNS regions, became a very useful drug for the treatment of essential hypertension and for some other indications.

With respect to the use of H_3 receptor *antagonists* as potential drugs, one has to realize that no endogenous tone (i.e. no activation by endogenous histamine) at the H_3 heteroreceptors involved in the inhibition of the release of glutamate[16] and monoamines (perhaps with the exception of the H_3 receptors on the serotoninergic neurones in the rat brain cortex)[11] is detectable. A different situation occurs for the cholinergic neurone. Both in the two in vitro studies in the rat entorhinal area[14,15] and in at least one[40] of the three in vivo studies in three brain regions of the rat[40-42] an endogenous tone was found. Thus, the H_3 heteroreceptors causing inhibition of ACh release may become also the target for H_3 receptor antagonists. This may, however, also hold true for H_3 heteroreceptors causing inhibition of monoamines, GABA and glutamate release under certain circumstances, e.g., under pathophysiological conditions. An example of an H_3 heteroreceptor activated by endogenous histamine under pathophysiological conditions only has been described in the peripheral nervous system[43]. Thus, the H_3 heteroreceptor, involved in the inhibition of NA release in the guinea-pig heart, is not activated by endogenous histamine under the usual experimental conditions. However, an endogenous tone of histamine develops at this receptor in reperfused hearts after 10-min global ischemia.

Even if H_3 heteroreceptors should not be involved in the main action of H_3 receptor ligands, they may be the substrate of side effects of H_3 receptor ligands but also of other drugs possessing, by accident, a high affinity for H_3 receptors. In this context, it is highly important to develop drugs which do not only possess a high selectivity towards H_3 receptors but, in addition, are capable of discriminating between H_3 receptor subtypes, provided that pharmacological differences, e.g., between H_3 auto- and heteroreceptors should be found.

5. CONCLUSIONS

Histamine H_3 receptors are not only involved in the inhibition of the release of histamine itself ("autoreceptors") but, in addition, also in the inhibition of the release of other neurotransmitters ("heteroreceptors"). Such an H_3 heteroreceptor-mediated inhibition of release has been shown for dopamine, 5-hydroxytryptamine (serotonin), noradrenaline, γ-aminobutyric acid, acetylcholine and glutamate in CNS preparations of various animals and for dopamine in guinea-pig retinal discs. Moreover, H_3 receptor-mediated inhibition of noradrenaline release has been identified in human cerebral cortex and hippocampal slices. H_3 receptors have been shown to be located *presynaptically* on dopaminergic (mouse striatum), serotoninergic (rat brain cortex), noradrenergic (guinea-pig, rat and mouse brain cortex) and glutamatergic neurones (rat dentate gyrus). Receptor interactions between H_3 heteroreceptors and other types of presynaptic receptors (e.g. α_2-adrenoceptors) have been identified; usually activation of one presynaptic receptor blunts the effect mediated via another presynaptic receptor which is activated subsequently. H_3 heteroreceptors might become targets for new drugs, i.e. for H_3 receptor agonists or, less likely, H_3 receptor antagonists.

REFERENCES

1 Arrang JM, Garbarg M, Schwartz JC (1983) Auto-inhibition of brain histamine release mediated by a novel class (H_3) of histamine receptor. Nature 302: 832-837.
2 Nowak JZ (1988) The isolated retina as a model of the CNS in pharmacology. Trends Pharmacol Sci 9: 80-82.
3 Schlicker E, Fink K, Detzner M, Göthert M (1993) Histamine inhibits dopamine release in the mouse striatum via presynaptic H_3 receptors. J Neural Transm [Gen Sect] 93: 1-10.
4 Schlicker E, Betz R, Göthert M (1988) Histamine H_3 receptor-mediated inhibition of serotonin release in the rat brain cortex. Naunyn-Schmiedeberg's Arch Pharmacol 337: 588-590.
5 Rodrigues AA, Jansen FP, Leurs R, Timmerman H, Prell GD (1995) Interaction of clozapine with the histamine H_3 receptor in rat brain. Br J Pharmacol 111: 1523-1524.
6 Smits RPJM, Mulder AH (1991) Inhibitory effects of histamine on release of serotonin and noradrenaline from rat brain slices. Neurochem Int 18: 215-220.
7 Taube HD, Starke K, Borowski E (1977) Presynaptic receptor systems on the noradrenergic neurones of rat brain. Naunyn-Schmiedeberg's Arch Pharmacol 299: 123-141.
8 Schlicker E, Fink K, Hinterthaner M, Göthert M (1989) Inhibition of noradrenaline release in the rat brain cortex via presynaptic H_3 receptors. Naunyn-Schmiedeberg's Arch Pharmacol 340: 633-638.
9 Rodrigues AA (1996) The histamine H_3 receptor in the rat brain. Pharmacological and (patho)physiological aspects. Thesis, Vrije Universiteit Amsterdam.
10 Schlicker E, Behling A, Lümmen G, Göthert M (1992) Histamine H_{3A} receptor-mediated inhibition of noradrenaline release in the mouse brain cortex. Naunyn-Schmiedeberg 's Arch Pharmacol 345: 489-493.
11 Schlicker E (1997) Histamine H_3 heteroreceptors in the CNS. In: Neurobiology of Histamine (G Prell ed) Humana Press, Totowa, NJ, USA, in press.
12 Celuch SM (1995) Possible participation of histamine H_3 receptors in the modulation of noradrenaline release from rat spinal cord slices. Eur J Pharmacol 287: 127-133.
13 Garcia M, Floran B, Arias-Montano JA, Young JM, Aceves J (1997) Histamine

H_3 receptor activation selectively inhibits dopamine D_1 receptor-dependent [^{3}H]GABA release from depolarization-stimulated slices of rat substantia nigra pars reticulata. Neuroscience 80: 241-249.

14 Clapham J, Kilpatrick GJ (1992) Histamine H_3 receptors modulate the release of [^{3}H]-acetylcholine from slices of rat entorhinal cortex: evidence for the possible existence of H_3 receptor subtypes. Br J Pharmacol 107: 919-923.

15 Arrang JM, Drutel G, Schwartz JC (1995) Characterization of histamine H_3 receptors regulating acetylcholine release in rat entorhinal cortex. Br J Pharmacol 114: 1518-1522.

16 Brown RE, Reymann KG (1996) Histamine H_3 receptor-mediated depression of synaptic transmission in the dentate gyrus of the rat in vitro. J Physiol (London) 496: 175-184.

17 Kathmann M, Schlicker E, Detzner M, Timmerman H (1993) Nordimaprit, homodimaprit, clobenpropit and imetit: affinities for H_3 binding sites and potencies in a functional H_3 receptor model. Naunyn-Schmiedeberg's Arch Pharmacol 348: 498-503.

18 Schlicker E, Kathmann M, Reidemeister S, Stark H, Schunack W (1994) Novel histamine H_3 receptor antagonists: affinities in an H_3 receptor binding assay and potencies in two functional H_3 receptor models. Br J Pharmacol 112: 1043-1048 + 113: 657 (erratum).

19 Schlicker E, Kathmann M, Bitschnau H, Marr I, Reidemeister S, Stark H, Schunack W (1996) Potencies of antagonists chemically related to iodoproxyfan at histamine H_3 receptors in mouse brain cortex and guinea-pig ileum: evidence for H_3 receptor heterogeneity? Naunyn-Schmiedeberg's Arch Pharmacol 353: 482-488.

20 Leurs R, Kathmann M, Vollinga RC, Menge WMPB, Schlicker E, Timmerman H (1996) Histamine homologues discriminating between two functional H_3 receptor assays. Evidence for H_3 receptor heterogeneity. J Pharmacol Exp Ther 276: 1009-1015.

21 Arrang JM, Garbarg M, Lancelot JC, Lecomte JM, Pollard H, Robba M, Schunack W, Schwartz (1987) Highly potent and selective ligands for histamine H_3-receptors. Nature 327: 117-123.

22 Howson W, Parsons ME, Raval P, Swayne GTG (1992) Two novel, potent and selective histamine H_3 receptor agonists. Bioorg Med Chem Lett 2: 77-78.

23 Garbarg M, Arrang JM, Rouleau A, Ligneau X, Dam Trung Tuong M, Schwartz JC, Ganellin CR (1992) S-[2-(4-imidazolyl)ethyl]isothiourea, a highly specific and potent histamine H_3 receptor agonist. J Pharmacol Exp Ther 263: 304-310.

24 Van der Goot H, Schepers MJP, Sterk GJ, Timmerman H (1992) Isothiourea analogues of histamine as potent agonists or antagonists of the histamine H_3-receptor. Eur J Med Chem 27: 511-518.

25 Ligneau X, Garbarg M, Vizuete ML, Diaz J, Purand K, Stark H, Schunack W, Schwartz JC (1994) [^{125}I]Iodoproxyfan, a new antagonist to label and visualize histamine H_3 receptors. J Pharmacol Exp Ther 271: 452-459.

26 West RE, Zweig A, Shih NY, Siegel MI, Egan RW, Clark MA (1990) Identification of two H_3-histamine receptor subtypes. Mol Pharmacol 38: 610-613.

27 Schwörer H, Reimann A, Ramadori G, Racké K (1994) Characterization of histamine H_3 receptors inhibiting 5-HT release from porcine enterochromaffin cells: further evidence for H_3 receptor heterogeneity. Naunyn-Schmiedeberg's Arch

Pharmacol 350: 375-379.
28 Leurs R, Smit MJ, Timmerman H (1995) Molecular pharmacological aspects of histamine receptors. Pharmac Ther 66: 413-463.
29 Starke K, Göthert M, Kilbinger H (1989) Modulation of neurotransmitter release by presynaptic autoreceptors. Physiol Rev 69: 864-989.
30 Schlicker E, Marr I (1997) Mutual interactions of the presynaptic histamine H_3 and prostaglandin EP_3 receptors on the noradrenergic terminals in the mouse brain. Neuroscience 79: 247-254.
31 Fink K, Schlicker E, Neise A, Göthert M (1990) Involvement of presynaptic H_3 receptors in the inhibitory effect of histamine on serotonin release in the rat brain cortex. Naunyn-Schmiedeberg's Arch Pharmacol 342: 513-519.
32 Limberger N, Späth L, Starke K (1988) Presynaptic α_2-adrenoceptor, opioid κ-receptor and adenosine A_1-receptor interactions on noradrenaline release in rabbit brain cortex. Naunyn-Schmiedeberg's Arch Pharmacol 338: 53-61.
33 Allgaier C, Daschmann B, Sieverling J, Hertting G (1989) Presynaptic κ-opioid receptors on noradrenergic nerve terminals couple to G proteins and interact with the α_2-adrenoceptors. J Neurochem 53: 1629-1635.
34 Molderings GJ, Göthert M (1990) Mutual interaction between presynaptic α_2-adrenoceptors and 5-HT_{1B} receptors on the sympathetic nerve terminals of the rat inferior vena cava. Naunyn-Schmiedeberg's Arch Pharmacol 341: 391-397.
35 Schlicker E, Behling A, Lümmen G, Malinowska B, Göthert M (1992) Mutual interaction of histamine H_3 receptors and α_2-adrenoceptors on noradrenergic terminals in mouse and rat brain cortex. Naunyn-Schmiedeberg's Arch Pharmacol 345: 639-646.
36 Fink K, Schlicker E, Göthert M (1994) N-Methyl-D-aspartate (NMDA)-stimulated noradrenaline (NA) release in rat brain cortex is modulated by presynaptic H_3-receptors. Naunyn-Schmiedeberg's Arch Pharmacol 349: 113-117.
37 Schwartz JC, Arrang JM, Garbarg M, Traiffort E (1995) Histamine. In: Psychopharmacology: The Fourth Generation of Progress (FE Bloom, DJ Kupfer eds) Raven, New York, pp 397-405.
38 Stark H, Schlicker E, Schunack W (1996) Developments of histamine H_3-receptor antagonists. Drugs Future 21: 507-520.
39 Arrang JM, Devaux B, Chodkiewicz JP, Schwartz JC (1988) H_3-Receptors control histamine release in human brain. J Neurochem 51: 105-108.
40 Mochizuki T, Okakura-Mochizuki K, Horii A, Yamamoto Y, Yamatodani A (1994) Histaminergic modulation of hippocampal acetylcholine release in vivo. J Neurochem 62: 2275-2282.
41 Blandina P, Giorgetti M, Bartolini L, Cecchi M, Timmerman H, Leurs R, Pepeu G, Giovannini MG (1996) Inhibition of cortical acetylcholine release and cognitive performance by histamine H_3 receptor activation in rats. Br J Pharmacol 119: 1656-1664.
42 Prast H, Fischer H, Tran MH, Grass K, Lamberti C, Philippu A (1997) Modulation of acetylcholine release in the ventral striatum by histamine receptors. Inflamm Res 46 (Suppl 1): S37-S38.
43 Imamura M, Poli E, Omoniyi AT, Levi R (1994) Unmasking of activated histamine H_3-receptors in myocardial ischemia: their role as regulators of exocytotic norepinephrine release. J Pharmacol Exp Ther 271: 1259-1266.

44 Schlicker E, Timm J, Göthert M (1996) Cannabinoid receptor-mediated inhibition of dopamine release in the retina. Naunyn-Schmiedeberg's Arch Pharmacol 354: 791-795.
45 Limberger N, Trendelenburg AU, Starke K (1995) Subclassification of presynaptic α_2-adrenoceptors: α_{2D}-autoreceptors in mouse brain. Naunyn-Schmiedeberg's Arch Pharmacol 352: 43-48.
46 Exner HJ, Schlicker E (1995) Prostanoid receptors of the EP_3 subtype mediate the inhibitory effect of prostaglandin E_2 on noradrenaline release in the mouse brain cortex. Naunyn-Schmiedeberg's Arch Pharmacol 351: 46-52.

R. Leurs and H. Timmerman (Editors)
The Histamine H_3 Receptor

H_3 receptor modulation of the release of neurotransmitters *in vivo.*

P. Blandina, L. Bacciottini, M.G. Giovannini and P.F. Mannaioni*

Dipartimento di Farmacologia Preclinica e Clinica, Universitá di Firenze, Viale G.B. Morgagni 65, 50134 Firenze, Italy

1. HISTAMINE AS A NEUROTRANSMITTER

Although the presence of histamine in nervous tissue was described more than 50 years ago [1], and much evidence for its neurotransmitter role has accumulated during the following two decades [2], only in the past 10 years has the evidence become persuasive [3-5]. Indeed, the delay in searching for a histaminergic neuronal system, in contrast to the exploration of other putative neurotransmitter systems, made it initially difficult to accept that histamine could have a specific transmitter role. Hence, the identification of a central histaminergic neuronal system, visualized immunocytochemically with antibodies against histidine decarboxylase [6] and histamine conjugates [7], has been a real breakthrough, providing new perspectives in histamine research. Histaminergic cell bodies are large, 20-30 µm in diameter, and confined to the tuberomammillary nucleus of the hypothalamus. Histamine in the brain is formed from L-histidine, which is almost exclusively decarboxylated by a specific histidine decarboxylase (E.C. 4.1.1.22), and only the tuberomammilary nucleus consistently showed the mRNA for histidine decarboxylase [8]. Electrophysiological studies indicated that histaminergic neurons fire spontaneously and regularly [9]. Histaminergic efferent fibers project predominantly ipsilaterally with multifold arborizations into the whole central nervous system, including most subcortical nuclei and the cerebral cortex [4]. In this respect histamine is like other biogenic amines, such as dopamine, norepinephrine and serotonin, in that it occurs in neural circuits that would allow centralized modulation of a wide area of the brain. Most of the histaminergic fibers are unmyelinated. Although they make relatively few contacts [10], the ability to form synaptic contacts was shown by electron microscopy [11]. These characteristics suggest that the histaminergic system is a regulatory center for whole-brain activity.

* This work was supported by grants 60% from M.U.R.S.T- Universitá di Firenze (Italy)

2. THE HISTAMINE H_3 AUTORECEPTOR

Soon after the identification of the histaminergic neuronal system, the histamine H_3 receptor was discovered. It was detected as an autoreceptor whose stimulation inhibited the potassium-evoked release of [^{3}H]-histamine from rat cortical slices preloaded with [^{3}H]-histidine [12]. Evidence suggests that H_3 receptors may act to restrict the influx of calcium ions, which is essential for histamine release [12-14]. Thus, histamine can reduce presynaptically its own release from axonal terminals, analogously to other aminergic neurotransmitters, such as noradrenaline through α_2 receptor activation. It is important to emphasize that these findings were related to endogenous histamine, e.g., histamine that the slices formed from histidine. On cerebral cortical slices that had been incubated with radiolabeled, exogenously preformed histamine, the effect of histamine on the depolarization-induced release was not seen [14]. There is much evidence that release of endogenous and exogenous amines do not always occur in parallel or in concert [15-17].

Originally, the H_3 receptor was postulated as a novel type of histamine receptor based on the deviating potencies of selective H_1 and H_2 receptor agents [12]. Eventually, confirmation of its existence was granted by selective H_3-receptor ligands, e.g. agonists such as R-α-methylhistamine (RAMH) [18], imetit [19-21], and immepip [22], and antagonists such as thioperamide [18], clobenpropit [20], iodoproxyfan [23], and GT-2016 [24]. The introduction of these agents has provided a valuable tool to assess the physiological functions of the H_3-receptor in living animals, and modulation of histamine release by H_3-receptor activation has been evidenced also *in vivo*. Indeed, systemic administration of thioperamide to rats resulted in reduced cortical levels of histamine [18]. This reduction presumably reflected an increased rate of histamine release, since thioperamide accelerated significantly the depletion of histamine in animals treated with α-fluoromethylhistidine [18, 25], an irreversible inhibitor of histidine decarboxylase [26]. Moreover, the activation of histamine turnover induced by thioperamide suggests that endogenous histamine exerts a tonic influence on presynptic H_3 receptors, but, since RAMH elicited an opposite response in the same experiments [18, 25], the activation by endogenous histamine is unlikely to be maximal. More recently, the *in-vivo* operativity of H_3 autoreceptors has also been demonstrated in the hypothalamus. The release of endogenous histamine from the hypothalamus of freely moving rats was measured by microdialysis coupled with high pressure liquid chromatography (HPLC) with fluorescence detection [27, 28]. Thioperamide, administered orally or intraperitoneally, caused an approximate three-fold increase in the histamine output [27, 29], whereas RAMH, injected intraperitoneally, decreased significantly the extracellular concentration of histamine [28]. Similar results were obtained with the push-pull superfusion technique: hypothalamic superfusion with RAMH inhibited, while superfusion with thioperamide enhanced the release of histamine from rat hypothalamus [30].

In addition to histamine release, histamine synthesis is also inhibited presynaptically by histamine, and the H_3 receptor is linked to this effect [31]. In the cerebral cortex of rats injected with the radiolabelled precursor of histamine, [^{3}H]L-histidine, [^{3}H]histamine synthesis decreased significantly in a dose-dependent manner after systemic administration of RAMH [25]. This inhibition could be reversed by intraperitoneal administration of thioperamide [18, 25]. Interestingly, administration of thioperamide alone produced a marked increase in histamine synthesis, indicative of the presence of a degree of tonic inhibition elicited by H_3-receptors [18].

The absence of a neuronal high-affinity uptake system for histamine [32] distinguishes the latter from other aminergic neurotransmitters whose synaptic activities are ended by re-uptake. Thus, histamine released from nerve endings is inactivated solely by metabolism. In the brain it is metabolized mainly, if not exclusively, by histamine methyltransferase [33]. The *tele*-methylhistamine that is formed is metabolized by monoamine oxidase-B [34, 35]. Since the brain rapidly converts histamine to *tele*-methylhistamine, the latter can provide a reliable index of brain histaminergic activity. Based on the accumulation of this metabolite after treatment with pargyline, a monoamine oxidase inhibitor, Oishi and coworkers [36] have demonstrated *in vivo* that RAMH potently inhibited, whereas thioperamide potently enhanced histamine metabolism in various brain regions of both rat and mouse. In the absence of a neuronal high-affinity uptake system, the modulation of histamine release and synthesis by presynaptic autoreceptors may play a pivotal role in the regulation of histamine synaptic levels. There are several functions that histamine may regulate in the brain [37, 38], and drugs selective for the H_3 autoreceptor, through the modulation of endogenous histamine release and synthesis, may influence all these functions.

3. MODULATION OF CATECHOLAMINES AND SEROTONIN RELEASE MEDIATED BY HISTAMINE H_3 HETERORECEPTORS

Autoradiographic studies have shown that the presence of H_3 receptors is not restricted to histaminergic neurons [39-41], and it has been suggested that H_3 receptors serve as general presynaptic heteroreceptors [42]. Consistently, *in-vitro* observations imply that H_3 receptors are involved in the inhibitory effects of histamine on electrically-evoked release of [^{3}H]-serotonin and [^{3}H]-noradrenaline from rat cortical and hypothalamic slices [43-45]. Moreover, H_3 receptor activation inhibited electrically-evoked release of [^{3}H]-dopamine from mouse striatal slices [46]. However, these findings were obtained by measuring the efflux of tritium after the tissue was preloaded with radiolabeled, exogenously preformed catecholamines or serotonin. As noted above, it cannot be assumed that the release of exogenous amines indicates the release of endogenous amines [47, 48]. Young et al. [49], measuring the release of endogenous neurotransmitters from slices of rat cerebral cortex, reported that histamine failed to modify the potassium-evoked release of endogenous noradrenaline and endogenous serotonin, thus leaving some doubts concerning the implication of H_3 receptors. We reported that histamine increased the release of endogenous noradrenaline from rat hypothalamic slices [50]. *In-vivo* studies have demonstrated that pharmacological manipulation of the H_3 receptor failed to alter the rate of monoamine turnover, thus implying that H_3 receptors are not involved in modulating catecholamines and indoleamine release or metabolism. More precisely, systemic administration of RAMH and thioperamide neither altered the levels of noradrenaline nor influenced the α-methyl-p-tyrosine-induced decline of noradrenaline levels in rat and mouse brain [51]. Similarly, they failed to alter serotonin and 5-hydroxyindoleacetic acid brain levels and to modify pargyline-induced serotonin accumulation [51]. Consistently, the blockade of H_3 receptors by systemic administration of clobenpropit has not affected basal rates of dopamine and serotonin turnover in several brain regions of the mouse [52]. On the other hand, clobenpropit slightly increased the noradrenaline turnover rate of noradrenaline, although only in the midbrain, the pons and the medulla oblungata, but not in the cerebral cortex nor in the diencephalon [52]. Thus reasonable doubts exist concerning the contribution of H_3 receptors to the modulation of catecholaminergic and serotonergic systems. Extrapolation of *in-vitro* observations must always be considered very cautiously.

4. EVIDENCE OF HISTAMINE H_3 RECEPTOR / ACh INTERACTIONS

High densities of H_3 receptors were found in ACh-rich areas such as the cerebral cortex [40]. Furthermore, two different laboratories reported that H_3 receptors are involved in the inhibitory effects of histamine on potassium-evoked release of [^{3}H]-acetylcholine (ACh) [53, 54]. However, H_3 receptor activation failed to inhibit [^{3}H]-ACh release from rat cortical synaptosomes [54]. Since, as noted above, prudence must always be exerted in *in-vivo* extrapolation of *in-vitro* observations, we investigated the effect of histamine on the release of ACh from the cortex of freely moving rats and characterized the underlying mechanisms [55-58].

4.1. Histamine moderates 100 mM potassium-evoked release of ACh from the cortex of freely moving rats.

Using microdialysis to simultaneously administer histamine and monitor changes in ACh release, we found that perfusion of the cortex with histamine does not alter spontaneous release of ACh, but results in a concentration-dependent inhibition of 100 mM potassium-evoked release of ACh [55], up to more than 60% (figure 1). This concentration of potassium is only apparently high, for the low recovery of potassium through the microdialysis membrane [59], and the rapid dilution of potassium in the extracellular space necessitate high concentrations in the perfusion fluid. Actually, 60 mM potassium has only a slight effect on ACh output during brain dialysis [60]. Finally, brain perfusion with 100mM potassium evokes an increase in ACh release similar to that obtained with incubation of cortical slices in 20 mM potassium [53].

4.2. H_3 receptor agonists mimic whereas H_3 receptor antagonists block the effect of histamine on 100 mM potassium-evoked release of ACh from the cortex of freely moving rats

Histamine interacts with its specific receptors, H_1 [61], H_2 [62] and H_3 [12], and with the polyamine-binding site of the NMDA receptor complex [63, 64]. The histamine-induced inhibition of ACh meets the proposed criteria of a H_3 receptor. The histamine agonists RAMH, imetit and immepip, selective for the H_3 receptor, mimic the effect of histamine (figure 1) with a slightly greater potency [55]. Consistently, the inhibition of 100 mM potassium-evoked ACh release produced by 100 μM histamine is completely antagonized by clobenpropit or thioperamide, both selective antagonists of the H_3 receptor (figure 1). These antagonists were added to the perfusion medium at concentrations that blocked H_3 receptor-mediated responses [18, 20, 65]. The antagonists alone were without effect on 100 mM potassium-evoked release of ACh [55, 56].

4.3. Neither H_1 nor H_2 receptors are involved in the histamine-induced inhibition of 100 mM potassium-evoked release of ACh from the cortex of freely moving rats

Inclusion of 2-thiazolylethylamine, a compound showing some selectivity for H_1 receptors [66], or dimaprit, a selective H_2 receptor agonist [67], in the perfusing Ringer solution produced no significant modification of 100 mM potassium-evoked ACh release [55, 56]. Consistently, neither triprolidine nor cimetidine, antagonists of the H_1 [68] and the H_2 [69] receptor, respectively, antagonized 100 μM histamine-elicited inhibition of 100 mM potassium-evoked ACh release [55]. The antagonists were added to the perfusion medium at concentrations more than 400 times their K_d for both the H_1 [68] and the H_2 [69] receptors.

In the cortical slice, mepyramine and ranitidine, antagonists at H_1 and H_2 receptors, caused a 10-fold increase of the potency of RAMH in inhibiting potassium-evoked release of [^{3}H]-ACh, without change in the maximal effect [53, 54]. We observed that the effect produced by 0.1 μM histamine, a concentration clearly submaximal, was increased significantly by H_1 and H_2 receptor blockade [55]. However, the inhibition elicited by 1 μM histamine, a concentration that triggered an almost maximal inhibition, was unchanged [55]. Although the lack of effectiveness of 2-thiazolylethylamine and dimaprit may preclude invoking either the H_1 or the H_2 receptor in promoting the release of ACh from the cortex, they might have been ineffective, because the release triggered by 100 mM potassium was already maximal. H_2 receptor activation stimulates neurotrans-

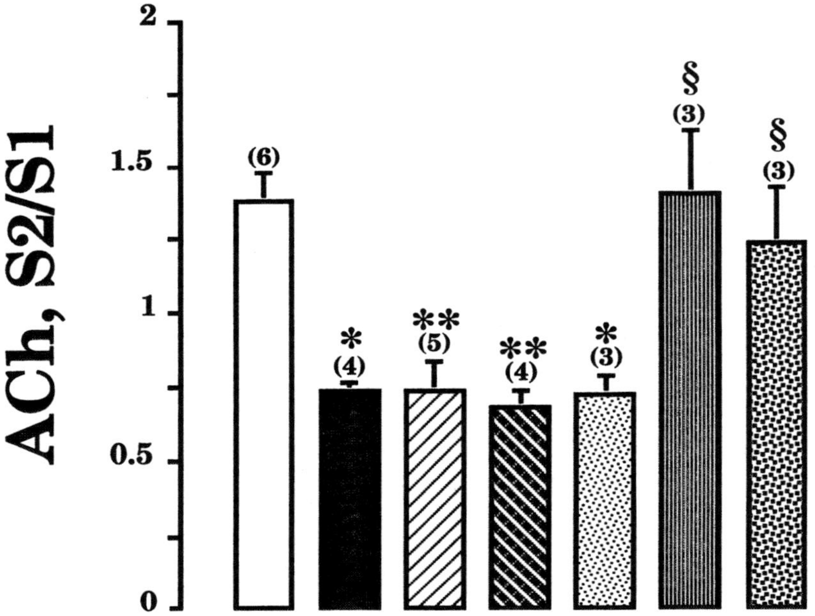

HISTAMINE	-	100	-	-	-	100	100	μM
RAMH	-	-	10	-	-	-	-	μM
IMETIT	-	-	-	10	-	-	-	μM
IMMEPIP	-	-	-	-	5	-	-	μM
THIOPERAMIDE	-	-	-	-	-	300	-	nM
CLOBENPROPIT	-	-	-	-	-	-	300	nM

Figure 1. Influence of local administration of histamine, alone and in the presence of thioperamide or clobenpropit, and histamine H_3 receptor agonists on 100 mM potassium-evoked release of ACh from the cortex of freely moving rats. Spontaneous ACh release was stable at 3.44 ± 0.16 pmol/10 min (n = 97). Perfusion through the dialysis fiber with a Ringer medium containing 100 mM potassium for 10 min (S_1) strongly stimulated the release of ACh (8.01 ± 0.32 pmol/10 min, n = 97). In a subset of 6 animals, a second identical 100 mM potassium perfusion (S_2), conducted 90 min after the end of the first (S_1), released similar amounts of ACh. The mean S_2/S_1 ratio was 1.37 ± 0.10. Histamine (100 μM), administered through the dialysis fiber for 30 min, failed to show any effect on the spontaneous release of ACh. Histamine receptor agonists and antagonists, alone or in association, were added 10 min before S_2 stimulation to the perfusion medium and remained throughout the S_2 stimulation. Shown are means ± S.E. of (n) experiments. The presence of significant treatment effects was determined by one-way analysis of variance followed by Scheffe's test. *$P < 0.05$ vs. control, **$P < 0.01$ vs. control, $^{\S}P < 0.05$ vs. histamine.

mitter release from other brain regions, such as ACh from rat hippocampus [70], and noradrenaline from rat hypothalamus [50].

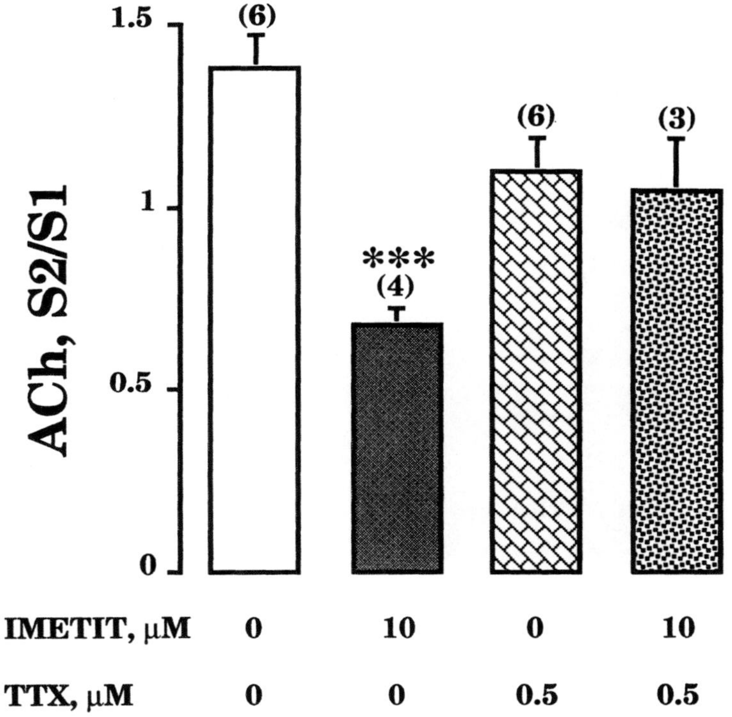

Figure 2. Effects of local administration of TTX, alone and in combination with 10 μM imetit, on 100 mM potassium-evoked release of ACh from the cortex of freely moving rats. At 40 (S_1) and 140 (S_2) min the perfusion medium was changed from 4 to 100 mM KCl for 10 min after equilibration. TTX was added 20 min, and imetit 10 min before S_2 stimulation to the perfusion medium. Both remained throughout the S_2 stimulation. Shown are means ± S.E. of (*n*) experiments. The presence of significant treatment effects was determined by one-way analysis of variance followed by Scheffe's test. ***$P < 0.001$ vs. control.

4.4. Tetrodotoxin-sensitivity of H_3 receptor-elicited inhibition of 100 mM potassium-evoked release of ACh from the cortex of freely moving rats

Tetrodotoxin (TTX) a voltage-dependent Na^+-channel blocker was infused directly into the prefrontal cortex through the dialysis fiber. TTX significantly decreased ACh spontaneous release by more than 40%, but the exposure to 100 mM potassium released similar amounts of ACh in the absence and in the presence of 0.5 μM TTX (figure 2). However, H_3 receptor-induced inhibition is completely abolished in cortices in which the traffic of action potentials was blocked by TTX (figure 2), thus H_3 receptors modulating ACh release are probably not located presynaptically on cholinergic nerve terminals, nor on non-cholinergic nerve endings impinging on the former. Therefore, it appears most likely that these H_3 receptors are somatodendritic receptors on interneurons, whose excitation produces Na^+-dependent action potentials that release an intermediary modulatory substance. Consistently, H_3 receptors decreased in the cerebral cortex after local infusion of neurotoxins [39, 40], and their stimulation failed to alter the potassium-evoked release of [^{3}H]-ACh from synaptosomes of the entorhinal cortex [54].

Table 1
The influence of bicuculline on immepip-induced inhibition of 100 mM potassium-evoked release of ACh from the cortex of freely moving rats. Bicuculline was infused into the prefrontal cortex through the dialysis fiber 20 min before S_2 and maintained during S_2 stimulation. Immepip was infused, alone or with bicuculline, 10 min before S_2 and maintained during S_2. Shown are the means ± S.E.

DRUGS	CONCENTRATION	ACh, S2/S1	n
NONE	-	1.26 ± 0.05	4
IMMEPIP	1 μM	0.65 ± 0.1***	3
IMMEPIP + BICUCULLINE	1 μM 0.1 μM	0.74 ± 0.11***	3
IMMEPIP + BICUCULLINE	1 μM 1 μM	0.92 ± 0.46	4
IMMEPIP + BICUCULLINE	1 μM 10 μM	1.17 ± 0.09	4
IMMEPIP + BICUCULLINE	1 μM 30 μM	1.11 ± 0.05	4
IMMEPIP + BICUCULLINE	1 μM 100 μM	1.55 ± 0.05	3

*** $p < 0.001$ vs control by one-way analysis of variance and Scheffe's test.

4.5. Bicuculline antagonized H_3 receptor-elicited inhibition of 100 mM potassium-evoked release of ACh from the cortex of freely moving rats

Since GABA is the most widespread inhibitory neurotransmitter in the brain [71], its role as hypothetical intermediary in the mechanism of H_3 receptor-mediated moderation of ACh release was investigated. Bicuculline, a $GABA_A$ receptor antagonist, reversed the inhibition of ACh release induced by immepip, an H_3 receptor agonist, in a concentration-dependent fashion (Table 1).

5. H_3 RECEPTOR ACTIVATION INCREASED 100 mM POTASSIUM-EVOKED RELEASE OF GABA FROM THE CORTEX OF FREELY MOVING RATS

Bicuculline, a $GABA_A$ receptor antagonist, reversed the inhibition of ACh release induced by immepip, an H_3 receptor agonist, thus suggesting a GABAergic involvement. More direct evidence of this involvement was provided by the observation that immepip, an H_3 receptor agonists, enhanced 100 mM potassium-evoked release of GABA from the cortex of freely moving rats by more than 50% [57]. Immepip was infused into the cortex through the dialysis fiber at a concentration that produced a maximal inhibition of potassium-evoked ACh release [55].

These findings strongly suggest that H_3 receptors, located postsynaptically on intrinsic perikarya, facilitate the release of GABA, which, in turn, inhibits ACh release. The simplest hypothesis is that these interneurons directly innervate the cholinergic presynaptic terminals and reduce ACh release. There is evidence that the cortical GABAergic system exerts a tonic inhibition of spontaneous release of ACh from the cortex, and that this inhibitory tone is maximal [72]. This could explain why neither histamine nor the H_3 receptor agonists altered spontaneous ACh release, much of which is TTX sensitive. Under resting conditions, the inhibition of ACh release caused by GABA being maximal, H_3 activation would have no effect on spontaneous ACh release. However, activation of H_3 receptors, by increasing the release of GABA, antagonizes the potassium-induced depolarization, thus depressing, at least partially, 100 mM potassium-evoked ACh release. Alternatively, another synaptic arrangement consonant with the lack of H_3 modulation of spontaneous release is that the activated interneuron inhibits the release of an excitatory presynaptic modulator of cholinergic terminals. If this excitatory pathway were not spontaneously active, H_3 activation would have no effect on spontaneous ACh release. In the presence of potassium, this excitatory modulator would be released and enhance the depolarization-induced release of ACh. Activation of H_3 receptors would remove this enhancement and partially, but not completely, depress potassium-evoked ACh release.

Cortical GABA interneurons control the activity of large populations of principal cells through their extensive axon arborization. Therefore, any pathway, even if relatively sparse such as the histaminergic pathway, may exert a powerful effect on the activity of the cortex if it modulates the activity of local GABA interneurons.

6. CONCLUSIONS

In the mammalian brain histamine is involved in the regulation of numerous physiological functions, including modulation of memory and synaptic plasticity [37, 73, 74]. Cholinergic systems have been closely linked to cognitive function [75], but ACh is unlikely to be the only neurotransmitter important for cognition. Disruption of cholinergic function is charateristic of aging and Alzheimer's disease, however, these changes typically occur within the context of system's alterations of other neurotransmitters, including noradrenaline, dopamine, serotonin, GABA, several neuropeptides, and histamine [38, 76]. Thus, cholinergic dysfunction might well interact with dysfunctions in other neurotransmitter systems to produce additive or even synergistic effects on cognition. Accordingly, the role of interactions between ACh and other neurotransmitters affecting cognition is of considerable interest. Since cognitive deficits might be related to reduced availability of ACh in the synaptic cleft [77], H_3 receptor activation, by moderating ACh release, would be expected to impair learning and memory. Indeed, systemic administration of doses of RAMH and imetit that reduced potassium-evoked cortical ACh release [55], also impaired rat performance in cognitive tests [55]. Another histamine H_3 receptor agonist, immepip, also impaired animal performance in the olfactory, social memory test, based on the recognition of a juvenile rat by a male, adult and sexually-experienced rat [78].

Conversely, H_3 receptor antagonists, such as thioperamide and clobenpropit, may provide a novel approach to restoring deficits in cognitive functions. Thioperamide exerts some procognitive activity in the olfactory, social memory test [78]. However, the procognitive effects of H_3 receptor antagonists may become fully evident only when behavioral deficits are pronounced. For example, while administration of scopolamine alone to rats impairs object recognition and a passive avoidance response, rats receiving thioperamide (5 mg/kg, i.p.) or clobenpropit (15 mg/kg, i.p.) in combination with scopolamine perform as well as saline-treated animals (Blandina, unpublished observations). However, both H_3 receptor antagonists lack any procognitive effect in control animals (Blandina, unpublished observations). Moreover, thioperamide improved learning and memory in senescence-accelerated mice (these animals showed a marked age-accelerated deterioration in learning tasks of passive avoidance), but it is ineffective in normal-rate aging mice [79]. Other studies report that administration of thioperamide or clobenpropit to scopolamine-impaired mice attenuated, although only slightly, scopolamine's amnestic effects in the elevated plus-maze test and the step-through passive avoidance test, [52, 80, 81].

Reduction of choline uptake into the brain of older adults may be a contributing factor to late life onset of neurodegenerative, particularly dementing, illnesses in which cholinergic neurons show special susceptibility to loss [82]. Conversely brain histaminergic activity is higher in the elderly [83, 84], and histamine content of the rat brain increases with age [85]. These observations could be readily integrated: decreased choline uptake in the elderly may be due to feed-back inhibition produced by decreased ACh release, to which high histaminergic activity may be contributing. Treatment strategies could take advantage of non-cholinergic drugs that potentiate cholinergic functions. H_3 receptor antagonists could relieve histaminergic inhibitory effects, hence facilitate ACh release, and correct the deficits resulting from cholinergic hypofunction. This property should be particularly considered since H_3 receptors showing pharmacological characteristics and functions closely similar to those of the H_3 receptor in rodent brain, have been identified in the human brain [86].

REFERENCES

1. H. Kwiatkowski, J. Physiol., 102 (1943) 32.
2. J.P. Green, Fed. Proc., 23 (1964) 1095.
3. J.C. Schwartz, J.M. Arrang, M. Garbarg, H. Pollard and M. Ruat, Physiol. Rev., 71 (1991) 1.
4. H. Wada, N. Inagaki, A. Yamatodani and T. Watanabe, Trends Neurosci., 14 (1991) 415.
5. J.C. Schwartz, J.M. Arrang, M. Garbarg and E. Traiffort, In: Psychopharmacology: the fourth generation of progress, F.E. Bloom and D.J. Kupfer (eds.), Raven Press Lts., New York (1995) 397.
6. T. Watanabe, Y. Taguchi, H. Hayashi, J. Tanaka, S. Shiosaka, M. Tohyama, H. Kubota, Y. Terano and H. Wada, Neurosci. lett., 39 (1983) 249.
7. P. Panula, H.Y. Yang and E. Costa, Proc. Natl. Acad. Sci. U.S.A., 81 (1984) 2572.
8. D.A. Bayliss, Y.M. Wang, C.A. Zahnow, D.R. Joseph and D.E. Millhorn, Mol. Cell. Neurosci., 1 (1990) 3.
9. H.L. Haas and P.B. Reiner, J. Physiol., 399 (1988) 633.
10. M. Tohyama, R. Tamiya, N. Inagaki and H. Tagaki, In: Histaminergic neurons: morphology and function, T. Watanabe and H. Wada (eds.), CRC Press, Boca Raton (FL) (1991) 107.
11. N. Inagaki, A. Yamatodani, K. Shinoda, Y. Shiotani, M. Tohyama, T. Watanabe and H. Wada, Brain Res., 418 (1987) 388.
12. J.M. Arrang, M. Garbarg and J.C. Schwartz, Nature, 302 (1983) 832.
13. S.J. Hill and R.M. Staw, Br. J. Pharmacol., 95 (1988) 1213.
14. J.M. Arrang, M. Garbarg and J.C. Schwartz, Neuroscience, 15 (1985) 553.
15. H. Herdon, J. Strupish and S. Nahorski, Brain Res., 348 (1985) 309.
16. J.P. Green, G.D. Prell, J.K. Khandelwal and P. Blandina, Agents Actions, 22 (1987) 1.

17. P. Blandina, J. Goldfarb, B. Craddock-Royal and J.P. Green, J. Pharmacol. Exp. Ther., 251 (1989) 803.
18. J.M. Arrang, M. Garbarg, J.C. Lancelot, J.M. Lecont, H. Pollard, M. Robba, W. Schunack and J.C. Schwartz, Nature, 327 (1987) 117.
19. M. Garbarg, J.-M. Arrang, A. Rouleau, X. Ligneau, M. Dam-Trung-Tuong, J.-C. Schwartzand C.R. Ganellin, J. Pharmacol. Exp. Ther., 263 (1992) 304.
20. H. VanderGoot, M.J.P. Schepers, G.J. Sterk and H. Timmerman, Eur. J. Med. Chem., 27 (1992) 511.
21. W. Howson, M.E. Parson, P. Raval and G.T.G. Swayne, Bioorg. Medicinal Chem. Lett., 2 (1992) 77.
22. R.C. Vollinga, J.P. de Koning, F.P. Jansen, R. Leurs, W.M.P.B. Menge and H. Timmerman, J. Med. Chem., 37 (1994) 332.
23. X. Ligneau, M. Garbarg, M.L. Vizuette, J. Diaz, K. Purand, H. Stark, W. Schunack and J.C. Schwartz, J. Pharmacol. Exp. Ther., 271 (1994) 452.
24. C.E. Tedford, S.L. Yates, G.P. Pawlowski, J.W. Nalwalk, L.B. Hough, M.A. Khan, J.G. Phillips, G.J. Durant and R.C.A. Frederickson, J. Pharmacol. Exp. Ther., 275 (1995) 598.
25. M. Garbarg, M.D. TrungTrong, C. Gros and J.C. Schwartz, Eur. J. Pharmacol., 164 (1989) 1.
26. M. Garbarg, G. Barbin, E. Rodergras and J.C. Schwartz, J. Neurochem., 35 (1980) 1045.
27. Y. Itoh, R. Oishi, M. Nishibori and K. Saeki, J. Neurochem., 56 (1991) 769.
28. Y. Itoh, R. Oishi, N. Adachi and K. Saeki, J. Neurochem., 58 (1992) 884.
29. T. Mochizuki, A. Yamatodani, K. Okakura, M. Takemura, N. Inagaki and H. Wada, Naunyn-Schmiedeberg's Arch. Pharmacol., 343 (1991) 190.
30. H. Prast, H.P. Fisher, M. Prast and A. Philippu, Naunyn-Schmiedeberg's Arch. Pharmacol., 350 (1994) 599.
31. J.M. Arrang, M. Garbarg and J.C. Schwartz, Neuroscience, 23 (1987) 149.
32. R.P.J.M. Smits, H.W.M. Steinbusch and A.H. Mulder, Neurochem. Int., 12 (1988) 193.
33. D.D. Brown, R. Tomchick and J. Axelrod, J. Biol. Chem., 234 (1959) 2948.
34. L.B. Hough and E.F. Domino, J. Pharmacol. Exp. Ther., 208 (1979) 422.
35. P.C. Waldmeier, J.J. Feldtrauer and L. Maitre, J. Neurochem., 29 (1977) 785.
36. R. Oishi, M. Nishibori and K. Saeki, J. Neurochem., 52 (1989) 1388.
37. H.L. Haas, P.B. Reiner and R.W. Greene, In: Histaminergic neurons: morphology and function, T. Watanabe and H. Wada (eds.), CRC Press, Boca Roton (FL) (1991) 196.
38. K. Onodera, A. Yamatodani, T. Watanabe and H. Wada, Progr. Neurobiol., 42 (1994) 685.
39. P. Cumming, C. Shaw and S.R. Vincent, Synapse, 8 (1991) 144.
40. H. Pollard, J. Moreau, J.M. Arrang and J.C. Schwartz, Neuroscience, 52 (1993) 169.

41. M.I. Martinez-Mir, H. Pollard, J.M. Arrang, M. Raut, E. Traiffort, J.C. Schwartz and J.M. Palacios, Brain Res., 526 (1990) 322.
42. J.F. Van Der Werf and H. Timmermann, Trends Pharmacol. Sci., 10 (1989) 159.
43. E. Schlicker, R. Betz and M. Göthert, Naunyn-Schmiedeberg's Arch. Pharmacol., 337 (1988) 588.
44. E. Schlicker, K. Fink, M. Hinterhaner and M. Göthert, Naunyn-Schmiedeberg's Arch. Pharmacol., 340 (1989) 633.
45. R.P.J.M. Smits and A.H. Mulder, Neurochem. Int., 18 (1991) 215.
46. E. Schlicker, K. Fink, M. Detzner and M. Göthert, J. Neural Transm., 93 (1993) 1.
47. I.J. Kopin, G.R. Breese, K.R. Krauss and V.K. Weise, J. Pharmacol. Exp. Ther., 161 (1968) 271.
48. H. Herdon and S.R. Nahorski, Naunyn-Schmiedeberg's Arch. Pharmacol., 335 (1987) 238.
49. C.S. Young, R. Mason and S.J. Hill, Biochem. Pharmacol., 37 (1988) 2799.
50. P. Blandina, P. Knott, L.K.H. Leung and J.P. Green, J. Pharmacol. Exp. Ther., 249 (1989) 44.
51. R. Oishi, M. Nishilori, Y. Itoh, S. Shishido and K. Sachi, Eur. J. Pharmacol., 184 (1990) 135.
52. S. Miyazaki, K. Onodera, M. Imaizumi and H. Timmerman, Life Sci., 61 (1997) 355.
53. J. Clapham and G.J. Kilpatrick, Br. J. Pharmacol., 107 (1992) 919.
54. J.M. Arrang, G. Drutel and J.C. Schwartz, Br. J. Pharmacol., 114 (1995) 1518.
55. P. Blandina, M. Giorgetti, L. Bartolini, M. Cecchi, H. Timmerman, R. Leurs, G. Pepeu and M.G. Giovannini, Br. J. Pharmacol., 119 (1996) 1656.
56. P. Blandina, M. Giorgetti, M. Cecchi, R. Leurs, H. Timmerman and M.G. Giovannini, Inflamm. Res., 45 (1996) S54.
57. M. Giorgetti, L. Bacciottini, L. Bianchi, M.G. Giovannini, M. Cecchi and P. Blandina, Inflamm. Res., 46 (1997) S3.
58. M. Giorgetti, M.G. Giovannini, L. Bacciottini, L. Bianchi and P. Blandina, In: Monitoring molecules in neuroscience, J.L. González-Mora, R. Borges and M. Mas (eds.), University of La Laguna, Tenerife (SP) (1996) 158.
59. B.H.C. Westerink and J.B. deVries, In: Monitoring molecules in neuroscience, J.L. Gonzáles-Mora, R. Borges and M. Msa (eds.), University of La Laguna, Tenerife (SP) (1966) 9.
60. B.H.C. Westerink, G. Damsa, H. Rolema, J.B. DeVries and A.S. Horn, Life Sci., 41 (1987) 1763.
61. A.S.F. Ash and H.O. Schild, Br. J. Pharmacol. Chemother., 27 (1966) 427.
62. J.W. Black, W.M. Duncan, C.J. Durant, C.R. Ganellin and E.M. Parson, Nature, 236 (1972) 385.
63. J.M. Bekkers, Science, 261 (1993) 104.

64. V.S. Vorobjev, I.N. Sharonova, I.B. Walsh and H.L. Haas, Neuron, 11 (1993) 837.
65. M. Kathmann, E. Schlicker, M. Detzner and H. Timmerman, Naunyn-Schmiedeberg's Arch. Pharmacol., 348 (1993) 498.
66. C.R. Ganellin, In: Pharmacology of Histamine Receptors, C. Ganellin and M.E. Parson (eds.), Wright, Bristol, England (1982) 10.
67. M.E. Parson, D.A.A. Owen, C.R. Ganellin and C.J. Durant, Agents Actions., 7 (1977) 31.
68. R.R. Ison and A.F. Casy, J. Pharm. Pharmacol., 23 (1971) 848.
69. G.J. Durant, J.C. Emmet and C.R. Ganellin, J. Med. Chem., 20 (1977) 901.
70. T. Mochizuki, K. Okakura-Mochizuki, A. Horii, Y. Yamamoto and A. Yamatodani, J. Neurochem., 62 (1994) 2275.
71. A. Nistri and A. Costanti, Prog. Neurobiol., 13 (1979) 117.
72. M. Giorgetti, L. Bacciottini, M. Giovannini and P. Blandina, Soc. Neurosci. Abs., 22 (1996) 1255.
73. K. Tasaka, New Advances in histamine research, Springer-Verlag, Tokyo, 1994.
74. J.P. Huston, U. Wagner and R.U. Hasenöhrl, Behav. Brain Res., 83 (1997) 97.
75. V. Haroutunian, E. Barnes and K.L. Davis, Psychopharmacology, 87 (1985) 266.
76. M.W. Decker and J.L. McGaugh, Synapse, 7 (1991) 151.
77. R. Quirion, A. Wilson, W. Rowe, I. Aubert, J. Richard, H. Doods, A. Parent, N. White and M.J. Meaney, J. Neurosci., 15 (1995) 1455.
78. H. Prast, A. Argyriou and A. Philippu, Brain Res., 734 (1996) 316.
79. K.-I. Meguro, K. Yanai, N. Sakai, E. Sakurai, K. Maeyama, H. Sasaki and T. Watanabe, Pharmacol. Biochem. Behav., 50 (1995) 321.
80. S. Miyazaki, M. Imaizumi and K. Onodera, Meth. Find. Exp. Clin. Pharmacol., 17 (1995) 653.
81. S. Miyazaki, M. Imaizumi and K. Onodera, Life Sci., 57 (1995) 2137.
82. B.M. Cohen, P.F. Renshaw, A.L. Stoll, R.J. Wurtman, D. Yurgelun-Todd and S.M. Babb, J.A.M.A., 274 (1995) 902.
83. G.D. Prell, J.K. Khandelwal, R.S. Burns, P.A. LeWitt and J.P. Green, Arch. Gerontol Geriatr., 12 (1991) 1.
84. G.D. Prell, J.K. Khandelwal, R.S. Burns, P.A. LeWitt and J.P. Green, Compr. Gerontol. A., 2 (1988) 114.
85. K. Onodera, L. Tuomisto, U. Tacke and M. Airaksinen, Meth. Find. Exp. Clin. Pharmacol., 14 (1992) 13.
86. J.M. Arrang, B. Devaux, J.P. Chodkievikz and J.C. Schwartz, J. Neurochem., 51 (1988) 105.

H3 receptor modulation of neuroendocrine responses to histamine and stress

Ulrich Knigge[a,b], Andreas Kjær[a,c], Henrik Jørgensen[a], Jørgen Warberg[a]

**[a]Department of Medical Physiology, The Panum Institute,
[b]Department of Surgery C, Rigshospitalet, and
[c]Department of Clinical Physiology and Nuclear Medicine, Frederiksberg Hospital, University of Copenhagen, Denmark[1]**

1. INTRODUCTION

The neurotransmitter histamine (HA) exerts several functions in the hypothalamus [1-2] including an involvement in the neuroendocrine regulation of pituitary hormone secretion [3]. HA has no effect directly at the level of the pituitary gland, but influences the secretion of anterior pituitary hormones either by an exerted e.g. in the paraventricular nucleus (PVN) on other central transmitters or hypothalamic regulating factors, which subsequently regulate the release of anterior pituitary hormones. In addition, HA acts on the supraoptic nucleus (SON) in the hypothalamus where the posterior pituitary hormones are synthesized and thereby exerts a direct effect on the release of the posterior pituitary hormones. Immunohistochemical studies have revealed that the histaminergic neurons, which originate in the tuberomammillary nuclei of the posterior hypothalamus, densely innervate most of the hypothalamic areas involved in the neuroendocrine control of pituitary hormone secretion [4-5]. Within the last two decades the effect of HA on pituitary hormone secretion have been explored in several studies and it has been

---

1. Correspondence to: Ulrich Knigge, M.D., Dr.Med.Sci., Department of Medical Physiology, Building 12.3., The Panum Institute, Blegdamsvej 3, University of Copenhagen, DK-2200 Copenhagen N, Denmark. Phone: +45 35327517; Fax: +45 35327546; E-mail: <Knigge@mfi.ku.dk>.
Acknowledgment: Our studies presented were supported by grants from the Commission of the European Communities (contract: CEE BMH1-CT 92-1087); the Danish Medical Research Council; the Lundbeck Foundation; the Ib Henriksen's Foundation; the Velux Foundation; the P. Carl Petersen's Foundation; the Danish Hospital Foundation for Medical Research, Regions of Copenhagen, The Faroe Islands and Greenland; NOVO's Foundation; Nordic Insulin Foundation Committee; Gerda and Aage Haensch Foundation; the Poul M. and Birthe Christiansen Foundation; the Danish Medical Association Research Fund; and the Foundation for the Advancement of Medical Research. Elsa Larsen and Jytte Oxbøl are thanked for skilled technical assistance. We thank professor H. Timmerman, Free University, Amsterdam, The Netherlands for imetit, professor J.C. Schwartz, Bioproject, France for R(α)methylHA and BP 2-94 and NOVO-Nordisk, Målev, Denmark for thioperamide. The materials for RIA of PRL and ACTH were kindly provided by the National Hormone and Pituitary Program of the NIDDK and by Dr. Mogens Fenger, Rigshospitalet, Copenhagen, Denmark.

attempted to establish a physiological role of the amine in this respect. Thus, histaminergic neurons have been found to participate in mediation of the hormone responses to stress, immune stimulation, suckling, estrogen surge and dehydration.

In order to clarify the role of HA in hormone regulation most studies have used central administration of HA either alone or in combination with blockade of postsynaptic H1 and H2 receptors or administration of compounds which affect HA synthesis or metabolism. The discovery of the presynaptic H3 receptor in 1983 [6-7] and the following development of specific agonists and antagonists for this receptor type, opened up new possibilities to study HA-neuroendocrine interactions. Since activation of the presynaptic H3 receptor inhibits HA synthesis and release [8], one should anticipate that H3 receptor agonists might exert effects on hormone secretion opposite to those exerted by HA itself or attenuate the effect of physiological reactions depending on neuronal HA. On the other hand H3 receptor antagonists, which augment neuronal HA release [8], is expected to enhance HA-mediated physiological reactions.

Autoradiographic studies have revealed that H3 receptors are located in most areas of the brain with the highest content in the basal ganglia and the cortex [9]. H3 receptors are distributed with moderate density in the hypothalamus, where the highest density of histaminergic nerve fibers are found [4-5], and were demonstrated in the anterior, medial and posterior hypothalamus [8]. However, their localization in hypothalamic areas primarily involved in the regulation of pituitary hormone secretion were not studied in detail. In the pituitary gland the H3 receptors were scarce.

Due to the identified localization of H3 receptors in the brain compared to the distribution of histaminergic nerve fibers this autoradiographic study suggests that H3 receptors might also be located on target cells different from histaminergic neurons [8]. This finding confirms previous results, which indicate that the H3 receptor in addition to their presence as autoreceptors on histaminergic neurons are also located presynaptically as heteroreceptors on serotonergic, noradrenergic, dopaminergic and cholinergic neurons [10-13]. The existence of heteroreceptors may complicate the use of H3 receptor ligands in the investigation of HA in neuroendocrine regulation, since the above mentioned transmitters are all involved in regulation of pituitary hormone secretion. In addition, the possible existence of H3 receptor subtypes and postsynaptically localized H3 receptors may further complicate the interpretation of available and future data. More detailed studies are required in order to distinguish between effects mediated via H3 auto- or H3 heteroreceptors, and proposed H3 receptor subtypes and postsynaptic H3 receptors.

When studying the role of various H3 receptor compounds on pituitary hormone secretion some points have to be considered: a) are the actions of the compounds related to H3 receptor binding or to a non-specific action, b) are the actions related to an auto- or heteroreceptor or eventually postsynaptic receptor effect of the compounds, and c) are the actions related to an effect known to involve histaminergic neurons. In the following, we will give a short description of the effect of HA in the neuroendocrine regulation of pituitary hormone secretion (Table 1) and focus on the role of H3 receptors in modulation of the secretion of the individual pituitary hormones based on previously published investigations and our own recent experiments.

2. THE H3 RECEPTOR AND ANTERIOR PITUITARY HORMONE SECRETION

2.1. ACTH and other POMC derived peptides

Several studies have indicated that HA in addition to other hypothalamic neurotransmitters [14] are involved in the neuroendocrine regulation of the proopiomelanocortin (POMC)-derived

Table 1

Effect of HA on hormone secretion from rat anterior, intermediate and posterior pituitary lobes

Hormone	Effect of HA	Postsynaptic HA receptor	Mediators and (Site of action of HA)	Physiological actions involving HA
ACTH **ß-END** (*Ant. pit.*)	stim.	H1/H2	CRH (*PVN*) AVP (*PVN, SON*) (*ME?*)	Stress, Suckling, Immune stim.
ß-END **α-MSH** (*Intermed. pit.*)	stim.	H1/H2	NE/E (*?*); DA* (*?*)	Stress
PRL (*Ant. pit.*)	stim.	H1/H2	DA* (*Arc. n.; ME*) 5-HT (*Raphe n., PVN*) AVP (*SON, PVN*)	Stress, Suckling E-induced PRL surge
TSH (*Ant. pit.*)	inhib.	H2	TRH (*Perivent. area*) (*PVN*)	Stress ?
GH (*Ant. pit.*)	inhib. stim.	H1 H1/(H2)	? GHRH	Stress? ?
LH (*Ant. pit.*)	stim.	H1/(H2)	GnRH (*Ant.hyp.*) (*Arc. n.*)	E-induced LH surge
FSH (*Ant. pit.*)	?			
AVP (*Post. pit.*)	stim.	H1/H2	- (*SON, PVN*) (*Post. pit.; ME*)	Osmotic stimuli
OT (*Post. pit.*)	stim.	H1/H2	- (*SON, PVN*) (*Post. pit.; ME*)	Suckling, Stress, Osmotic stimuli

*: DA inhibits hormone secretion and HA may inhibit that effect.

DA: dopamine; 5-HT: serotonin; TRH: thyrotropin-releasing hormone; GHRH: growth hormone-releasing hormone; GnRH: gonadotropin-releasing hormone; Ant. pit.: anterior pituitary lobe; Intermed. pit.: intermediate pituitary lobe; Post. pit.: posterior pituitary lobe; SON: supraoptic nucleus; PVN: paraventricular nucleus; Arc. n.: arcuate nucleus; Raphe n.: raphe nucleus; ME: median eminence; Perivent.: periventricular; Ant. hyp.: anterior hypothalamus; E-induced: estrogen-induced.

peptides adrenocorticotrophic hormone (ACTH), β-endorphin (β-END) and α-melanocyte-stimulating hormone (α-MSH) in rats and other species [for review see 3]. Central administration of HA has been found to stimulate the secretion of ACTH, β-END and α-MSH dose-dependently via activation of postsynaptic H1 and H2 receptors [15-16] and to stimulate hormone synthesis as indicated by an increase in POMC mRNA in the anterior pituitary lobe [17].

The action of HA is indirect, mediated primarily via activation of corticotropin-releasing hormone (CRH) originating in parvocellular neurons in the PVN and secondly via vasopressin (AVP) originating in parvo- and magnocellular neurons in the PVN and in the SON [18-24]. The effect of CRH is predominantly mediating in character (i.e. HA releases CRH which subsequently stimulates ACTH secretion) [18] while the effect of AVP seem to be mediating as well as permissive in character (i.e. AVP has to be present in order for HA to exert its effect on ACTH secretion) [25]. Besides these two important mediators, prostaglandins are involved in HA-induced release of the POMC-derived peptides from the anterior lobe [26], whereas catecholamines, oxytocin (OT) and serotonin (5-HT) do not participate [27-28, Willems et al. (unpublished observations)].

A physiological role of hypothalamic histaminergic neurons have been established in a number of studies, which has shown that HA participates in the mediation of the ACTH and β-END responses to stimuli such as stress, immune system activation and suckling. This conclusion is based on findings, which show that inhibition of neuronal HA synthesis by α-fluoromethylhistidine (α-FMH), blockade of postsynaptic H1 or H2 receptors and lesion of the tuberomammillary nuclei reduced or prevented the hormone response to these stimuli [16,29-32, Søe-Jensen (unpublished observations)].

It has been reported that the murine anterior pituitary tumor cell line AtT-20 possesses H3 receptors since the ligand [^{3}H]N$^\alpha$meHA binds with high affinity to membrane preparations from this cell line [33-34]. The binding is decreased by steroids. Furthermore, the H3 receptor agonist R(α)methylhistamine (RmHA) increased the release of ACTH from this cell line [34], an effect which was blocked by the H3 receptor antagonist thioperamide (THIOP) but not by H1 or H2 receptor antagonists. The H2 receptor agonist dimaprit and HA were less potent. The finding may explain the previously reported stimulatory effect of systemically administered HA on ACTH and corticosteroid secretion *in vivo*, although this effect was reported to occur via H1 or H2 receptors [35-36]. In contradiction to a direct H3 receptor mediated effect of HA on ACTH secretion are the findings that H3 receptors were scarcely distributed in the normal pituitary gland of rats and guinea pigs in autoradiographic studies [9,37] and that HA had no effect on ACTH secretion from isolated pituitary cells *in vitro* [38]. Furthermore, systemic administration of different H3 receptor agonists or the antagonist THIOP had no effect on basal or stimulated ACTH secretion. Thus, a physiological role of H3 receptors located in the normal pituitary gland is doubtful.

In some studies H3 receptor compounds were used in order to evaluate the involvement of histaminergic neurons in regulation of ACTH and β-END secretion *in vivo*. Systemic administration of the H3 receptor agonist RmHA (10 mg/kg at -180 and -60 min) had no effect on basal hormone secretion but reduced by approximately 50% the ACTH and β-END responses to 5 min of restraint stress [31]. The inhibitory effect of RmHA, which was comparable to the inhibitory effect of the HA synthesis inhibitor α-FMH (200 μg icv), was reversed by pretreatment with the H3 receptor antagonist THIOP (5 mg/kg ip). However, when administered alone THIOP (5 or 10 mg/kg ip) had no effect on basal or restraint stress-stimulated ACTH and β-END secretion. Lower doses of RmHA (2.5 or 5 mg/kg ip at -180 and 60 min) had no effect on basal or stress-induced hormone release, indicating a rather steep dose response curve [31]. In these experiments activation of H3 receptors had no effect on hypothalamic content of HA, while THIOP increased the content of tele-methylHA, which is in agreement with an increased release and turnover of neuronal HA. However, as mentioned above, this increased release of neuronal HA did not result in an enhancement of the ACTH or β-END response to stress.

Similar to the effect on restraint stress, RmHA (10 mg/kg x 2 ip) inhibited the ACTH and β-END response to insulin-induced hypoglycemia, which increases neuronal HA turnover [30-31]. This inhibitory effect was completely or partly reversed by THIOP. Likewise, the responses of the POMC-derived peptides and corticosterone to immune stimulation with the *E.coli* lipopolysaccharide (LPS) endotoxin, which augmented neuronal HA turnover, were reduced by RmHA pretreatment [39]. The effect of RmHA was equal to that of HA synthesis inhibition by α-FMH [39]. In lactating female rats suckling-induced ACTH secretion was reduced by pretreatment with RmHA (Fig. 6) as well as by α-FMH [32].

In order to assess the role of H3 receptors further, we compared the effect of different H3 receptor agonists administered either alone or in combination with THIOP on the ACTH and β-END responses to various stimuli [Knigge (unpublished observations)]. In these experiments conscious male Wistar rats were exposed to restraint stress (the animal was fixed on its back for 5 minutes immediately before decapitation), hypoglycemia stress (injection 45 minutes before decapitation of 3 IU/kg insulin actrapid was ip, which reduced plasma glucose from approximately 7 to 2 nmol/l) or endotoxin stress (injection 120 minutes before decapitation of 10 μg/kg LPS ip). The animals were pretreated ip 180 and 60 minutes before decapitation with the H3 receptor agonists RmHA (10 mg/kg) [7], the prodrug BP 2-94 ((*R*)-(-)-2[[N-[1-(1*H*-imidazol-4-yl)-2-propyl]imino]phenylmethyl] phenol; 25 mg/kg) [40] or imetit (2 mg/kg) [41-42] immediately preceded by ip injection of saline or the H3 receptor agonist THIOP (5 mg/kg).

Restraint stress-induced stimulation of ACTH and β-END secretion was inhibited almost 40% by each of the three H3 receptor agonists (Figs. 1 and 2). Prior administration of THIOP completely prevented the inhibitory effect of the H3 receptor agonists. The H3 receptor agonists exerted a more profound inhibition of the β-END and in particular the ACTH responses to insulin-induced hypoglycemia causing an inhibition up to 75% (Figs. 1 and 2). The inhibitory effect of the H3 receptor agonists on hypoglycemia-induced ACTH and β-END secretion was only in part reestablished by THIOP. The inhibitory effect of the H3 receptor agonists on LPS-induced ACTH secretion varied considerably, since RmHA only reduced the response about 50% whereas imetit almost completely inhibited it (Fig. 1). THIOP prevented the inhibitory effect of RmHA but only partially reduced the inhibitory effect of BP 2-94 and imetit. Basal secretion of ACTH and β-END were not affected by any of the three H3 receptor agonists or by THIOP.

These studies indicate that activation of H3 receptors by RmHA, the prodrug BP 2-94 or imetit is capable of reducing the response of POMC-derived peptides from the anterior pituitary gland to a number of physiological stimuli. The effect of the three agonists seems to be related to H3 receptor activation since their inhibitory effect are completely or partially reversed by the H3 receptor antagonist THIOP. Since exogenous HA stimulates the secretion of the POMC-derived peptides [15] and since blockade of postsynaptic H1 or H2 receptors or HA synthesis exert a similar effect as the three H3 receptor agonists [16,29], the above results seem to be explained by an effect of the agonists exerted on presynaptic H3 receptors located on histaminergic neurons.

The inhibitory efficiency of the H3 receptor agonists varies with the particular form of stress applied, although some of this is likely to be explained by the magnitude of the hormone response induced by the stressor. Thus, the agonists inhibited the ACTH response to hypoglycemia about 75% but only reduced the response to restraint stress by 50%. In accordance with these results

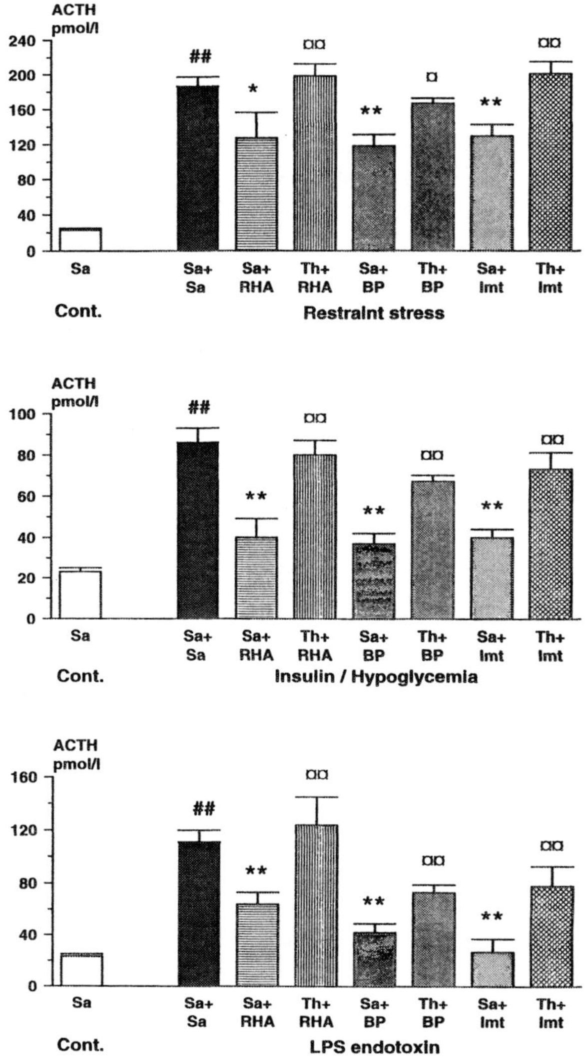

Figure 1. Effect of the H3 receptor agonists R(α)methylHA (RHA; 10 mg/kg), BP 2-94 (BP; 25 mg/kg) or imetit (Imt; 2 mg/kg) administered ip at -180 and -60 minutes in combination with saline (Sa) or the H3 receptor antagonist thioperamide (Th; 5 mg/kg ip at -180 min) on the ACTH response to restraint stress (5 minutes), insulin-induced hypoglycemia (3 IU/kg ip -45 minutes) or LPS endotoxin (10 µg/kg ip at -120 minutes). The rats were decapitated at 0 minutes. ##: $p<0.01$ vs. saline control; * or **: $p<0.05$ or $p<0.01$ vs. stimulus alone; and ¤ or ¤¤: $p<0.05$ or $p<0.01$ vs. stimulus and H3 receptor agonist.

THIOP completely prevented the inhibitory effect of the H3 receptor agonists to restraint stress but only in part antagonized the inhibitory effect of the agonists to hypoglycemia stress. Furthermore, imetit prevented the ACTH response to LPS while the response was reduced 75% by BP 2-94 and only 50% by RmHA, which indicates a differentiation of the potency of the agonists in that response compared to the almost equal potencies shown in the other stress responses. The findings may suggest an existence of heterogeneity in the H3 receptor population and different involvement of H3 auto- and heteroreceptors.

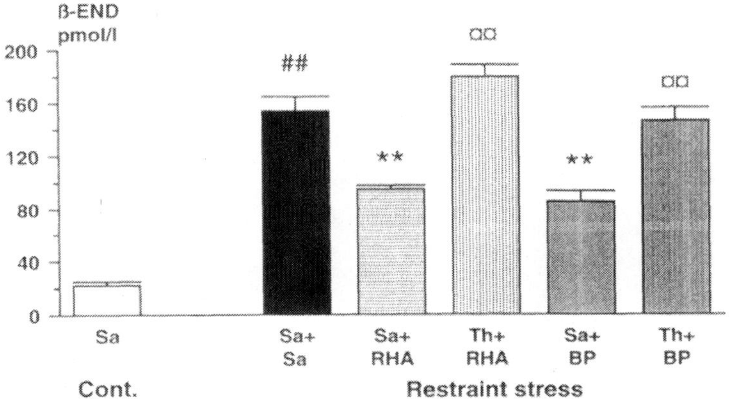

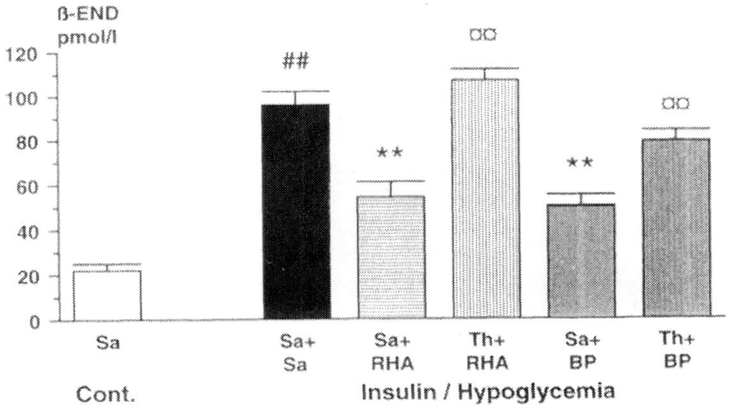

Figure 2. Effect of the H3 receptor agonists R(α)methylHA (RHA) or BP 2-94 (BP) administered in combination with saline (Sa) or the H3 receptor antagonist thioperamide (Th) on the β-END response to restraint stress or insulin-induced hypoglycemia. Further informations are given in legend to figure 1.

As mentioned above, the existence of H3 receptors located presynaptically as heteroreceptors on other aminergic neurons, such as serotonergic, noradrenergic and dopaminergic neurons has been suggested [10-13]. However, in a series of experiments we found that neither RmHA nor THIOP affected the ACTH response to serotonergic activation induced by administration of the 5-HT precursor 5-hydroxytryptophan in combination with the 5-HT re-uptake inhibitor fluoxetine [27].

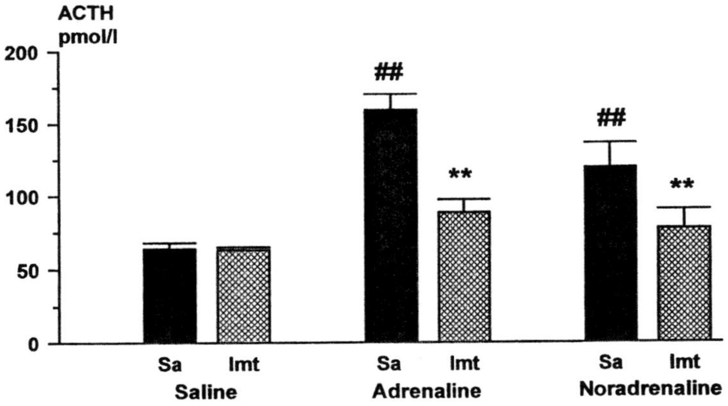

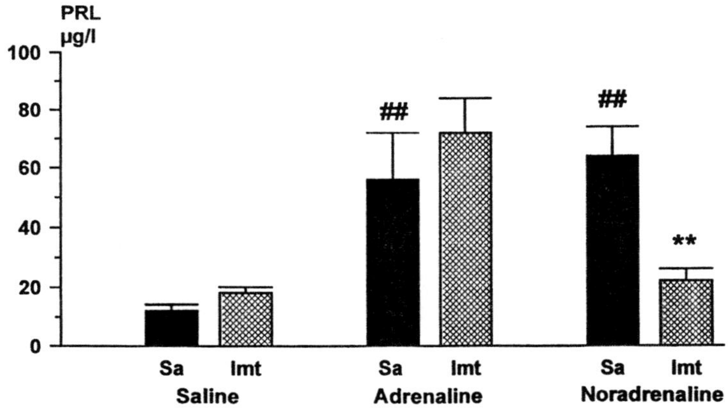

Figure 3. Effect of the H3 receptor agonist imetit (Imt; 2 mg/kg ip at -180 minutes) on the ACTH and PRL responses to icv administration of adrenaline (8.3 µg/rat) or noradrenaline (8.3 µg/rat) administered ip at -15 minutes.The rats were decapitated at 0 minutes. ##: $p<0.01$ vs. saline control; **: $p<0.01$ vs. stimulus alone.

In contrast to this finding, we recently found that imetit almost blocked the ACTH response to adrenaline and noradrenaline [Willems et al. (unpublished observations)] (Fig. 3). Since blockade of postsynaptic H1 and H2 receptors had no effect on the ACTH response to stimulation by catecholaminergic compounds, the results may suggest an effect of imetit on H3 heteroreceptors located on catecholaminergic neurons. However, to exclude a non-specific effect of the H3 receptor agonist THIOP should be administered prior to injection of imetit in additional experiments. Furthermore, the catecholaminergic receptors were stimulated by exogenous administration of catecholaminergic compounds and not by an increase in endogenous catecholamines. Thus, experiments are needed where endogenous adrenergic and noradrenergic neurons are activated during administration of H3 receptor agonists administered alone and in combination with H3 receptor antagonists.

In summary, H3 receptors play an important role in the histaminergic regulation of the secretion of ACTH and other POMC-derived peptides since their activation inhibits the hormone response to various physiological stimuli, which is in accordance with the finding that an inhibition of the histaminergic system by blockade of synthesis or postsynaptic receptors exert a similar effect as activation of H3 receptors.

2.2. Prolactin

HA is involved in the central regulation of PRL secretion and stimulates PRL release via activation of H1 and in particular H2 receptors [43-46]. The effect of HA is indirect, exerted in the hypothalamus by affecting factors, which directly regulate PRL secretion. Both serotonin and catecholamines as well as AVP seem to mediate the effect of HA [27,47-49, Willems (unpublished observations)] whereas the involvement of dopamine, which exerts a tonic inhibitory action on PRL secretion, is more controversial [50-53]. The hypothalamic sites where HA affect these factors are presently unknown, but the PVN may be involved, at least concerning AVP as a mediator. A number of studies have indicated that HA participate in the mediation of the effect of some physiological stimuli of PRL secretion. Thus, blockade of HA synthesis or postsynaptic H1 or H2 receptors reduce the PRL response to restraint and ether stress, suckling and the proestrus estrogen surge [32,43, 54-57].

The selective H3 receptor agonists RmHA has been investigated in relation to PRL secretion. Systemic administration of RmHA in doses of 10 mg/kg or lower did not affect basal PRL secretion [31], but reduced the PRL response to 5 minutes of restraint stress in male rats [31] and to suckling in female rats (Fig. 6) [32]. The inhibitory effect of RmHA on restraint stress-induced PRL release was antagonized completely by THIOP. The H3 receptor antagonist did not affect basal [31,58] or stress-induced PRL secretion [31]. However, THIOP enhanced the PRL response to morphine [58], while an expected inhibitory effect of RmHA on morphine-stimulated PRL secretion was not observed [58].

In recent experiments we investigated the effect of different H3 receptor agonists on basal or restraint stress-stimulated PRL secretion [Knigge et al. (unpublished observations)]. The H3 receptor compounds RmHA, BP 2-94 or imetit were administered alone or in combination with THIOP before exposure of male rats to restraint stress (Fig. 4). The study design was performed as described above. Five minutes of restraint stress caused a more than 10-fold increase in plasma PRL level. The effect of stress was inhibited by 50 to 80% by RmHA, BP 2-94 or imetit. Concomitant administration of THIOP prevented the inhibitory effect of the H3 receptor agonists.

The histaminergic neuronal system may interact with serotonergic and catecholaminergic systems in the regulation of PRL secretion and H3 receptors may be located presynaptically on serotonergic and catecholaminergic neurons (see above). However, we found that administration of RmHA or THIOP did not affect the PRL response to increased serotonergic neuronal activity induced by administration of the 5-HT precursor 5-hydroxytryptophan in combination with the re-uptake inhibitor fluoxetine [27]. In a recent study (Fig. 3) we found that the H3 receptor agonist imetit did not affect the PRL response to adrenaline but significantly decreased the response to noradrenaline [Willems et al. (unpublished observations)]. This effect of imetit does not suggest an inhibitory effect of the compound on H3 heteroreceptors located presynaptically on catecholaminergic neurons, since the neuronal catecholaminergic system was not activated by exogenous administration of the two catecholamines.

In all studies mentioned above the effect of the H3 receptor compounds were investigated following systemic administration. However, in a recent study RmHA administered centrally (1-5 μg) caused a dose-dependent and long lasting stimulation of PRL secretion [59]. The effect was prevented by THIOP but not by H1 or H2 receptor antagonists or by pretreatment with the HA synthesis inhibitor α-FMH. It was suggested that RmHA acted on postsynaptic H3 receptors, the existence of which have recently been suggested. However, it is difficult to understand how the H3 receptor agonist acts via presynaptic H3 receptors following systemic administration and via postsynaptic H3 receptors following central administration. Further studies are needed to explore these differences.

The available data indicate that systemic administration of H3 receptor agonists oppose the stimulatory effect of endogenous HA on PRL secretion and that this effect, which is prevented by concomitant blockade of the H3 receptors, occurs on presynaptic histaminergic neurons rather than on other aminergic neurons. At least, the effect is not exerted via activation of H3 heteroreceptors on serotonergic neurons.

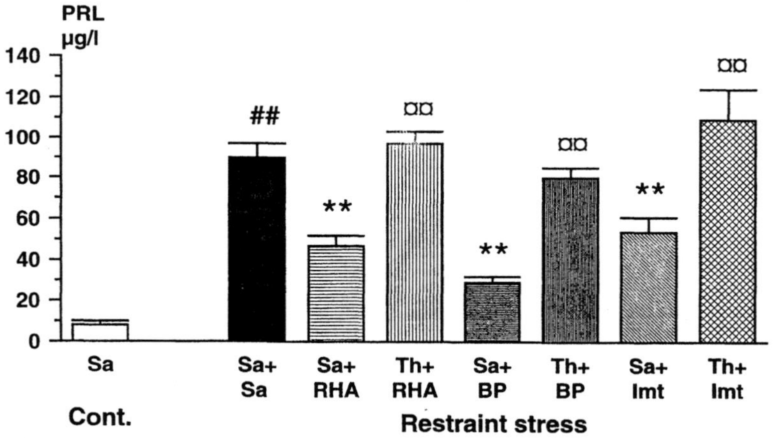

Figure 4. Effect of the H3 receptor agonists R(α)methylHA (RHA), BP 2-94 (BP) or imetit (Imt) administered in combination with saline (Sa) or the H3 receptor antagonist thioperamide (Th) on the PRL response to restraint stress. Further informations are given in legend to figure 1.

2.3. Growth hormone

Few studies have evaluated the effect of HA on GH secretion. Central administration of HA inhibits basal and stimulated GH release predominantly via activation of H1 receptors [58,60-61], whereas systemic injection of HA seem to stimulate GH secretion either via H1 or H2 receptors [62-63]. Following both administration routes, the effect of HA is indirect presumably mediated by activation of GHRH or other GH releasing factors but presumably not via somatostatin [63-64].

Only two studies have evaluated the involvement of H3 receptors in regulation of GH secretion. In accordance with an inhibitory effect of centrally administered HA on GH secretion, it was found that activation of presynaptic H3 receptors by systemic administration of RmHA slightly increased basal GH secretion and enhanced GH secretion stimulated by morphine or the α_2 receptor agonist clonidine [58,61]. The H3 receptor antagonist THIOP decreased the stimulatory effect of morphine without affecting basal GH levels [58]. Thus, these results indicate that endogenous HA exerts a tonic inhibitory action on basal and stimulated GH secretion, since presynaptic H3 receptor-induced inhibition of HA release by RmHA stimulates GH secretion and since presynaptic H3 receptor induced stimulation of HA release by THIOP inhibited enhanced GH secretion.

2.4. Thyrotropin

Studies have indicated that HA participates in the hypothalamic regulation of thyrotropin (TSH) secretion by an inhibitory action [for review see 3,14]. However, to our knowledge data have not been presented concerning the effect of H3 receptor manipulation of thyrotropin secretion.

2.5. Gonadotropins

HA has been found to stimulate the secretion of the gonadotropin LH, probably via activation of hypothalamic gonadotropin-releasing hormone (GnRH) neurons [for review see 3]. Whether H3 receptors play any role in HA-induced activation of the hypothalamic-pituitary-gonadal axis is obscure. However, it was recently reported that THIOP, in contrast to the inhibitory effect of H1 receptor antagonists, had no effect on the HA-induced release of GnRH from GT1-1 cells *in vitro* [65]. Further studies are required in order to investigate a role of H3 receptors in regulation of the hypothalamic-gonadal axis.

3. THE H3 RECEPTOR AND POSTERIOR PITUITARY HORMONE SECRETION

3.1. Vasopressin

Vasopressin is synthesized in magnocellular neurons in the SON as well as in magnocellular and parvocellular neurons in the PVN. In general, AVP originating in magnocellular neurons is axonally transported to the posterior pituitary lobe and released to the peripheral circulation while parvocellular AVP is released to the long pituitary portal vessels from nerve endings in the median eminence and is involved in regulation of anterior pituitary hormones. Vasopressin was one of the first pituitary hormones found to be affected by HA. Administered into the SON or into the brain ventricles HA has been found to induce antidiuresis and to increase plasma levels of AVP dose-dependently [48,66-68]. This indicates that HA stimulates magnocellular neurons in the SON and the PVN. This is further substantiated by the findings that the SON and PVN are densely innervated by histaminergic nerve fibers projecting from the posterior hypothalamic area

[4-5,69], that supraoptic neurons are excited by central administration of HA as well as by activation of histaminergic neurons in the tuberomammillary nuclei [70-71] and that centrally infused HA stimulates AVP neuronal activity and synthesis by increasing immunoreactivity of c-*fos* protein as well as c-*fos* and AVP mRNA in AVP-immunoreactive magnocellular neurons of the PVN and SON [20,21]. In addition to its effect on magnocellular AVP neurons, HA may also augment the activity in parvocellular AVP neurons [20,21]. The effect of HA on AVP release seem to be mediated via H1 and H2 receptors, since central administration of agonists to these postsynaptic receptors mimic the response of HA while concomitant administration of H1 and H2 receptor antagonists inhibit the response to HA [72].

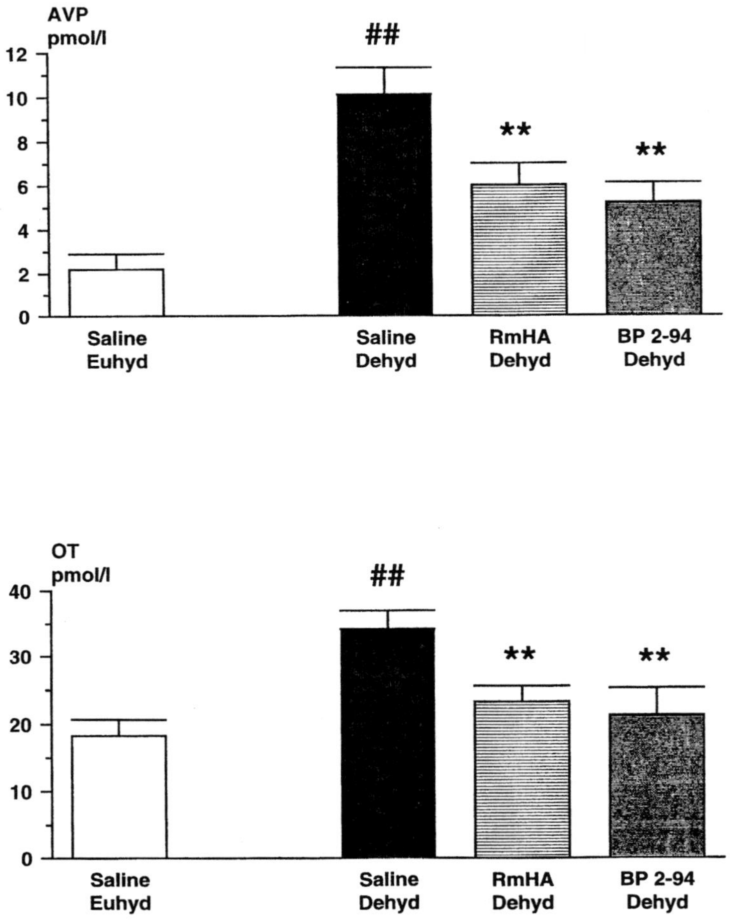

Figure 5. Effect of the H3 receptor agonists R(α)methylHA (RHA; 10 mg/kg) or BP 2-94 (BP; 25 mg/kg) administered ip at -24, -18, -6 and -2 hours on the AVP and OT responses to 24 hours of dehydration. The animals were decapitated at time 0. Isotonic saline served as control solution. ##: $p<0.01$ vs. euhydrated rats. **: $p<0.01$ vs. saline injected dehydrated rats.

Hyperosmolality induced by dehydration stimulates expression of AVP mRNA in the SON and PVN as well as AVP secretion from the posterior pituitary lobe [73]. These effects are inhibited by blockade of HA synthesis by α-FMH or by blockade of H1 or H2 receptor antagonists [72-73]. Furthermore, dehydration increases mRNA for the HA synthesizing enzyme histidine decarboxylase in the tuberomammillary nuclei [73] and augments neuronal HA turnover in the hypothalamus [72]. These findings strongly indicate that histaminergic neurons are involved in the mediation of the AVP response to dehydration-induced hyperosmolality.

In accordance with these results we found that the H3 receptor agonist RmHA inhibited the AVP response to dehydration-induced hyperosmolality (Fig. 5) similar to the effect observed by α-FMH [72]. In other studies we found that the prodrug BP 2-94 exerted an inhibitory action comparable to that observed for RmHA causing a more than 60% inhibition of the AVP response to 24 hours of dehydration (Fig. 5). A further involvement of H3 receptors in histaminergic activation of AVP neurons was recently indicated by the finding that the H3 receptor antagonist THIOP increased c-*fos* mRNA expression and *Fos*-like immunoreactivity in magnocellular neurons in the SON and PVN [74], indicating that an increased neuronal HA release induced by THIOP augment neuronal AVP activity. Whether this effect may result in an increased plasma level of AVP is presently unknown.

In summary, the present findings indicate that activation of presynaptic H3 receptors inhibits AVP secretion induced by hyperosmolality - an effect which is similar to and in accordance with that of blockade of HA synthesis or postsynaptic H1 and H2 receptors.

3.2. Oxytocin

Until recently an involvement of neuronal HA in the hypothalamic regulation of oxytocin (OT) secretion was unknown. However, within the recent years it has been reported that HA administered centrally stimulates c-*fos* expression and c-*fos* immunoreactivity in oxytocinergic neurons and OT mRNA in the SON and PVN [21]. In accordance with this, central infusion of HA stimulates OT secretion, an effect which occurs via activation of H1 and H2 receptors [28]. Although HA releases OT, this neuropeptide does not seem to be involved in the mediation of the HA-induced release of ACTH and PRL [28].

Histaminergic neurons may be involved in the mediation of the OT response to physiological stimuli. Thus, blockade of the histaminergic system by α-FMH or H1 or H2 receptor antagonists inhibited the dehydration-induced increase in OT mRNA in the SON and peripheral OT release in male rats [73, Kjær et al. (unpublished observations)] and the suckling-induced OT release in lactating female rats [32]. Furthermore, suckling increased histidine decarboxylase mRNA in the tuberomammillary nuclei [32].

H3 receptor agonists have been used to investigate the role of HA in regulation of OT secretion in relation to the stimuli of dehydration and suckling. Thus, the OT response to 24 hours of dehydration was almost prevented by pretreatment with the H3 receptor agonists RmHA and the pro-drug BP 2-94 administered ip (Fig. 5). Likewise the suckling-induced OT release was prevented by RmHA pretreatment (50 μmol/kg ip twice) (Fig. 6) [32]. These results indicate that H3 receptor activation, thereby reducing neuronal HA release, affect OT responses to physiological stimuli in male and female rats.

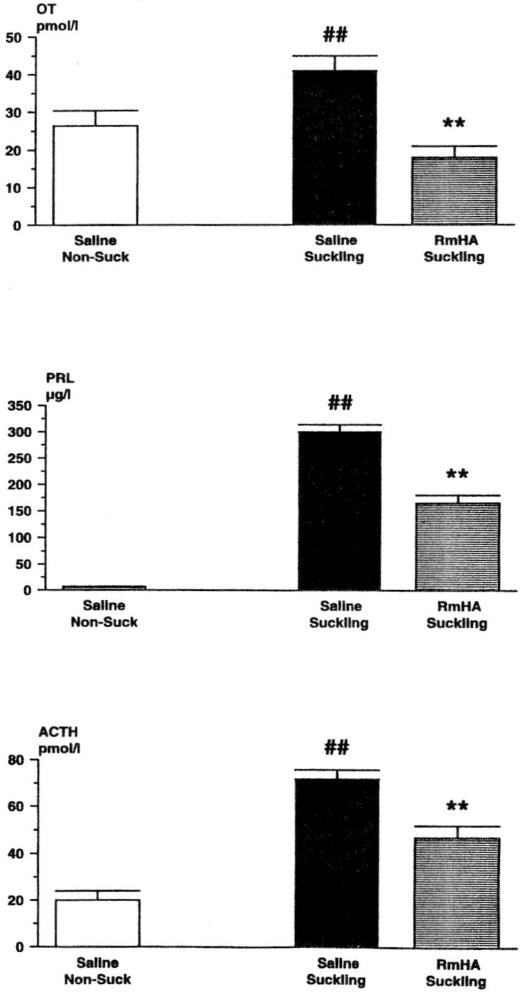

Figure 6. Effect of the H3 receptor agonist R(α)methylHA (RHA; 10 mg/kg ip at -180 and -60 minutes) on the OT, PRL and ACTH responses to 20 minutes of suckling in lactating rats. The animals were decapitated at 0 minutes. ##: $p<0.01$ vs. lactating rats not exposed to suckling; ** $p<0.01$ vs. suckling exposed lactating rats pretreated with saline.

4. CONCLUDING REMARKS

The discovery of the presynaptic H3 receptor and the development of specific H3 receptor agonists and antagonists have provided new opportunities to investigate the role of histaminergic neurons in neuroendocrine regulation. As the H3 receptor is located presynaptically on histaminergic neurons (in addition to its putative location on other neurons), the effect on hormone secretion of H3 receptor agonists by inhibition of HA release and synthesis may be useful when studying the involvement of histaminergic neurons in the mediation of the pituitary hormone responses to physiological stimuli. In this respect it should be considered that the effect of the H3 receptor agonists are comparable to that of the histidine decarboxylase inhibitor α-FMH. This similarity between different H3 receptor agonists and α-FMH has also been found in most studies. Furthermore, the specificity of the effect of the H3 receptor agonists have in most studies been indicated by the ability of the H3 receptor antagonist THIOP to prevent the effect of the various agonists. In different studies H3 receptors have been implicated in the participation of the histaminergic regulation of ACTH, β-END, PRL and growth hormone secretion from the anterior pituitary lobe and of AVP and OT secretion from the posterior pituitary lobe, since activation of the receptor has been found to inhibit the responses of these hormones to physiological stimuli involving HA (Table 2). The effects on hormone secretion caused by H3 receptor activation is in accordance with an inhibition of neuronal HA synthesis and release and suggest a role of H3 autoreceptors located on histaminergic neurons.

However, receptor autoradiography and *in vitro* studies have suggested that H3 receptors are located on other aminergic neurons in the brain. Since amines such as serotonin and catecholamines are involved in the regulation of pituitary hormone secretion, it is obvious that an action of the H3 receptor compounds may be exerted via these H3 heteroreceptors. Only few studies have evaluated this heteroreceptor action. It has been excluded that the effect of the H3 receptor agonists is due to an effect on H3 receptors located on serotonergic neurons, while an effect on catecholaminergic neurons has yet not been excluded.

Table 2
Effect of H3 receptor compounds on pituitary hormone secretion.

	ACTH	β-END	PRL	GH	TSH	LH	AVP	OT
H3-ago	inhib.	inhib.	inhib.	stim. stim.#	?	?	inhib.	inhib.
H3-anta	no	no pre*	no pre*	no/inhib.¤ pre*	?	?	?	?

#: H3 receptor agonist infused icv stimulated PRL secretion.
¤: The H3 receptor antagonist THIOP had no effect on basal but inhibited stimulated hormone secretion.
*: The H3 receptor antagonist THIOP prevented the inhibitory effect of H3 receptor agonist

The suggestion of the existence of postsynaptic H3 receptors have not been clarified in relation to pituitary hormone secretion, although the finding in a single study of a stimulatory effect of the H3 receptor agonist RmHA infused centrally on PRL may suggest such an effect - in as much as the effect was prevented by THIOP.

In summary, the H3 receptor agonists are useful tools to investigate pituitary hormone secretion when a reduction of the activity of histaminergic neurons is required. Conversely, H3 receptor antagonists may be useful to study hormone secretion during increased histaminergic activity. However, an increased activity induced by THIOP have not been shown to have any marked effect on either basal or stimulated hormone secretion.

REFERENCES

1. J.-C. Schwartz, J.M. Arrang, M. Garbarg, H. Pollard, M. Ruat, Physiol. Rev., 71 (1991) 1.
2. G.D. Prell, J.P. Green, Annu. Rev. Neurosci., 9 (1986) 209.
3. U. Knigge, J. Warberg, Acta Endocrinol. Copenh., 24 (1991) 609.
4. P. Panula, U. Pirvola, S. Auvinen, M.S. Airaksinen, Neuroscience, 28 (1989) 585.
5. T. Watanabe, Y. Taguchi, S. Shiosaka, J. Tanaka, H. Kubota, Y. Terano, M. Tohyama, H. Wada, Brain Res., 295 (1984) 13.
6 J.M. Arrang, M. Garbarg, J.-C. Schwartz, Nature, 302 (1983) 832
7. J.M. Arrang, M. Garbarg, J. C. Lancelot, J.M. Lecomte, H. Pollard, M. Robba, W. Schunack, J.-C. Schwartz, Nature, 327 (1987) 117.
8. M. Garbarg, M.D. Tuong, C. Gros, J.-C. Schwartz C. Eur. J. Pharmacol., 164 (1989) 1.
9. H. Pollard, J. Moreau, J.M. Arrang, J.-C. Schwartz, Neuroscience, 52 (1993) 169.
10. E. Schlicker, R. Betz, M. Gothert, Naunyn Schmiedebergs Arch. Pharmacol, 337 (1988) 588.
11. E. Schlicker, B. Malinowska, M. Kathmann, M. Gothert, Fundam. Clin. Pharmacol., 8 (1994) 128.
12. E. Schlicker, K. Fink, M. Hinterthaner, M. Gothert, Naunyn Schmiedebergs Arch. Pharmacol., 340 (1989) 633.
13. E. Schlicker, K. Fink, M. Detzner, M. Gothert, J. Neural. Transm. Gen. Sect., 93 (1993) 1.
14. J. Tuomisto, P. Mannisto, Pharmacol. Rev., 37 (1985) 249.
15. U. Knigge, F.W. Bach, S. Matzen, P. Bang, J. Warberg, Acta Endocrinol. Copenh., 119 (1988) 312.
16. U. Knigge, S. Matzen, T. Hannibal, H. Jørgensen, J. Warberg, Neuroendocrinology, 54 (1991) 646.
17. U. Knigge, A. Kjær, P.J. Larsen, H. Jørgensen, F.W. Bach, M. Møller, J. Warberg, Neuroendocrinology, 62 (1995) 319.
18. A. Kjær, U. Knigge, P.M. Plotsky, F.W. Bach, J. Warberg, Neuroendocrinology, 56 (1992) 851.
19. A. Kjær, U. Knigge, J. Warberg, Eur. J. Endocrinol., 131 (1994) 391.
20. A. Kjær, P.J. Larsen, U. Knigge, J. Warberg, Endocrinology, 135 (1994) 1171.
21. A. Kjær, P.J. Larsen, U. Knigge, M. Møller, J. Warberg, Endocrinology, 134 (1994) 482.
22. A. Kjær, U. Knigge, F.W. Bach, J. Warberg, Neuroendocrinology, 57 (1993) 1035.
23. A. Kjær, U. Knigge, H. Vilhardt, F.W. Bach, J. Warberg, Neuroendocrinology ,57 (1993) 503.
24. A Kjær, U. Knigge, F.W. Bach, J. Warberg, Neuroendocrinology, 56 (1992) 419.

25. A. Kjær, U. Knigge, F.W. Bach, J. Warberg, Neuroendocrinology, 58 (1993) 588.
26. M. Anthonisen, U. Knigge, A. Kjær, J. Warberg, Neuroendocrinology, 66 (1997) 68.
27. H. Jørgensen, U. Knigge, A. Kjær, J. Warberg, Neuroendocrinology, 64 (1996) 329.
28. A. Kjær, U. Knigge, J. Warberg, Neuroendocrinology, 61 (1995) 704.
29. U. Knigge, S. Matzen, F.W. Bach, P. Bang, J. Warberg, Acta Endocrinol. Copenh., 120 (1989) 533.
30. A. Kjær, U. Knigge, E.L. Madsen, P. Søe-Jensen, F.W. Bach, J. Warberg, Endocrinology, 132 (1993) 2213.
31. P. Søe-Jensen, U. Knigge, M. Garbarg, A. Kjær, A. Rouleau, F.W. Bach, J.-C. Schwartz, J. Warberg, Neuroendocrinology, 57 (1993) 532.
32. F.H. Schagen, U. Knigge, A. Kjær, P.J. Larsen, J. Warberg, Neuroendocrinology, 63 (1996) 550.
33. R.E. West Jr., J. Myers, A. Zweig, M.I. Siegel, R.W. Egan, M.A. Clark, Eur. J. Pharmacol., 267 (1994) 343.
34. M.A. Clark, A. Korte, J. Myers, R.W. Egan, Eur. J. Pharmacol., 210 (1992) 31.
35. Y. Morita, K. Koyama, Jpn. J. Pharmacol., 29 (1979) 59.
36. M.A. Reilly, E.B. Sigg, J. Pharmacol. Exp. Ther., 222 (1982) 583.
37. A. Korte, J. Myers, N.Y. Shih, R.W. Egan, M.A. Clark, Biochem. Biophys. Res. Commun., 168 (1990) 979.
38. K. Hashimoto, S. Yunoki, J. Takahara, T. Ofuji, Endocrinol. Jpn., 26 (1979) 103.
39. U. Knigge, A. Kjær, H. Jørgensen, M. Garbarg, C. Ross, A. Rouleau, J. Warberg, Neuroendocrinology, 60 (1994) 243.
40. A. Rouleau, M. Garbarg, X. Ligneau, C. Mantion. P. Lavie, C. Advenier, J.-M. Lecomte, M. Krause, H. Stark, W. Schunack, J.-C. Swartz, J. Pharmacol Exp. Ther., 281 (1997) 1085.
41. M. Garbarg, J.M. Arrang, A. Rouleau, X. Ligneau, M.D. Trung Tuong, J.-C. Schwartz, C.R. Ganellin, J. Pharmacol. Exp. Ther., 263 (1992) 304.
42. M. Kathmann, E. Schlicker, M. Detzner, H. Timmerman, Naunyn Schmiedebergs Arch. Pharmacol., 348 (1993) 498.
43. U. Knigge, S. Matzen, J. Warberg, Neuroendocrinology, 47 (1988) 68.
44. U. Knigge, S. Matzen, J. Warberg, Neuroendocrinology, 44 (1986) 41.
45. A.O. Donoso, M.B. Zarate, A. Seltzer, Neuroendocrinology, 36 (1983) 436.
46. C. Netti, V. Sibilia, F. Guidobono, I. Villa, P. Franco, A. Pecile, Acta Endocrinol. Copenh., 119 (1988) 488.
47. U. Knigge, I. Sleimann, S. Matzen, J. Warberg J. Neuroendocrinology, 48 (1988) 527.
48. A. Kjær, U. Knigge, L. Olsen, H. Vilhardt, J. Warberg, Endocrinology, 128 (1991) 103.
49. A. Kjær, U. Knigge, H. Vilhardt, J. Warberg, Neuroendocrinology, 57 (1993) 314.
50. D.M. Gibbs, P.M. Plotsky, W.J. de Greef, J.D. Neill, Life Sci., 24 (1979) 2063.
51. U. Knigge, S. Matzen, J. Warberg, Neuroendocrinology, 48 (1988) 167.
52. A.E. Fleckenstein, K.J. Lookingland, K.E. Moore, Life Sci., 51 (1992) 741.
53. A.M. Seltzer, A.O. Donoso, J. Neural. Transm., 65 (1986) 115.
54. E.O. Alvarez, J. Endocrinol., 95 (1982) 409.
55. E.O. Alvarez, Brain Res. Bull., 12 (1984) 11.
56. C. Netti, F. Guidobono, V. Sibilia, I. Villa, E. Cazzamalli, A. Pecile, Agents Actions, 20 (1987) 236.
57. A.M. Seltzer, A.O. Donoso, E. Podesta, Physiol. Behav., 36 (1986) 251.
58. C. Netti, F. Guidobono, V. Sibilia, F. Pagani, I. Villa, A. Pecile, Agents Actions, 33 (1991)

147.
59. C.E. Navarro, R. Otoya, A.O. Donoso, Neuroendocrinology, 57 (1993) 654.
60. C. Netti, F. Guidobono, V.R. Olgiati, V. Sibilia, A. Pecile, Horm. Res., 14 (1981) 180.
61. C. Netti, V. Sibilia, F. Guidobono, A. Pecile, Neuroendocrinology, 57 (1993) 1066.
62. F. Casanueva, R. Betti, D. Cocchi, T. Chieli, P. Mantegazza, E.E. Muller, Endocrinology, 108 (1981) 157.
63. J. Tanaka, A. Yamatodani, H. Niigawa, A. Miyake, K. Murakoshi, K. Tada, S. Hariguchi, H. Wada, T. Nishimura, R. Cacabelos, Endocrinol. Jpn., 34 (1987) 151.
64. C. Netti, F. Guidobono, V. Sibilia, V.R. Olgiati, F. Pagani, A. Pecile, Horm. Res., 19 (1984) 12.
65. G. Noris, D. Hol, C. Clapp, G. Martinez de la Escalera, Endocrinology, 136 (1995) 2967.
66. C.T. Bennett, A. Pert, Brain Res., 78 (1974) 151.
67. J. Dogterom, T.B. van Wimersma Greidanus, D. De Wied, Experientia, 32 (1976) 659.
68. L. Tuomisto, L. Eriksson, F. Fyhrquist, Agents Actions, 14 (1984) 558.
69. M.L. Weiss, Q.Z. Yang, G.I. Hatton, Neuroscience, 31 (1989) 299.
70. W.E. Armstrong, C.D. Sladek, Neuroscience, 16 (1985) 307.
71. G.I. Hatton, Q.Z. Yang, J. Neurosci., 16 (1996) 123.
72. A. Kjær, U. Knigge, A. Rouleau, M. Garbarg, J. Warberg, Endocrinology, 135 (1994) 675.
73. A. Kjær, P.J. Larsen, U. Knigge, J. Warberg, Endocrinology, 136 (1995) 2189.
74. M.L. Vizuete, V. Dimitriadou, E. Traiffort, N. Griffon, A. Heron, J.-C. Schwartz, Neuroreport, 6 (1995) 1041.

R. Leurs and H. Timmerman (Editors)
The Histamine H_3 Receptor

Functional role of histamine H_3 receptors in peripheral tissues

G. Bertaccini, G. Coruzzi and E. Poli

Institute of Pharmacology, University of Parma, School of Medicine, Via Gramsci 14, I-43100 Parma, Italy.

1. INTRODUCTION

The discovery of three different subtypes of histamine receptors has clarified many aspects of the physiologic role of histamine in different tissues. The first subtype (H_1) identified by Bovet and Staub (1937), has shown to be of fundamental importance in the phenomena related to allergy and inflammation; the second subtype (H_2) (Black et al., 1972) caused a real revolution in the understanding of the physiologic pathways of gastric secretion and, consequently, in the treatment of acid-related diseases. Finally, the third most recently discovered subtype (H_3) (Arrang et al. 1983), has received considerable attention because of its importance in the regulation of central functions. Subsequent studies, however, have shown that H_3 receptors are also extensively distributed in peripheral tissues, where they negatively modulate histamine synthesis and release as well as the release of different neurotransmitters (Leurs and Timmerman, 1992; Bertaccini and Coruzzi, 1995).

The availability of selective agonists and antagonists of this receptor subtype unravelled the role of histamine in various physiologic functions. Binding experiments with radiolabelled H_3 agonists ($[^3H](R)\alpha$-methylhistamine and $[^3H]N^{\alpha}$-methylhistamine) paved the way for the study of the distribution of H_3 receptors in different body areas. According to a study by Korte et al., (1990) on the guinea pig, who considered both central and peripheral distribution, the highest density of H_3 receptors was found in the brain; a considerable amount, however, was detected in the large intestine, in the ileum and in the pancreas, whereas very small amounts were measured in the other areas of the digestive, cardiovascular, genitourinary and respiratory systems, as well as in the skeletal muscle. However, these data were not entirely confirmed in rat tissues (Arrang et al., 1987). This finding leads us to conclude that there may be marked quantitative differences in the tissue distribution of H_3 receptors.

In this chapter we will review the location and functional role of H_3 receptors in the peripheral tissues, paying particular attention to a) the digestive, b) the cardiovascular, c) the respiratory, d) the genitourinary and e) the immune systems of various animal species including humans.

2. DIGESTIVE SYSTEM

In the digestive system of all the species we considered H_3 receptors were found to be relatively abundant considering the different peripheral tissues. These receptors were shown to have a certain, if not fundamental, role in the regulation of gastric acid secretion, gastroprotection, intestinal motility and absorption.

2.1. Gastric acid secretion

It is well known that gastric acid secretion is a very complex process, which is regulated by the interplay of nervous, hormonal and paracrine mediators. A fundamental role in this process, however, is played by histamine, which is produced by histaminocytes (ECL and/or mast cells according to the species) and directly stimulates parietal cells through the activation of membrane H_2 receptors. The recently discovered H_3 receptor (Arrang et al., 1983) led investigators to reconsider the still unsolved problems remaining from the H_2 antagonists era. During the last few decades many papers by various groups have appeared in the literature which dealt with the role played by H_3 receptors in acid secretion control (for review, see Bertaccini and Coruzzi, 1995). Most of these studies suggest an inhibitory role of H_3 receptors at this level. However, a few studies suggest the possibility that at least in some species, H_3 receptors mediate an increase in acid production, due to the inhibition of somatostatin release. Finally, according to other studies, this new receptor subtype is not involved in the regulation of acid secretion. At least some of these discrepancies are due to the use of different species and/or experimental conditions (in vivo *vs* in vitro models).

Table 1
Effect of (R)α-methylhistamine on gastric acid secretion (in vivo data)

Species	Technique	Effect	Reference
Conscious cat	Gastric fistula	↓ (2-DG and BBS) 0 (dimaprit and Pgas)	Coruzzi et al., 1991a
Conscious cat	Gastric fistula and Heidenhain pouch	↓ (food and Pgas) 0 (basal and histamine)	Badò et al., 1991
Conscious dog	Gastric fistula	↓ (2-DG, Pgas and BBS 0 (basal and histamine)	Soldani et al., 1993;1994
Conscious rat	Pylorus-ligated	0 (i.v.) ↓ (i.c.v.)	Coruzzi et al., 1992 Barocelli et al., 1995
Anaesthetized rat	Lumen-perfused stomach	0 (basal, histamine) and 2-DG)	Coruzzi et al., 1992

2-DG = 2-deoxy-D-glucose; Pgas = Pentagastrin; BBS = bombesin, ↓ = decrease; 0 = no effect.

A synopsis of the most important data concerning H_3 receptors and gastric acid secretion is reported in Tables 1 and 2. Table 3 reports the effects of H_3 ligands on hormonal or paracrine secretions which are directly or indirectly connected with acid secretion.

2.1.1 In vivo experiments

Early in vivo experiments, that were carried out in conscious cats with gastric fistula and/or Heidenhain pouch, showed that H_3 receptor activation could inhibit acid secretion induced by indirect stimuli (2-deoxy-D-glucose, pentagastrin, bombesin and food); however, this activation could not modify basal or directly (histamine and dimaprit) stimulated secretion (Badò et al., 1991; Coruzzi et al., 1991a). This inhibition of acid secretion was not paralleled by concomitant reduction in serum gastrin levels (Badò et al., 1990; Coruzzi, unpublished data). Moreover, the same results were obtained in the conscious dog fitted with gastric fistula (Soldani et al., 1993). In these experiments, the selective H_3 agonist (R)α-methylhistamine caused a maximum inhibition of 50-60% which was counteracted by thioperamide, thus indicating the involvement of specific H_3 receptors. No effect on serum gastrin level was found either in basal conditions or following stimulation with bombesin or 2-deoxy-D-glucose in cats (Figure 1) and in dogs (Soldani et al., 1994). In a more recent study carried out in

Table 2
Effect of H_3 receptor activation on gastric acid secretion (in vitro data)

Species	Technique	Agonist	Effect	Reference
Mouse	Whole stomach	(R)α-MHA	↑	Vuyyuru and Schubert, 1993
Mouse	Gastric glands	(R)α-MHA	0	Muller et al., 1993
Rat	Gastric fundus	(R)α-MHA	0	Coruzzi et al., 1992
Rat	Vascularly-perfused stomach	(R)α-MHA	0	Sandvik et al., 1989
Guinea pig	Gastric fundus	(R)α-MHA	0	Coruzzi, unpublished
Rabbit	Fundic glands	(R)α-MHA	↓*	Badò et al., 1992a
Rabbit	Parietal cells	N^{α}-MHA	0	Beales and Calam, 1997

(R)α-MHA = (R)α-methylhistamine; N^{α}-MHA = N^{α}-methylhistamine; * = thioperamide increased basal acid secretion; ↑ = increase; 0 = no effect; ↓ = decrease.

anaesthetised dogs (Soldani et al., 1996), (R)α-methylhistamine was found to reduce the increase of histamine release induced by pentagastrin. This effect was reversed by thioperamide, which *per* se was able to dose-dependently enhance the increase of histamine induced by very low doses of pentagastrin, thus demonstrating a definite role for H_3 receptors in the regulation of histamine release from ECL cells. In the rat, though peripheral (i.v.) administration of (R)α-methylhistamine was ineffective, -the increase in acid secretion observed at high doses is due to H_2 receptor activation (Coruzzi et al., 1992)- central (i.c.v.) administration of this agonist in pylorus-ligated rats caused a 40-50 % reduction of total acid output (Barocelli et al., 1995).

2.1.2 In vitro experiments

Results obtained by in vitro experiments are not as homogeneous as in vivo experiments. According to most of these studies, in various isolated preparations of rats and guinea pigs, H_3 receptor activation had no effect on acid secretion (Table 2). However, it is noteworthy to

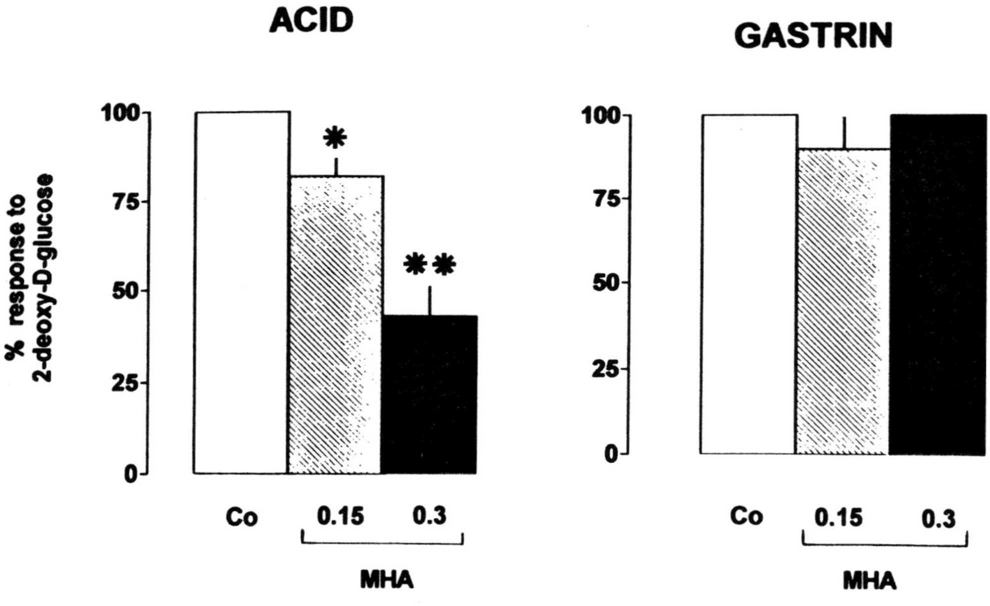

Figure 1. Conscious cat with gastric fistula. Effect of (R)α-methylhistamine (MHA, μmol/kg i.v.) on acid secretion and plasma gastrin levels stimulated by 2-deoxy-D-glucose (100 mg/kg i.v.) (Co). On the ordinate: percent responses in comparison with maximum responses considered arbitrarily as 100. Mean values ± SEM from 5 cats. * $P < 0.05$ and ** $P < 0.01$ *vs* control values.

mention that in the rat, H_3 agonists caused a remarkable inhibitory effect on both histamine and somatostatin release (Table 3). It can thus be hypothesised that at least in this species, these inhibitory effects, which have opposite consequences on acid secretion, are responsible for the lack of activity observed in gastric secretory studies.

In the rabbit fundic glands, (R)α-methylhistamine was reported to inhibit acid secretion induced by histamine and carbachol, whereas thioperamide had the opposite effect. This finding led Badò et al., (1992a; 1995) to hypothesise the presence of inhibitory H_3 receptors on the parietal cells. Moreover, recent studies (Badò et al., 1995) demonstrated that H_3 agonists also reduced acid secretion stimulated by carbachol. However, since these agonists did not affect binding of [^{3}H]-N-methylscopolamine to parietal cells, an inhibitory H_3 receptor that counteracts M_3 receptor stimulation through post-receptor mechanism, has been hypothesised. Contrastingly, a subsequent study in isolated cultured rabbit parietal cells did not support this hypothesis, since N^{α}-methylhistamine, when used in the presence of H_2 antagonists, did not inhibit basal or carbachol or forskolin-stimulated aminopyrine (AP) accumulation (Beales and Calam 1997). Additionally, in rabbit fundic glands (Badò et al., 1992a) (R)α-methylhistamine decreased both basal and thioperamide-induced histamine release. This finding suggests that histamine H_3 receptors occur as autoreceptors on the histamine-containing cells. Moreover, histamine synthesis also proved to be modulated by histamine H_3 receptors (Hollande et al., 1993).

In the mouse, whereas no evidence of H_3 receptors was found in isolated gastric glands (Muller et al., 1993), in the whole stomach, (R)α-methylhistamine actually increased, and thioperamide decreased acid secretion, thus indicating a definite stimulatory role for H_3 receptors in this species (Table 2). Apparently, this excitatory effect, which contrasts with the observations obtained in other models, was due to an inhibitory effect on somatostatin release from fundic D cells (Schubert et al., 1993; Vuyyuru and Schubert 1993). Also, an inhibitory effect on somatostatin secretion mediated by H_3 agonists was observed in other species (rat and dog). However, contrarily to what might have been expected, in these species, the inhibitory effect on somatostatin is not followed by an increase in acid secretion, but it is instead followed by a decrease, owing to the predominant H_3-mediated inhibition on the release of excitatory mediators (histamine, acetylcholine) from other sites (ECL, cholinergic nerve terminals).

In the human gastric tumoral cell line HGT1, (R)α-methylhistamine has shown to inhibit in a nanomolar range basal and carbachol-stimulated inositol phosphate formation (Cherifi et al., 1992). This suggests the presence of H_3 receptors on ECL and/or on parietal cells, that are coupled with phosphatidyl inositol turnover.

In conclusion, histamine H_3 receptors have multiple locations in the gastric mucosa, occurring in different cell types, according to the different animal species; however, the final effect on acid secretion seems to be an inhibitory one, although in a particular experimental model of the mouse, a stimulatory effect seems to be predominant. A scheme representing the multiple locations of H_3 receptors in the gastric mucosa from available data is reported in Figure 2.

2.1.3. H_3 receptors and *Helicobacter pylori*

Recently, a possible role of H_3 receptors in the changes of gastric acid secretion produced by *Helicobacter pylori* (*H.p.*) was advanced (Courillon-Mallet et al., 1995). In fact *H.p.*, besides producing histamine (Velasquez et al., 1996), was found to possess the enzyme N-methyltransferase, which transforms histamine into its methylated derivative N^{α}-

methylhistamine. This compound, described as as a naturally occurring imidazole derivative by Bertaccini and Vitali as early as 1964, was found to be slightly weaker (about 80-90%) than histamine at H_1 receptors. It was also described as more active than histamine in stimulating dog gastric secretion (Bertaccini et al., 1971) and nearly as active as the parent substance in anaesthetised rats with lumen-perfused stomach (Bertaccini et al., 1973), thus indicative of good activity at histamine H_2 receptors. Moreover, N^{α}-methylhistamine was reported to be a good -although non-selective- agonist at H_3 receptors (Trzeciakowski, 1987; Coruzzi et al., 1991b). This activity was considered to be responsible for the decrease in somatostatin

Table 3
Effect of H_3 receptor agonists on hormonal and paracrine mediators of acid secretion process

Species	Agonist	Effect	Reference
Mouse in vitro	(R)α-MHA	↓ somatostatin	Schubert et al., 1993
Rat in vitro	(R)α-MHA	↓ histamine	Prinz et al., 1993 Sandvik et al., 1989 Vuyyuru et al., 1997
Rat in vitro	Imetit	↓ histamine	Modlin et al., 1995 Kidd et al., 1996
Rat in vitro	(R)α-MHA	↓ somatostatin	Badò et al., 1992b; 1994 Vuyyuru et al., 1995; 1997
Rabbit in vitro	(R)α-MHA	↓ histamine	Badò et al., 1992a
Cat in vivo	(R)α-MHA	0 gastrin	Badò et al., 1990 Coruzzi, unpublished
Dog in vivo	(R)α-MHA	0 gastrin ↓ histamine	Soldani et al., 1993; 1994 Soldani et al., 1996
Dog in vitro	(R)α-MHA	↓ somatostatin	Vuyyuru et al., 1995
Human in vitro	N^{α}-MHA*	↓ somatostatin	Courillon-Mallet et al., 1995
	(R)α-MHA	↓ somatostatin	Vuyyuru et al., 1995

(R)α-MHA = (R)α-methylhistamine; N^{α}-MHA = N^{α}-methylhistamine; ↓ = decrease; ↑ = increase; 0 = no effect; * = derived from *Helicobacter pylori*

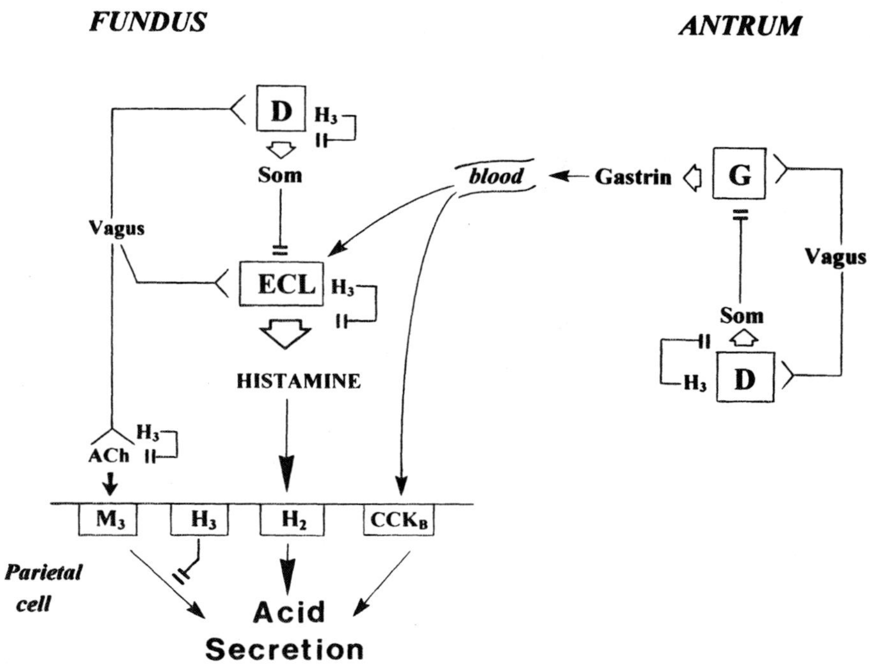

Figure 2. Scheme illustrating the neural, hormonal and paracrine pathways regulating acid secretion from the parietal cell. The model includes all the possible localizations of H_3 receptors, based on the available data from the literature concerning different experimental preparations. The inhibitory function (—||) of H_3 receptors gives rise to different final effects on acid secretion according to the cell involved; this may depend on the animal species, as described in the text.

production by D cells, with consequent hypergastrinemia and hyperchlorhydria in patients infected with *H.p.* (Courillon-Mallet et al., 1995; McGowan et al., 1996). Of course the acid hypersecretion by N^{α}-methylhistamine could also be the consequence of the above mentioned H_2 receptor activity. However, we must consider that activation of H_3 receptors by this amine can cause a decrease in histidine decarboxylase (HDC) activity (the histamine forming enzyme), thus reducing the amount of histamine available and consequently acid secretion. Indeed, a reduced HDC activity was observed in both the fundus and antrum of *H.p.* infected patients and it returned to normal levels after eradication therapy (Courillon-Mallet et al.,

1995). A scheme illustrating the possible mechanisms by which N^{α}-methylhistamine produced by *H.p.* may influence parietal cell function is shown in Figure 3. This scheme clearly shows that N^{α}-methylhistamine produced by *H.p.* could change the regulation of acid secretion in two opposite ways: its effect on histamine formation in the fundus could reduce acid secretion, whereas the effect on somatostatin in the antrum could induce hypergastrinemia and thus increase acid secretion. One could speculate that the final effect of this bacterium on acid production would depend on its distribution in the stomach.

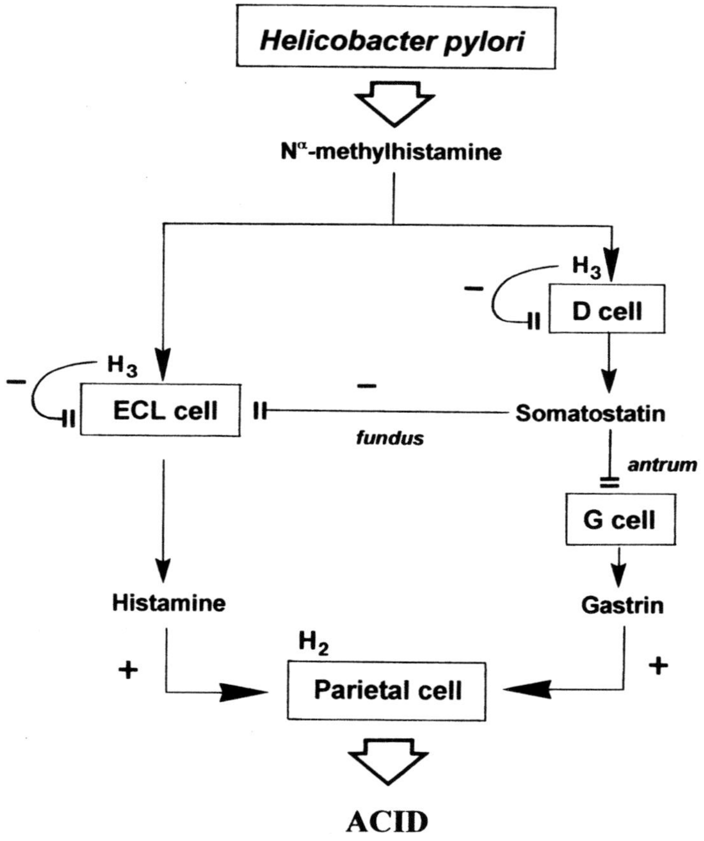

Figure 3. Effect of N^{α}-methylhistamine produced by *Helicobacter pylori* on acid secretion. On the one hand, this compound may reduce acid secretion by inhibiting, *via* H_3 receptor activation, histamine synthesis and release from ECL cells; on the other hand, H_3 receptor activation on D cells, with consequent inhibition of somatostatin release, may increase acid secretion. Additionally, direct activation of H_2 receptors on parietal cell by N^{α}-metylhistamine must also be considered (this mechanism is not shown in the scheme).

2.2. Gastroprotection

Despite a number of reports on this topic, the role of histamine in the formation and restitution of gastric mucosal lesions, is still unclear. In this regard, histamine was considered both as an ulcerogenic and a gastroprotective agent (Andersson et al., 1990; Aures et al., 1982; Galli et al., 1987; Arakawa et al., 1988; Takeuchi et al., 1987; Nishiwaki et al., 1989). Also the role of histamine H_3 receptors is not yet completely understood, even though many reports seem to indicate an involvement of this histamine receptor subtype in the protection against gastric damage induced by different noxious stimuli (Palitzsch et al., 1989; Bertaccini et al., 1995; Morini et al., 1995).

Table 4
Effect of (R)α-methylhistamine (MHA, 100 mg/kg intragastrically) on adherent and intracellular mucus of rat gastric mucosa after treatment with saline or absolute ethanol (0.1 ml)

	Adherent mucus		Intracellular mucus
	μg Alcian Blue/ g wet tissue	Thickness of mucus layer (μm)	Alcian Blue/PAS stained area (μm²)
Saline	171.4	92.3	3438.5
Ethanol	194.5	89.3	2769.5
MHA/ethanol	148.8	93.1	10905.5*
MHA/ethanol	272.3*	200.0**	33612.3**

* $P<0.05$ and ** $P<0.01$ *vs* saline; PAS = Periodic acid Schiff.

In the rat, (R)α-methylhistamine dose-dependently inhibited gastric lesions induced by ethanol administration, an effect which was reversed by thioperamide, although to a lesser extent. Damage induced by aspirin, indomethacin or stress, was also inhibited. Histological studies showed that (R)α-methylhistamine facilitates a process of re-epithelization (Figure 4); these studies also indicated a remarkable parallel increase in mucus granules (Table 4), with a complete disappearance of necrotic haemorrhagic lesions (Morini et al., 1995; Morini et al., 1996a;1996b). Scanning electron micrographs of gastric mucosa showed that ethanol induced a complete surface desquamation with consequent exposure of the lamina propria. In addition, pre-treatment with (R)α-methylhistamine caused a marked migration of pit cells and a restitution of the mucosal surface covered with a continuous layer of the surface epithelial cells (Figure 5). Apart from mucus production, the increase in HCO_3^- secretion (Coruzzi et al., 1996) may also be responsible for the gastroprotection induced by (R)α-methylhistamine. Contrastingly, an antisecretory effect of (R)α-methylhistamine can be excluded, since this compound has demonstrated to be ineffective in the rat after peripheral administration (Coruzzi et al., 1992). Interestingly, the same results were obtained by stable azomethine prodrugs of (R)α-methylhistamine (Morini et al., 1997).

The precise role that H_3 receptors play in the gastroprotection against noxious stimuli is still unknown. On the one hand, the remarkable effect of (R)α-methylhistamine -which is one of the most employed H_3 agonists- and the reversal effect of thioperamide and clobenpropit,

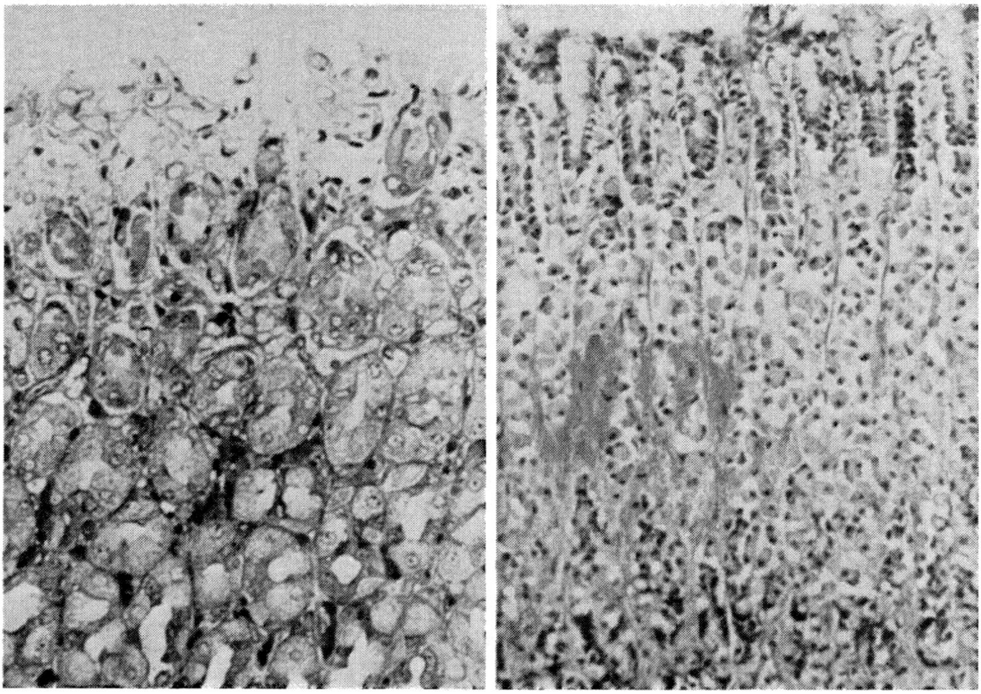

Figure 4. Light micrographs of the rat gastric mucosa 1 h after the administration of ethanol (*on the left*): necrotic lesions and destroyed luminal and gastric pit cells are evident. *On the right*: pretreatment with (R)α-methylhistamine 100 mg/kg intragastrically caused a marked reepithelization with surface and gastric pit cells almost normal (magnification x 420). (from Morini et al., 1995).

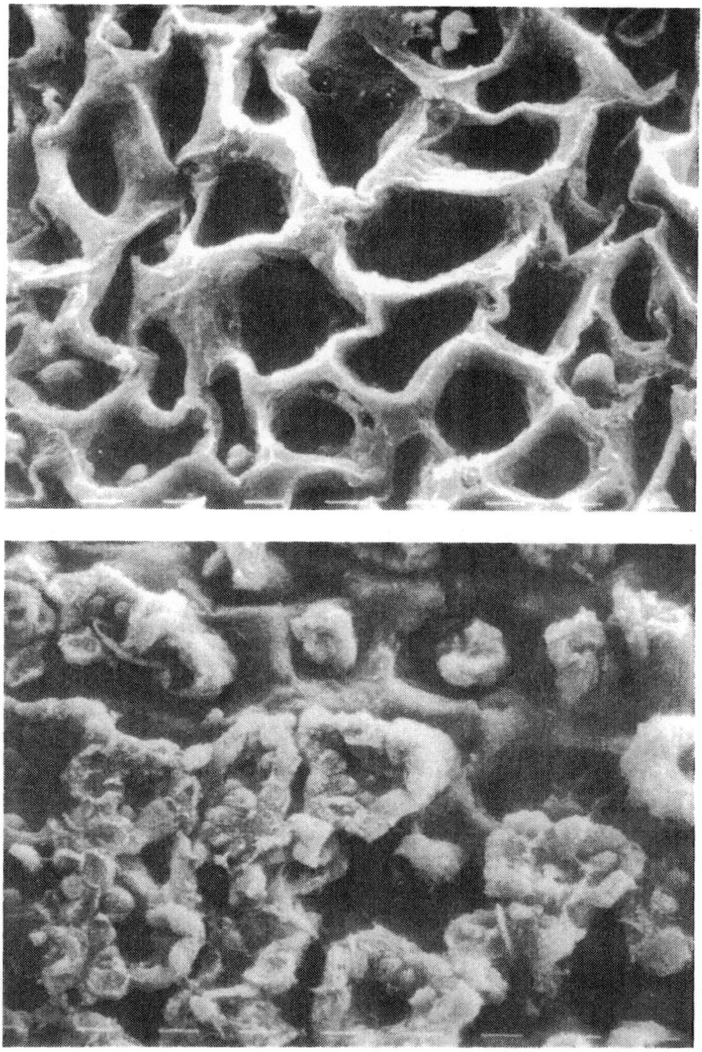

Figure 5. Scanning electron micrographs of rat gastric mucosa 1 h after administration of ethanol. *On the left*: loss of the interpit cells with complete surface desquamation and exposure of the lamina propria; *on the right*: pretreatment with (R)α-methylhistamine 100 mg/kg intragastrically caused the migration of pit cells with restitution of the mucosal surface (magnification x 640).

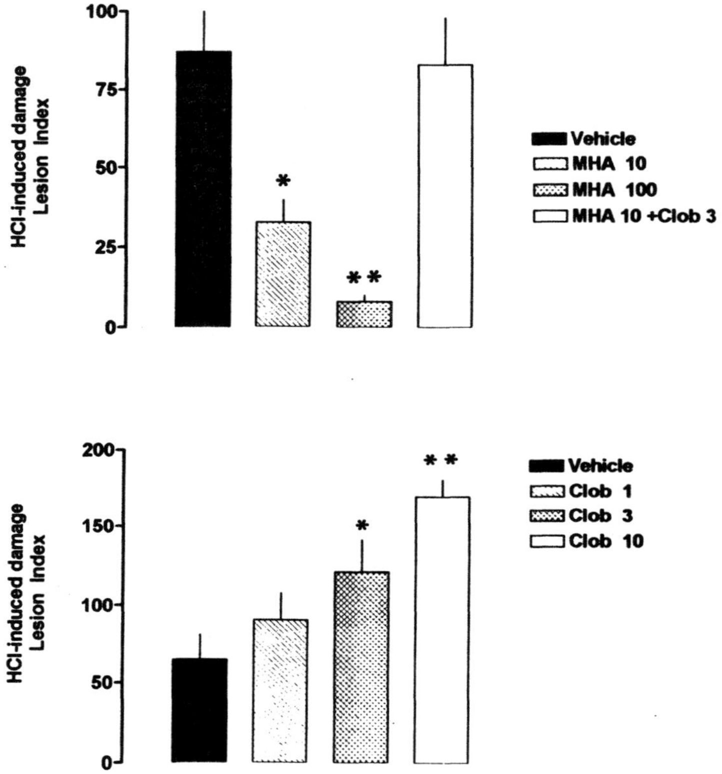

Figure 6. Effect of (R)α-methylhistamine (MHA, mg/kg intragastrically) and clobenpropit (Clob, mg/kg intragastrically) on the gastric damage induced in the conscious rat by HCl 0.6 N (■). On the ordinata lesion index. Values are mean ± SEM from 8-10 experiments. * $P<0.05$ and ** $P<0.01$ *vs* vehicle.

seem to indicate that histamine H_3 receptors are involved. Moreover, when the H_3 antagonists are administered alone, they markedly increase gastric damage induced by various stimuli (Figure 6). Furthermore, the S isomer of (R)α-methylhistamine, a compound which is about 100 fold weaker at histamine H_3 receptors (Arrang et al., 1987), has a very weak protective effect (Morini et al., unpublished data).

On the other hand, other data raise doubts about the involvement of H_3 receptors in gastroprotection: 1) imetit and immepip, the most recently described H_3 agonists are not effective; 2) the H_3 antagonists are unable to overcome the effect of very high doses of (R)α-methylhistamine and 3) contrary to the effect of (R)α-methylhistamine which ranged from 20 to 60% in all the experimental models, in the gastroprotection experiments, high doses of the agonist caused as much as 100% protection against noxious stimuli. To date, we cannot exclude that the protective effect of (R)α-methylhistamine (at least at the highest doses) is related to secondary actions and/or to interactions with other receptors, which can further complicate the equilibrium of different endogenous transmitters. We must not forget, however, that gastroprotection is a multifactorial phenomenon where histamine receptor subtypes are no more than minute tesserae of a complex mosaic whose details have not yet been completely understood.

2.3. Gastrointestinal motility

The effects of histamine on gastrointestinal motility are very complex and consist of neurally-mediated as well as direct effects (for review see Bertaccini and Coruzzi, 1992). Different receptor subtypes are involved: H_1 receptors mediate a direct excitatory response to histamine; H_2 receptors mediate both muscle relaxation and neurally-induced contractions; furthermore, the existence of so-called "anomalous" H_2 receptors sensitive to the agonists but not to the antagonists has been reported (Bertaccini and Zappia, 1981). The last discovered (Arrang et al., 1983) H_3 receptor subtype was found to be widely distributed along the gastrointestinal tract, where it negatively modulates the release of acetylcholine and other neurotransmitters from the myenteric plexus (Bertaccini et al., 1991). As a consequence, H_3 receptor activation caused an attenuation of neurogenic contractions of the smooth muscle (Figure 7). So far, conversely from what was reported for other tissues (vascular and bronchial), there is no evidence for the presence of H_3 receptors in the smooth muscle cells of the gastrointestinal tract.

The presence of H_3 receptors in the small and large intestine of the guinea pig reported in several papers (see Table 5) was assessed both in functional studies with electrically-stimulated preparations or isolated myenteric neurones and in tissues incubated with [^{3}H] choline in which the release of radiolabelled acetylcholine was measured (Poli et al., 1991). A summary of the inhibitory effects of two H_3 agonists, namely N^α-methylhistamine and the more selective (R)α-methylhistamine, on the release of acetylcholine or NANC neurotransmitter from different areas of the guinea pig intestine is reported in Table 5. This table clearly shows that the two compounds have approximately the same potency in the different experimental models. However, it was also shown in the gastrointestinal tract that (R)α-methylhistamine has a higher degree of selectivity towards H_3 receptors, and whereas the potency ratio H_3/H_1 for N^α-methylhistamine is about 10 fold, the potency ratio for (R)α-methylhistamine is about 1000 fold (Coruzzi et al., 1991b).

In spite of the well documented inhibitory effects mediated by H_3 receptors in electrically-stimulated preparations of the guinea pig intestine, the activation of this receptor subtype does not influence the reflex-evoked peristaltic motility of the guinea pig ileum (Poli et al., 1997; Poli and Pozzoli 1997). Since this experimental model reproduces peristalsis in quasi-physiologic conditions (Holzer, 1989), H_3 receptors apparently play a minor role when compared to that of the other prejunctional receptors, such as α_2-adrenoceptors and adenosine A_1-receptors, in the control of the physiologic motility of the gut (Figure 8).

Conversely from what was found in the guinea pig, H_3 receptors were not found in the rabbit colon (Pozzoli et al., 1997) either in vitro or in vivo experimental models. In this regard, (R)α-methylhistamine, immepip, thioperamide and clobenpropit were unable to modify spontaneous motility and neurogenic contractions elicited by field stimulation. In these aspects, the rabbit resembles the rat- another rodent for which H_3 receptors were not detected along the whole gastrointestinal tract. Preliminary experiments, carried out on human colonic specimens

Table 5
Inhibitory effect (pD_2 values) of H_3 receptor agonists on peripheral nervous tissues of the guinea pig intestine

Tissue	Neurotransmitter	N^{α}-MHA	(R)α-MHA	Reference
Duodenum	ACh	7.17	7.76	Coruzzi et al., 1991b
	NANC	NT	8.09	Leurs et al., 1991
Jejunum	NANC	NT	8.18	Leurs et al., 1991
Ileum	ACh	8.77	NT	Trzeciakowski, 1987
	ACh	NT	7.80	Hew et al., 1990
	ACh	NT	7.76	Schlicker et al., 1994b
	ACh	NT	7.10	Rizzo et al., 1995
	NANC	7.70	7.60	Taylor and Kilpatrick, 1992
	NANC	NT	8.10	Leurs et al., 1991
	NANC	8.40	8.30	Menkveld and Timmerman, 1990
Colonic submucous neurons	ACh	10 μM	NT	Cooke and Wang, 1994
Colon	NANC	NT	8.27	Leurs et al., 1991
Myenteric ganglia	ACh	8.06	NT	Tamura et al., 1988
Longitudinal muscle-myenteric plexus	ACh	NT	7.68	Poli et al., 1991
	ACh	NT	(10^{-9}-10^{-6} M)	Yau and Youther, 1993

N^{α}-MHA = N^{α}-methylhistamine; (R)α-MHA = (R)α-methylhistamine; ACh = acetylcholine; NANC = nonadrenergic, noncholinergic; NT = not tested.

removed during surgery, showed that (R)α-methylhistamine slightly reduced electrically-evoked contractions at 0.1-10 μM concentrations (Bertaccini et al., unpublished data).

Experiments carried out in the conscious rat demonstrated that H_3 receptors contribute only to the central, and not the peripheral, regulation of intestinal motility (Fargeas et al., 1989). In addition, the concomitant ineffectiveness on acid secretion of peripherally administered H_3 ligands seems to exclude a significant role of H_3 receptors in the peripheral regulation of gastrointestinal functions. Lastly, in the mouse H_3 receptors do not seem to have any appreaciable influence on gastrointestinal motility (Oishi et al., 1993)

Histamine H_3 receptors were also found to occur in porcine small intestine where they inhibit the release of 5-hydroxytryptamine from enterochromaffin cells (Schwörer et al., 1992).

In all the experimental models considered, the inhibitory effects of H_3 agonists on intestinal motility were antagonised by a series of selective H_3 receptor antagonists (Table 6).

With regard to the signal transducing mechanisms coupled to histamine H_3 receptors, we performed an extensive investigation in our lab on the guinea pig duodenum. Poli et al., (1993)

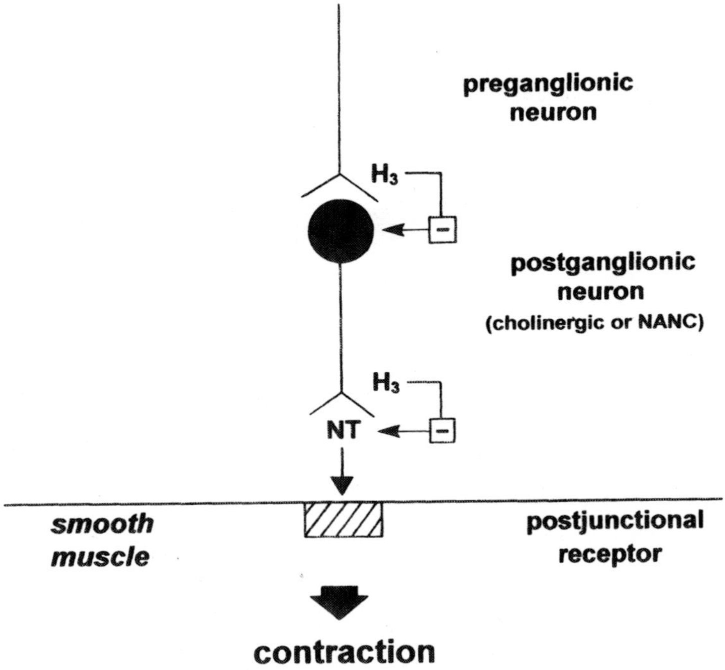

Figure 7. Scheme illustrating the location and functional role of the histamine H_3 receptor in the intestine; NT = neurotransmitter.

observed that H_3 receptor-mediated effects in the guinea pig duodenum are not related to changes in intracellular levels of cyclic AMP or GMP. Contrastingly, these effects were enhanced in low- and markedly reduced in high Ca^{++} medium or when intracellular Ca^{++} concentration was increased by compound Bay K 8644 (Poli et al., 1994a). Thus, in the guinea pig duodenum H_3-mediated effects seem to be associated with a restriction of Ca^{++} ion access into the axon terminal, with consequent decreased availability of Ca^{++} for the stimulation-secretion coupling. This same mechanism has been reported to explain H_3

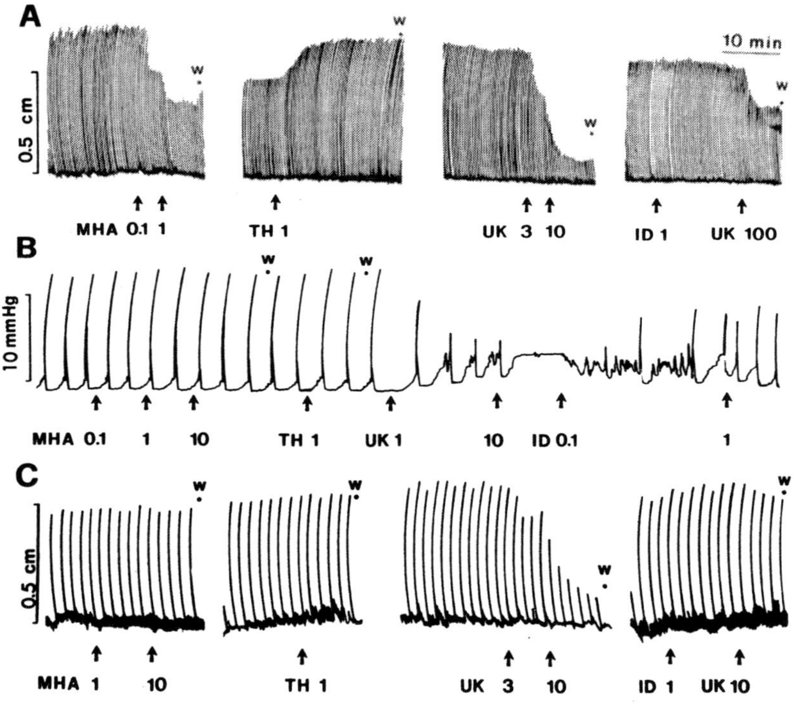

Figure 8. Original tracings showing the effects of (R)α-methylhistamine (MHA, μM), thioperamide (TH, μM), the selective α_2 adrenoceptor agonist UK-14304 (UK, nM) and the α_2-adrenoceptor antagonist idazoxan (ID, μM) on different in vitro assays from the guinea pig ileum. A) Electrically-evoked longitudinal contractions of the whole ileum; B) Peristaltic waves of the perfused ileum; C) Reflex-evoked circular muscle contractions. Vertical calibrations represent (A and C) centimeters of isotonic contractions or (B) changes in perfusion pressure. Horizontal calibration is the chart speed. w = washing of the preparation.

receptor-mediated inhibition of noradrenaline release in the guinea pig heart (Endou et al., 1994) and in the mouse brain cortex (Schlicker et al., 1994a). In this context, we suggested the involvement of neuronal N-type Ca^{++} channels since nerve-mediated contraction in the guinea pig duodenum are blocked by ω-conotoxin (Poli et al., 1994a). In preliminary experiments on the guinea pig duodenum, the effect of (R)α-methylhistamine was not modified in low K^+ medium, thus indicating that K^+ channels are unlikely to be involved in the H_3 mediated effects on cholinergic neurotransmission (Poli et al., 1994a). The histamine H_3 receptor at peripheral cholinergic nerve endings in the gut seems to be coupled to G proteins, since pertussis toxin attenuated the effect of H_3 receptor agonists (Poli et al., 1993).

In conclusion, available data indicate that H_3 receptors can mediate inhibitory effects on intestinal motility. However, it is still unclear whether these effects have a physiologic significance. On the one hand, the lack of involvement of histamine H_3 receptors in the control

Table 6
Activities of some histamine H_3 receptor antagonists on various in vitro functional models of the guinea pig intestine

	Duodenum	Jejunum	Ileum	Colon
Burimamide	NT	NT	7.0[1] 7.3[2]	NT
Impromidine	NT	NT	7.0[1] 7.1[2] 7.2[3] 7.6[4] 7.2[5]	NT
Thioperamide	7.99[6] 8.09[7]	8.13[7]	8.2[1] 8.7[2] 8.5[3] 8.15[7] 8.79[8] 8.96[5]	8.36[7]
Clobenpropit	*	9.9[9]	9.95[10]	NT
Impentamine	8.53[11]	8.4[12]	NT	NT

1= Taylor and Kilpatrick, 1992; 2 = Rizzo et al., 1995; 3 = Menkveld and Timmerman, 1990; 4 = Trzeciakowski, 1987; 5 = Hew et al., 1990; 6 = Coruzzi et al., 1991b; 7 = Leurs et al., 1991; 8 = Schlicker et al.,1994b; 9 = Leurs et al., 1995a; 10 = Barnes et al., 1993; 11 = Bertaccini, unpublished data; 12 = Vollinga et al., 1995. NT = not tested; * unsurmountable antagonism, starting at 0.1 nM (Poli, unpublished data).

of intestinal peristalsis tend to minimize their role in the regulation of intestinal motility and would therefore suggest that this receptor subtype simply represents one of the numerous prejunctional mechanisms (such as α_2 adrenergic, opioid, adenosine A_1, etc) which modulate intestinal neurotransmission. On the other hand, the finding that H_3 antagonist administration can increase the release of acetylcholine from the myenteric plexus, indicates the existence of a negative control which functions also in physiologic conditions.

Recent evidence shows that H_3 receptors are heterogeneous: the study done by West et al., (1990) proposed for the first time the existence of H_3 receptor subtypes, termed H_{3A} (high affinity) and H_{3B} (low affinity), that is based on the biphasic displacement of N^{α}-methylhistamine binding to rat brain homogenates by thioperamide and burimamide. According to this classification, guinea pig ileum, mouse and rat brain cortex receptors belong to the H_{3A} subtype (Schlicker et al., 1992; 1994b; 1996; Kathman et al., 1993; Jansen et al., 1994) whereas those in the guinea pig sympathetic postganglionic nerve terminals belong to the H_{3B} subtype (Hey et al., 1992a). In a subsequent paper (Leurs et al., 1996) the new H_3 receptor antagonist impentamine was found to discriminate between two functional models. Indeed impentamine acts as a competitive antagonist at the guinea pig jejunum but as a partial agonist in the mouse brain, thus suggesting the existence of H_3 receptor subtypes, which seem to be different from the previously suggested classification for H_{3A} and H_{3B}. At present, it is unclear whether or not this discrepancy is related to different sub-subtypes, species differences or may simply reflect differences in the efficiency of receptor coupling. In this regard, the H_3 receptor mediating inhibition of 5HT release from porcine enterochromaffin cells has shown to have a particular pharmacological profile, thus indicating that this receptor does not belong to either H_{3A} or H_{3B} subtype (Schwörer et al., 1994).

2.4. Intestinal electrolyte transport

It is well known that histamine plays a role in the intestinal ion and water transport (Cooke and Reddix, 1994); however, species differences can modify these processes. Moreover, histamine was shown to act both directly on the transporting epithelial cells and indirectly, by modulating enteric nervous activity or by releasing substances like eicosanoids. In some species a single receptor subtype is involved, whereas in others, different subtypes are responsible for the complex mechanisms observed. In the dog, only H_1 and H_2 receptors appear to regulate the transport of chloride ions, whereas H_3 receptors are functionally absent (Rangachari and Prior, 1994). In the guinea pig distal colon H_2 receptors located in the submucosal neurons stimulate chloride secretion, though H_3 receptors modulate the amplitude of recurrent cycles of secretion. In this regard, the H_3 agonist N^{α}-methylhistamine reduced the amplitude of dimaprit-evoked response, while the H_3 antagonist burimamide had the opposite effect (Cooke and Wang, 1994). Intracellular recordings from submucous neurons in the guinea pig colon suggested that H_3 receptors were located on presynaptic terminals and modulated acetylcholine release at nicotinic synapses (Wang and Cooke, 1990). Therefore, histamine released from mast cells or other non neuronal cells into the milieu surrounding the synaptic circuits, inhibits nicotinic transmission *via* H_3 receptors and functions as a brake to prevent excessive excitation of neural pathways regulating ion transport.

2.5. Gallbladder

Although histamine is distributed in the gallbladder, its functional effects have been identified only following activation of H_1 and H_2 receptors which mediate muscle contraction and relaxation, respectively (Impicciatore 1978). The H_3 receptor agonist (R)α-

methylhistamine was found to be inactive in modifying electrically-induced contractions of the guinea pig and human gallbladder. This finding indicates that no major role exists for prejunctional H_3 receptors in this tissue (Poli, unpublished data). These findings in the guinea pig have been confirmed by other studies (Jennings et al., 1995).

2.6. Pancreas

Studies on the role of histamine in the regulation of pancreatic secretion are obviously not so numerous as those concerning acid secretion. Most of these data refer to the effects mediated by H_1 and H_2 receptors (for review see Bertaccini and Coruzzi 1992). As far as H_3 receptors are concerned, binding studies (Korte et al., 1990) performed with $[^3H]$ N^{α}-methylhistamine showed the presence of this histamine receptor subtype in the guinea pig pancreatic tissue, in amounts comparable to those found in the gastrointestinal tract, but definitely lower amounts than those found in the CNS. Functional studies carried out in the guinea pig with (R)α-methylhistamine and its antagonist thioperamide showed that low concentrations of the agonist (0.01-10 μM) elicited a concentration-dependent decrease in the release of α amylase induced by electrical field stimulation, which was totally reversed by thioperamide. Furthermore, concentrations of (R)α-methylhistamine above 10 μM resulted in an increase of this secretion, that however, was not related to H_3 receptor activation, since it was blocked by mepyramine. In addition, approximately the same results were obtained in vivo, and this allowed the authors (Jennings et al., 1996) to conclude that activation of histamine H_3 receptors in the pancreas, leads to decreased fluid and enzyme secretion induced by acetylcholine release from intrinsic pancreatic nerves.

3. CARDIOVASCULAR SYSTEM

In the cardiovascular system, histamine induces marked effects, such as hypotension and tachycardia, which result from complex interactions with specific receptors located on the vascular and cardiac cell membranes, and with components of the autonomic nervous system. Receptor subtypes mediating cardiovascular responses to the amine have been pharmacologically classified in the past, by using specific agonists and antagonists of the "classical" H_1- and H_2-receptors (see for review Levi et al., 1982; 1991). Histamine H_2 receptors exert a predominant control on the heart, where they mediate positive inotropic, chronotropic and batmotropic activity, although species differences there may exist (Levi et al., 1982; McNeil, 1984). However, at the vascular level, H_1 and H_2 receptors are involved in the regulation of vascular permeability and vasodilation (Owen et al., 1980).

The occurrence of histamine H_3-receptors on the cardiovascular system and their role in the actions of histamine were observed for the first time on the guinea pig mesentry (Ishikawa and Sperelakis, 1987). This study revealed the presynaptic localisation of these receptors on adrenergic terminals surrounding the vascular beds. More recently, binding studies performed in homogenized guinea pig tissues using N^{α}-methylhistamine as a specific ligand of histamine H_3-receptors (Korte et al., 1990), suggested that these receptors occurs also in the heart and in the aorta. The density of these receptors in the heart and in aorta seems to be low (< 1 pmol/mg of proteins), compared to the their density in central nervous system (Korte et al., 1990). Hovever, this factor does not "a priori" minimise the importance of these receptors, considering that the heart and the endothelial system are sites of active histamine synthesis and storage (Levi et al., 1991) and large amounts of the endogenous ligand are available for receptor stimulation at these levels, thus compensating for the relatively low number of receptor moyeties. On the other hand, we cannot say when cardiovascular H_3 receptors are activated by endogenous histamine.

3.1. Heart

Since histamine H_3-receptors on the heart are characterised by marked species differences, the effects mediated by these receptors will be considered in relation to the different animal models where these receptors have been investigated.

3.1.1. Guinea pig

Early evidence that prejunctional histamine H_3-receptors may modulate the sympathetic nerve activity on the heart was provided by Luo et al., (1991). These authors clearly stated that the selective H_3-agonist (R)α-methylhistamine attenuates the inotropic response induced by transmural stimulation of the adrenergic nerve terminals in the isolated right atrium, without affecting basal contractile force of the preparation or the positive inotropic effect elicited by exogenous noradrenaline. The effect of (R)α-methylhistamine, which is not modified by H_1 and H_2-receptor blockade, was reversed by the specific H_3-receptor antagonist thioperamide, at concentrations which do not influence the inhibitory activity mediated by other presynaptic receptors, like α_2-adrenoceptors.

Other authors showed that (R)α-methylhistamine, together with the inotropic and chronotropic adrenergic response from transmurally-stimulated atrial preparations, also inhibits the release of noradrenaline, thus providing direct evidence that histamine H_3-receptors negatively modulate the cardiac sympathetic activity at a presynaptic site of action (Endou et al., 1994). In addition, it was demonstrated that (R)α-methylhistamine, at concentrations greater than 1 μM, produces further antiadrenergic activity by acting at inhibitory presynaptic α_2-adrenoceptors. This result is supported by the fact that yohimbine reverses this effect.

As far as the inhibitory mechanism of H_3-receptors is concerned, the effect of (R)α-methylhistamine is decreased by pretreatment with pertussis toxin and, like the α_2-adrenoceptor- and A_1-adenosine receptor-mediated effects, it is potentiated by ω-conotoxin GVIA, a blocker of the N-type Ca^{++} channel (Endou et al., 1994). One may conclude from these findings that presynaptic histamine H_3-receptors are probably coupled to a pertussis toxin-sensitive Gi/Go protein, which exerts a negative control on the neuronal Ca^{++}-currents, that are responsible for the exocytotic release of noradrenaline.

The above described antiadrenergic activity elicited by H_3-receptor activation can be also seen on the isolated hearts with intact sympathetic innervation (Imamura et al., 1995) and in anesthetised animals, where basal heart rate (McLeod et al., 1993) or the tachycardia evoked by electrical stimulation of the medulla oblongata (Hey et al., 1992a) were considered as parameters of the sympathetic nerve activity. In these studies, (R)α-methylhistamine inhibits basal heart rate or the tachycardia evoked by central nervous system stimulation (McLeod et al., 1993; Hey et al., 1992a). Also, tachycardic response and noradrenaline overflow elicited by electrical stimulation of the sympathetic nerve trunks arising from stellate ganglia are diminished (Imamura et al., 1994). Moreover, thioperamide, which antagonises the inhibition elicited by (R)α-methylhistamine, does not by itself produce any modification of either neurogenic tachycardia or noradrenaline release (Imamura et al., 1994). It is evident that endogenous histamine, known to be released in response to nerve stimulation, does not exert any H_3-receptor mediated modulation of the sympathetic nerve activity under "physiologic" conditions. Conversely, when sympathetic nerve activity is enhanced by short (10 min) periods of global ischemia, H_3-receptors are fully activated by an endogenous ligand, probably histamine. This activation is suggested by the facilitatory effect of thioperamide on the noradrenaline release (Imamura et al., 1994). These observations are compatible with a protective role of H_3-receptors in the ischemic myocardium. Accordingly, histamine H_3-receptors modulate not only the exocytotic noradrenaline release, but also the carrier-mediated noradrenaline spillover, which characterises the ischemic injury to the heart, following prolonged (20 min) coronary flow interruption. This protective effect, which is also produced by the activation of adenosine A_1-receptors, seems to be due to an interaction with the Na^+-H^+ transporter, an ion exchanger which operates to compensate for the H^+ accumulation in sympathetic nerve endings, when metabolic deprivation of ion pumps occurs (Schoming, 1990; Dart and Du, 1993). In the ischemic heart, the Na^+-H^+ transporter activation accumulates Na^+ in the axoplasm, thus causing a reversal of the (cocaine- and desmethylimipramine-sensitive) uptake-1 for catecholamines, so that noradrenaline is actively pumped out of the sympathetic nerve endings. Indirect evidence that the Na^+-H^+ transporter may be involved is the synergistic inhibitory activity of imetit and the compound 5-(*N*-Ethyl-*N*-Isopropyl)-amiloride, which is a specific inhibitor of the transporter (Gupta et al., 1989). The main consequence of the inhibition of the carrier-mediated noradrenaline release is the marked reduction in the incidence and duration of ventricular fibrillation during reperfusion. This suggests a protective action against the disfunctional consequences of prolonged myocardial ischemia. Interestingly, α_2-adrenoceptor activation potentiates the carrier-mediated noradrenaline release, while their blockade inhibits it. This difference should be emphasised, since α_2 adrenoceptors and histamine H_3-receptors usually mediate similar effects and, in addition, they share a common postreceptor mechanism at the cellular level (Poli et al., 1994a; Endou et al., 1994).

In addition to the adrenergic (efferent) component of the cardiac innervation, histamine H_3-receptors also modulate the release of neuromediators from the afferent system. In the isolated guinea pig heart, the tachycardic and calcitonin gene-related peptide (CGRP)-releasing effects elicited by C-fiber stimulation with capsaicin are decreased by the H_3-agonist imetit. Similarly, the H_3-agonist reduces both the CGRP release and the chronotropic response evoked by bradykinin on the isolated atria. The presynaptic localisation of this effect is confirmed by the inability of the H_3-agonist to modify the chronotropic activity of exogenous CGRP (Imamura et al., 1996b). Since CGRP released from sensory nerves, in turn, releases histamine from mast cells, one can hypothesize a cross talk between the afferent and the histaminergic system. As shown in Figure 9, the histamine released from mast cells in response to the antidromic stimulation decreases the release of CGRP and, moreover, it indirectly modulates its own release. This mechanism can be demonstrated by the facilitatory activity of thioperamide on the release of CGRP and on tachycardia induced by antidromic stimulation of C-fibers.

Additionally, cardiac function can also be modulated by centrally-located H_3-receptors. The intracerebroventricular administration of (R)α-methylhistamine in conscious animals is associated with a marked reduction of the heart rate, an effect which is antagonised by thioperamide (McLeod et al., 1991). The peripheral administration of ipratropium, a muscarinic antagonist which does not cross the blood-brain barrier, prevents such inhibition, demonstrating that the activation of centrally-located histamine H_3-receptors leads to an increase in the vagal tone to the heart, rather than to a facilitatory effect on the sympathetic efferent fibers.

The results obtained in the guinea pig heart preparations are summarized in Table 7 and in Figure 9.

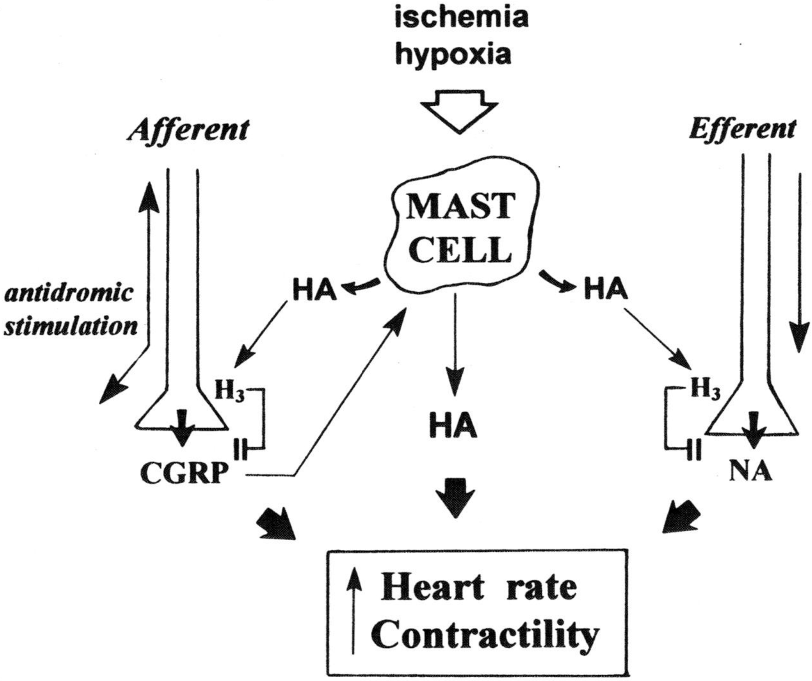

Figure 9. Scheme illustrating the main locations and functional role of H_3 receptors in the heart. Histamine H_3-receptors located on sympathetic post-ganglionic nerve endings negatively modulate the release of noradrenaline (NA) and, thus, the associated response at the post-junctional level. Histamine H_3-receptors may also occur on nerve terminals of the afferent fibers (C-fibers), where they contribute to the negative feed-back loop linking C-fiber terminals to mast cells. In this loop, calcitonin gene-related polypeptide (CGRP), released for antidromic stimulation of C-fibers, causes histamine release from mast cells, which, in turn, activates prejunctional histamine H_3-receptors on C-fibers (as well as H_2 receptors on cardiac cells). H_3 receptors negatively modulate the release of CGRP and, as a consequence, of histamine (HA) itself from mast cells. All these phenomena involving H_3-receptors could contribute to the "protective effect" of endogenous histamine, in conditions like hypoxia or ischemia, where the hyperactivity of intrinsic nerves represents one of the main causes of tissue damage.

3.1.2. Rat

Differently from that observed in the guinea pig, contrasting findings regarding both the occurrence and the role of H_3-receptors in the heart have been obtained in the rat. It has been reported that the activation of histamine H_3-receptors by (R)-methylhistamine in pithed animals attenuates the rise in heart rate induced by electrical stimulation of the preganglionic nerve fibers (Malinowska and Schlicker, 1991; 1993a, 1993b), as well as the nicotine-induced rise in heart rate in anaesthetised animals. In both models, the tachycardic effect of exogenous noradrenaline was not modified, hence excluding a postjunctional control of the cardiovascular functions. These findings suggest that inhibitory histamine H_3-heteroreceptors, occurring on postganglionic adrenergic nerve endings, are involved, and that these receptors negatively modulate noradrenaline release, in a fashion similar to that seen in the guinea pig (Malinowska and Schlicker, 1993a; Oudart et al., 1995). Compatible with this assumption is the evidence that (R)α-methylhistamine attenuates the electrically-evoked noradrenaline release from portal vein nervous plexuses in freely moving rats (Smit et al., 1997). Moreover, thioperamide, *per se* inactive, reverses this inhibitory effect (Smit et al., 1997). The ability of the NO-synthase inhibitor L-monomethyl arginine methyl ester (L-NAME) to prevent the (R)α-methylhistamine-induced inhibition suggests that the L-arginine/NO pathway, operating presynaptically on post-ganglionic sympathetic nerve fibers, is involved (Oudart et al., 1995). It is worth noticing that NO also plays a role in the negative control of vascular contractility by endothelial H_3-receptors (see 3.2.2).

The above described inhibitory activity mediated by H_3-receptors was not confirmed by others in anaesthetised animals with intact innervation (Coruzzi et al., 1995), or in the pithed rat (Hegde et al., 1994), These authors found tachycardic response by (R)α-methylhistamine, that was not correlated to the activation of H_3-receptors. The blockade of β-receptors by propranolol significantly reduced such tachycardic effects (Hegde et al., 1994; Coruzzi et al., 1995), suggesting that (R)α-methylhistamine interacts with the sympathetic nervous system by releasing catecholamines or by acting at the β-adrenergic receptor level.

Similarly to the in vivo data, studies performed on isolated heart preparations gave equivocal results. Histamine H_3-receptors are not apparently involved in the control of noradrenaline release from isolated hearts perfused with the histamine releaser 48/80. In this model, endogenous histamine enhances the carrier-mediated noradrenaline overflow through the activation of histamine H_2-receptors (Fuder et al., 1994). Conversely, H_3 receptors do not seem to play any role in the control of noradrenaline release. In the same model, a facilitatory activity mediated by H_3-receptors on the acetylcholine release has been described (Fuder et al., 1994). Apparently, the compound 48/80 enhances tritiated choline overflow elicited by vagus nerve stimulation and thioperamide prevents this effect, thus suggesting that endogenous histamine may activate H_3-receptors. Some doubts about the involvement of histamine H_3-receptors should be raised, since the authors did not test any specific H_3-agonist (or exogenous histamine after H_1 and H_2-receptor blockade), which would reproduce the putative H_3 receptor-mediated effect of endogenous histamine. Moreover, the possibility that thioperamide may interact also with 5-HT receptors (Leurs et al., 1995b) was not taken into account. This latter point is very important since, as stated by the authors, the compound 48/80 induces a release of 5-HT (together with histamine) and, like thioperamide, the $5\text{-}HT_3/5\text{-}HT_4$ antagonist tropisetron enhances the acetylcholine release. Therefore, further research is necessary to establish what is the involvement of histamine H_3 receptors in the control of acetylcholine release from cardiac vagal nerves.

Table 7
Cardiac effects mediated by histamine H_3 receptors.

Species	Preparation	Localisation	Parameter investigated	Effect	Reference
Guinea pig	Right atrium	Adrenergic nerves	Neurogenic inotropic response	inhibition	Luo et al., 1991
	Right atrium	Adrenergic nerves	Neurogenic chronotropic response	inhibition	Endou et al., 1994
	Left atrium		Neurogenic inotropic response	inhibition	
			NA release (ES-evoked)	inhibition	
	Isolated heart	Adrenergic nerves	Tachycardia (ES-evoked)	inhibition	Imamura et al., 1994
			NA release (ES-evoked)	inhibition	
		Adrenergic nerves	NA release (ischemia-induced)	inhibition	
	Left atrium	Afferent neurons	BK-evoked inotropic response	inhibition	Imamura et al., 1996b
			BK-evoked CGRP release	inhibition	
	Isolated heart	C-fibers	Capsaicin-evoked tachycardia	inhibition	Imamura et al.,1996b
			Capsaicin-evoked CGRP release	inhibition	
	Isolated heart	Adrenergic nerves	NA release (carrier-mediated)	inhibition	Imamura et al., 1996a
			Ischemia-reperfusion arrhythmias	antiarrhythmic effect	
	Isolated heart	not explored	Hypoxia-induced cardiac depression	cardioprotection	Akagi et al., 1995
	Conscious animal	CNS	Heart rate	inhibition	McLeod et al., 1991
	Anaesthetised animal	Adrenergic nerves	Heart rate (basal)	inhibition	McLeod et al., 1994
	Anaesthetised animal	Adrenergic nerves	Tachycardia (nicotine-evoked)	inhibition	Ea-Kim et al., 1996
	Pithed animal	Adrenergic nerves	Tachycardia (spinal stimulation)	inhibition	Hutchison and Hey, 1994

	Anaesthetised animal	Sympathetic nerves	Heart rate (basal)	inhibition	McLeod et al., 1993
	Anaesthetised animal	Sympathetic nerves	Tachycardia (bulbar stimulation)	inhibition	Hey et al., 1992a
	Anaesthetised animal	Sympathetic nerves	Heart rate (basal)	inhibition	Ea-Kim et al., 1993
Rat	Isolated heart	Not investigated	Arrhythmias (coronary ligation)	proarrhythmic effect	Banning and Curtis, 1994
	Isolated heart	Cholinergic nerves? Not found	ACh release NA release	increase? no effect	Fuder et al., 1994
	Pithed animal	Sympathetic nerves	Tachycardia (ES-evoked)	inhibition	Malinowska and Schlicker 1993a
	Anaesthetised animal	Not found	Heart rate	no effect	Oudart et al., 1995
	Anaesthetised animal	Not found	Heart rate (basal)	no effect	Coruzzi et al., 1995
	Anaesthetised animals	Not found	Heart rate (basal)	no effect	Hegde et al., 1994
Man	Pectinate muscle Synaptosomes	Adrenergic nerves Membrane	Positive inotropism (ES-evoked) K^+-evoked NA release	inhibition inhibition	Imamura et al., 1995
	Atrial specimen	Adrenergic nerves	Carrier-mediated NA release	inhibition	Hatta et al., 1996, 1997

CGRP = Calcitonin gene-related polypeptide; ES = electrical stimulation; NA = noradrenaline; ACh = acetylcholine; BK = bradykinin.

In contrast with observation on the guinea pig, histamine H_3-receptors do not modulate noradrenaline release evoked by bilateral stimulation of stellate ganglia (Levi, personal communication), while their activation by endogenous histamine in an ischemia-reperfusion model resulted in a proarrhythmic effect, antagonised by thioperamide (Banning and Curtis, 1994). Also on this study an exogenous agonist was not tested.

It is evident, therefore, that the role of histamine H_3-receptors in the control of rat cardiac function is still an open question, since results were obtained in different in vitro and in vivo conditions and were related to the methods employed (see Table 7), as well as perhaps to the side effects of some histamine H_3 receptor ligands.

3.1.3. Man

Studies on humans are scanty and, for obvious reasons, were obtained in in vitro models. Only right atrium preparations, obtained from surgical specimens of patients undergoing reconstructive heart surgery, have been employed. Similarly to guinea pig results, histamine H_3-receptors are present on sympathetic nerve endings in the human heart, where they function to modulate the adrenergic responses by decreasing noradrenaline exocytosis. This point was clearly stated in K^+-stimulated cardiac synaptosomes, which were freshly prepared from atrial tissues. Also, differently from the response to exogenous noradrenaline, the inotropic response of the pectinate muscle evoked by transmural stimulation of the adrenergic nerves was attenuated by H_3-agonists like (R)α-methylhistamine and imetit, an effect reversed by thioperamide (Imamura et al., 1995).

As far as model of protracted myocardial ischemia using right atrium fragments is concerned, the activation of histamine H_3-receptors by imetit causes a 40% decrease of the carrier-mediated noradrenaline release, possibly by way of the same mechanism operating in the guinea pig myocardium (Hatta et al., 1996, 1997). Moreover, this effect of imetit is prevented by thioperamide, thus confirming that histamine H_3-receptors are involved (see Table 7). Altogether, these results suggest that the human heart contains histamine H_3-receptors which, similarly to the guinea pig heart, modulate the exocytotic and carrier-mediated noradrenaline release, and the end-organ responses associated with sympathetic receptor activation. On the other hand, however, an involvement of these receptors in the control of sinus node and ventricular activity as yet cannot be established.

3.2. Vessels

In contrast to that observed in the heart, histamine H_3-receptors in vessels may be located at neuronal, as well as non-neuronal cell membranes. Three different localisations, prejunctional, endothelial and post-junctional (muscular cell) level have been found. A synopsis of the results obtained in vessels is depicted in Table 8.

3.2.1. Presynaptic histamine H_3-receptors

As we previously stated, Ishikawa and Sperelakis (1987) described an inhibitory effect mediated by presynaptic histamine H_3-receptors on the sympathetic vasoconstriction in the guinea pig mesenteric arteries, using histamine as agonist and burimamide (H_2/H_3-antagonist) or impromidine (H_2-agonist/H_3-antagonist) as antagonists. The pA_2 values for these antagonists were well correlated with values against the histamine-induced inhibition of histamine release measured in the rat brain (Arrang et al., 1983), and thus suggestive of the involvement of the H_3-receptor. The presynaptic localisation of, and the antiadrenergic activity mediated by these receptors on sympathetic nerves, was later confirmed in the same tissue in vitro by the use of the more specific antagonist thioperamide (Beyak and Vanner, 1995).

Histamine H_3-receptors can be also found in in vivo guinea pigs. Here the activation of histamine H_3-receptors by (R)α-methylhistamine decreased basal arterial pressure (Ea-Kim et al., 1992a, 1993; McLeod et al., 1993, 1994). Furthermore, the H_3 agonist attenuates the hypertensive effect evoked by the electrical stimulation of spinal (Huchinson and Hey, 1994) or bulbar (Hey et al., 1992a) sympathetic nerves, as well as the pressor response to nicotine (Ea-Kim et al., 1996). Pretreatment with thioperamide completely prevented the (R)α-methylhistamine-induced antiadrenergic effects, thus suggesting that H_3-receptors are involved in this antiadrenergic activity.

The effects induced by the H_3-receptor agonist can also be prevented by NO-synthase blockade with L-NAME, thus indicating the partecipation of the L-arginine-NO pathway (Ea-Kim et al., 1993, 1996). This pathway is probably located at the neuronal ending level, since (R)α-methylhistamine does not modify the hypertensive effects induced by exogenous noradrenaline (Ea-Kim et al., 1996), thus excluding the involvement of endothelial receptors (see below).

Differently from guinea pig results, contrasting findings have been obtained in other species, such as rabbit and rat (McLeod et al., 1994). According to some authors, histamine H_3-receptor activation decreases basal arterial pressure in rabbit (McLeod et al., 1994). In this species, however, H_3-receptor activation causes tachycardia, an effect which could be due to central reflexes or to vagal withdrawal. Moreover, basal arterial pressure is not modified in normotensive rats (McLeod et al., 1994; Hegde et al., 1994; Coruzzi et al., 1995), as well as in spontaneously-hypertensive rats (McLeod et al., 1994). In this species, the H_3 receptor agonist (R)α-methylhistamine causes a rise in arterial pressure, prevented by yohimbine, thus suggesting the involvement of vascular (excitatory) α_2-adrenoceptors. This effect remembers the α_2-adrenoceptor-mediated inhibition of the adrenergic response by (R)α-methylhistamine described in the heart (Endou et al., 1994).

Contrastingly, it was found that presynaptic histamine H_3-receptor activation lowers the hypertensive effect evoked by nicotinic stimulation of sympathetic ganglia (Oudart et al., 1995), or by electrical stimulation of the spinal sympathetic nerves (Malinowska and Schlicker, 1991, Godlewski et al., 1997a). In spontaneously hypertensive rats, these receptors are probably activated by endogenous histamine, since thioperamide increases arterial pressure. Therefore, histamine H_3-receptors appear to be operative in hypertension, where histamine could play a modulatory role in the control of sympathetic system hyperreactivity (Godlewski et al., 1997a).

A possible explanation for the differences observed in rat could be the use of animals under basal conditions *vs* electrically-driven systems and/or different receptor densities for different strains. Certainly, histamine H_3-receptors are not homogeneously distributed along the rat vascular system, as they were found in some districts, like pulmonary microvessels (Danko et al., 1994), portal vein (Smit et al., 1997) or tail artery (Godlewski et al., 1997b), but not in vena cava (Schneider et al., 1991).

As far as man is concerned, a presynaptic inhibition of the noradrenaline release by H_3 receptors was only studied in the isolated human saphenous vein. In this tissue, these receptors, similarly to α_2-adrenoceptors, decrease the ^{3}H-noradrenaline overflow elicited by electrical field stimulation (Molderings et al., 1992). Therefore, the role of histamine H_3-receptors in the global control of the blood pressure in humans cannot be stated at present.

Table 8
Vascular effects mediated by histamine H_3-receptors.

Species	Preparation	Localisation	Parameter investigated	Effect	Reference
Guinea pig	Mesenteric artery	prejunctional	Sympathetic transmission	inhibition	Ishikawa and Sperelakis, 1987
	Mesenteric artery	prejunctional	Sympathetic transmission	inhibition	Beyak and Vanner, 1995
	Aorta	endothelial	Muscle contraction	inhibition	Rosic et al., 1991
	Basilar artery	endothelial	Muscle contraction	inhibition	Ea Kim et al., 1990
	Anaesthetised animal	prejunctional (symp. nerves)	Microvascular leakage	increase	Danko et al., 1994
	Anaesthetised animal	vagal afferents	Microvascular leakage	inhibition	Ichinose et al., 1990
	Anaesthetised animal	prejunctional	Arterial pressure (basal)	decrease	McLeod et al., 1993
	Anaesthetised animal	prejunctional	Pressor response to nicotine	reduction	Ea-Kim et al., 1996
	Anaesthetised animal	prejunctional	Pressor response (ES-evoked)	inhibition	Hey et al., 1992a
	Anaesthetised animal	not defined	Basal blood pressure	decrease	Ea-Kim et al., 1993
	Pithed animal	prejunctional	Pressor response (ES-evoked)	inhibition	Hutchison and Hey, 1994
	Conscious animal	central nervous system	Mean blood pressure (basal)	decrease	McLeod et al., 1991
Rabbit	Saphenous artery	postjunctional (sarcolemma)	Ca^{++}-conductance	facilitation	Oike et al., 1992

	Cerebral artery	endothelial	Muscle contraction	inhibition	Ea-Kim and Oudart, 1988; Ea-Kim et al., 1992b
Rat	Isolated aorta	endothelial	Muscle contraction	inhibition	Djuric and Andjelkovic, 1995; Djuric et al., 1996
	Isolated heart	endothelial?	Coronary flow	increase	Kostic and Jakovljevic, 1996
	Freely moving animal	prejunctional	Noradrenaline overflow	inhibition	Smit et al., 1997
	Pithed animal	prejunctional	Vasopressor response to ES	inhibition	Malinowska and Schlicker, 1991; 1993a; Godlewski et al., 1997a
	Anaesthetised animal	not found	Arterial pressure (basal)	no effect	Coruzzi et al., 1995; Hegde et al., 1994; McLeod et al., 1994
	Anaesthetised animal	prejunctional	Pressor response to nicotine	inhibition	Oudart et al., 1995
	Vena cava "in situ"	not found	Noradrenaline release	no effect	Schneider et al., 1991
Cow	Oviductal artery	post-junctional (muscular)	Muscle contraction	contraction	Martinez et al., 1997
Pig	Lymph vessels	not found	Muscle contraction	no effect	Reeder et al., 1996

ES = electrical stimulation

Similarly to what was described for the heart (Imamura et al., 1996b), histamine H_3-receptors may also modulate the function of sensory nerves in vessels. In fact, these receptors control the release of neuropeptides, like CGRP and tachykinins, from C-fibers in rat meningeal (Matsubara et al., 1992), cutaneous (Ohkubo et al., 1995) and pulmonary microvessels (Ichinose et al., 1990). These peptides, released by antidromic stimulation of C-fibers, cause neurogenic inflammation, vasodilatation and plasma protein extravasation, together with degranulation of mast cells and consequent release of active substances. In these studies, (R)α-methylhistamine attenuates protein extravasation (Matsubara et al., 1992), immunoreactive substance P release (Ohkubo et al., 1995) and neurogenic leakage (Ichinose et al.,1990), which are induced by electrical stimulation of trigeminal, sciatic and vagus nerves, respectively (see also Respiratory System section). These effects of (R)α-methylhistamine were antagonised by thioperamide. Furthermore, the inability of (R)α-methylhistamine to modify the vascular effects of exogenous substance P was indicative of the presynaptic localisation of histamine H_3-receptors.

Much like what was described in heart and lung, neuropeptides may release histamine from mast cells, which in turn may act on presynaptic H_3-receptors (Matsubara et al., 1992). In this regard, endogenous histamine exerts a two faceted influence on neurogenic inflammation, since this substance may also cause vasodilatation and rise in vascular permeability through the H_1-receptor/NO-pathway (Levi et al., 1991; Lassen et al., 1995).

Since neurogenic inflammation may be important in the pathophysiology of vascular headache (Moskowitz, 1991; Göthert et al., 1995), the use of specific H_3 receptor agonists could be useful in the treatment of such disease (Mansfield, 1990).

Histamine H_3 receptor stimulation is not involved in the direct effects of histamine in the permeability response in in vivo rats, (Pile and Smaje., 1992). In this last study, H_1 and H_2 receptors seem to have a dominant role. However, the H_3-receptor antagonist thioperamide prevents the effects induced by histamine and by the H_1-agonist 2-thiazolylethylamine, thus suggesting non-specific interactions of these histamine receptor ligands.

3.2.2. Endothelial histamine H_3-receptors.

Vessels also contain non-neuronal histamine H_3-receptors, which are located on the endothelial layer, whose activation promotes vasodilatation with a mechanism independent of the adrenergic system inhibition. Ea-Kim et al., were the first to demonstrate the existence of these receptors in the isolated rabbit basilar artery precontracted with K^+, where both histamine and (R)α-methylhistamine induce a vasorelaxant effect, prevented by the H_2-agonist/H_3-antagonist impromidine - in presence of H_2-receptor blockade by cimetidine (Ea-Kim and Oudart, 1988) - or by thioperamide (Ea-Kim et al., 1992b). The effect of the histamine H_3-receptor agonists is completely lost in preparations denuded of the endothelium (Ea-Kim et al., 1992b), thus suggesting the partecipation of endothelium-derived relaxing factor(s) (EDRF). This vasorelaxant behaviour is similar to that described for acetylcholine and other indirectly-acting vasodilators (Furchgott and Zawadsky, 1980; Furchgott, 1983; Luscher, 1994). Moreover, the vasorelaxant activity of (R)α-methylhistamine was partially suppressed by the NO-synthase inhibitor N^G-nitro-L-arginine methyl ester (L-NAME) and/or by the cyclooxygenase inhibitors tranylcypromine, indomethacin or dexamethasone, while a combination of L-NAME and tranylcypromine almost completely prevented this activity. It is evident from these studies that the activation of endothelial H_3-receptors is followed by the release of NO and a prostanoid(s), possibly prostacyclins, which cooperate for the vasorelaxant activity at the muscular level.

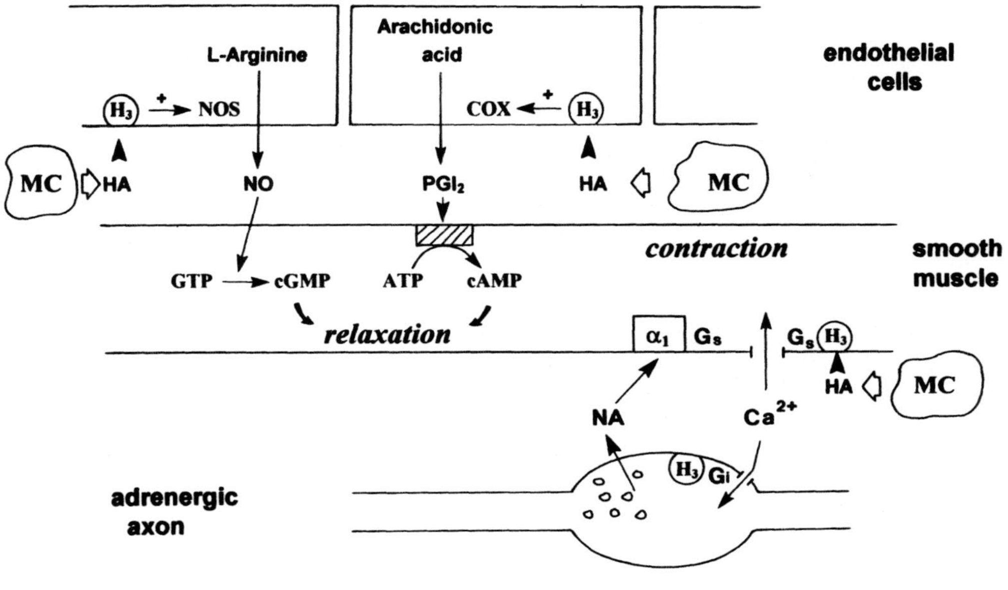

Figure 10. Scheme illustrating the main locations and functional role of H_3 receptors in the vessels. H_3 receptors coupled with inhibitory G proteins (G_i) occur as prejunctional receptors in the adrenergic varicosities, where they negatively modulate noradrenaline (NA) release. Moreover, their activation in endothelial cells can induce muscle relaxation, by the release of inhibitory factors, such as nitric oxide (NO) and prostacyclin (PGI_2). In some districts, excitatory H_3 receptors were found in muscle cells and they mediate muscle contraction. MC = mast cell; NOS = NO synthase; COX = cyclooxygenase.

In this regard, endothelial histamine H_3 receptors closely resemble the H_1-subtype, which is known to operate with a similar mechanism on the resistance vessels to produce vasodilatation, increased vascular permeability and hypotensive effects, by way of a mechanism involving the constitutive NO-synthase pathway (Beyak and Vanner, 1993, 1995; Levi et al., 1991; Luscher, 1994; Lassen et al., 1995). Such endothelium-dependent relaxation, mediated by H_3 receptors, can be also observed in rat aorta (Djuric and Andjelkovic, 1995; Djuric et al., 1996) and in coronaries (Kostic` and Jakovljevic`, 1996), as well as in guinea pig aorta (Rosic et al., 1991) and basilar artery ((Ea-Kim et al., 1990). In coronaries, H_3 receptors cooperate with H_2 receptors in the inhibition of vasomotility. For rat, an endothelium-derived hyperpolarising factor (EDHF) seems to be involved in the vasodilatation, while no contribution of prostacyclin is evident (Djuric et al., 1996). EDHF is a poorly understood factor, which causes hyperpolarisation of the cell membranes by activating an ATP-dependent K^+-channel (Luscher, 1994), thus preventing contractile stimuli of endogenous vasoconstrictors. The role of EDHF in the H_3-mediated response is suggested by the functional antagonism observed with ouabain, an inhibitor of the Na-K ATPase (Djuric et al., 1996). Accordingly, the inhibition of this enzyme shifts the membrane potential to a less negative level, thereby counteracting the hyperpolarisation associated with the K^+-channel opening.

Similarly to what identified in the rat, histamine H_3 receptor activation does not contribute to prostacyclin production in coltured human endothelial cells (Bull et al., 1992). There are no findings explaining any possible involvement of these receptors in the control of endothelial functions in humans.

3.2.3. Post-junctional (muscular) histamine H_3-receptors.

Recent studies using the patch-clamp method showed that histamine H_3-receptors may be also located on post-junctional membranes of dispersed myocites of the rabbit saphenus artery (Oike et al., 1992). On these cells, histamine or (R)α-methylhistamine accelerate the voltage-dependent Ca^{++}-channel conductance, in a fashion similar to that observed for the vasoconstrictors noradrenaline and angiotensine II, by way of the respective receptor activation. This facilitatory effect is followed by a rise in the citoplasmatic free Ca^{++} concentration, a phenomenon which may be predictive of a contractile effect in the whole tissue. Direct evidence of this has not been so far demonstrated for (R)α-methylhistamine or for other H_3-agonists. In this regard, some authors recently described an H_3-receptor-mediated contraction in bovine oviductal arteries (Martinez et al., 1997), an effect which suggests possible implications of these postjunctionally-located receptors in the control of contractility in intact vessels.

In dispersed myocites of rabbit saphenous vein, thioperamide antagonised the facilitatory effect of (R)α-methylhistamine, at concentrations (1-10 μM) 100 times higher than those usually requested to block H_3-receptors in other tissues, thus showing a pA_2 value 7.54. This value is relatively low, compared to the values (from 8.00 to 9.30) usually obtained for presynaptic or endothelial H_3-receptors (Ea-Kim and Oudart., 1988; Coruzzi et al., 1991b; Hey et al., 1992a). Nevertheless, it seems premature to postulate the existence of receptor subclasses among peripheral H_3-receptors. However, the rabbit saphenous artery H_3-receptor also differentiates from presynaptic H_3-receptors in the post-receptor events. In fact, the facilitatory effect evoked by (R)α-methylhistamine in saphenous artery is prevented by non-hydrolysable GTP analogs, like GTPγS, but not by pertussis toxin. This finding suggests possible involvement of PTX-insensitive G-protein, instead of the Gi/Go protein found in the heart or in the intestine (Endou et al., 1994; Poli et al., 1994a). The involvement of different G-proteins could explain the different coupling (inhibitory *vs* excitatory) of the H_3-receptors with the Ca^{++}-channels in different cell types. The investigation of other H_3-antagonists would elucidate whether or not these differences in the transductional system are associated with differences in the recognition sites of the respective receptor moyeties.

A scheme representing the multiple locations and functional role of H_3 receptors in the vessels is reported in Figure 10.

4. RESPIRATORY SYSTEM

The localization and function of histamine H_3 receptors in the respiratory tract was excellently reviewed in 1992 by Barnes. H_3 receptors were detected in lungs by binding studies (Arrang et al., 1987) and following these, in an attempt to elucidate the possible role of H_3 receptors in the airway system, a series of functional studies were performed in different species and by different experimental models. A synopsis of the results obtained by several authors is summarized in Table 9.

It is evident from the table that the bulk of the results were obtained in the guinea pig airway system. H_3 receptors are located both on cholinergic ganglia and on postganglionic cholinergic and NANC nerve endings, where they exert an inhibitory control on neurotransmitter release. In contrast with results obtained in isolated tissues of the guinea pig, no evidence for an H_3-mediated inhibition of cholinergic neurotransmission was found in vivo (Hey et al., 1992b). In this study, (R)α-methylhistamine did not modify neurally mediated cholinergic bronchoconstrictor response to electrical stimulation of medulla oblongata. Regardless of nerve location, H_3 receptors also seem to be present in the tracheal muscle where their activation evokes a relaxation of precontracted trachea (Cardell and Edvinsson, 1994). In addition, according to a more exhaustive paper (Burgaud and Oudart, 1993a) H_3 receptors are also present in bronchial epithelium and their activation causes a relaxation of precontracted muscle, which is counteracted by thioperamide. Moreover, pretreatment with indomethacin prevents the above effects, thus suggesting an involvement of prostanoids. However, results obtained in the guinea pig trachea were not in line with these data, since thioperamide prevents the histamine-induced epithelium-dependent relaxation, while (R)α-methylhistamine was completely ineffective (Guc et al., 1988).

In isolated perfused rabbit lung, imetit, -a novel selective H_3 agonist- was found to inhibit the effect of acetylcholine and capsaicin on capillary filtration (Delaunois et al., 1995). The effect of imetit was reversed by carboperamide, a novel H_3 receptor antagonist. On the basis of these data, the authors conclude that H_3 receptors could protect the lung against acetylcholine and capsaicin-induced pulmonary edema, via a prejunctional modulatory effect on the C fibers. Moreover, since the response to exogenous substance P was also inhibited by imetit, the presence of H_3 receptors at a postsynaptic level -probably on mast cells- was also suggested.

Another interesting effect related to H_3 receptors is its possible interference in the airway microvascular leakage (AML) induced by a variety of stimuli. Activation of H_3 receptors inhibits NANC-induced leakage by reducing neuropeptides release from sensory fibers (Ichinose et al., 1990). Hence, H_3 receptors may act as a safety device in asthmatic diseases to prevent, not only bronchoconstriction, but also increased airway permeability. On the other hand, it was also demonstrated that H_3 receptor activation inhibits sympathetic tone, which, in turn, reduces antigen-induced AML (Danko et al., 1994). Therefore, the authors conclude that the blockade of H_3 receptors may have therapeutic potential in inflammatory pulmonary

Table 9.
Histamine H_3 receptor-mediated effects in airways

Species	Tissue	Effect	Reference
Rat	Lung (in vitro)	↓ Neuropeptide release from sensory nerves	Dimitriadou et al., 1994
Guinea pig	Trachea (in vitro)	↓ ACh release	Ichinose et al., 1989
	Trachea (in vitro)	Relaxation	Cardell and Edvinsson, 1994
	Bronchi (in vitro)	↓ NT release from NANC nerves	Burgaud and Oudart, 1993a
	Bronchi (in vivo)	↓ NT release from NANC nerves	Ichinose and Barnes, 1989a
	Bronchi (in vivo)	No effect on ACh release	Hey et al., 1992b
	Bronchioles (in vitro)	Epithelium-dependent relaxation	Burgaud and Oudart, 1993b
	Lung (in vitro)	Histamine forming capacity	Allen et al., 1996
	Vessels (in vivo)	↓ Neuropeptide release from sensory nerves [↓ leakage]	Ichinose et al., 1990
	Vessels (in vivo)	↓ Sympathetic nerve transmission [↑ leakage]	Danko et al., 1994
Rabbit	Lung (in vitro)	↓ Endogenous and exogenous SP	Delaunois et al., 1995
Human	Bronchi (in vitro)	↓ ACh release	Ichinose and Barnes, 1989b
Human (asthmatic patients)	Bronchi (in vivo)	No effect	O'Connor et al., 1993

NT = neurotransmitter; NANC = nonadrenergic, noncholinergic; ACh = acetylcholine; SP = substance P; ↓ = decrease; ↑ = increase.

disorders. The above considerations indicate that the role of H_3 receptors in the control of microvascular leakage requires further studies. In a thorough investigation concerning porcine tracheobronchial lymph vessels, no important effects concerning H_3 receptor activation were reported (Reeder et al., 1996). It is true, however, that histamine may contribute to the regulation of lymphatic vascular smooth muscle tone, but this function is apparently only controlled by histamine H_1 receptors.

An hypothesis was advanced speculating that H_3 receptors are located in guinea pig lung mast cells, having a negative effect on histamine synthesis and/or release (Barnes, 1992). In this regard, in the guinea pig bronchi the H_3 receptor antagonist thioperamide increased the allergen-induced histamine forming capacity, while (R)α-methylhistamine was ineffective. As hypothesized in other studies, the ineffectiveness of the exogenous agonist might be possibly due to the fact that H_3 receptors are already maximally stimulated by mast cell derived histamine, following allergen exposure (Allen et al., 1996) Moreover, in an extensive study performed in rats (Dimitriadou et al., 1994) it was shown that H_3 receptors are absent from lung mast cells, but they occur in sensory fibers where they inhibit neuropeptide release. It has been suggested that a short regulatory feed-back loop exists between mast cells and adjacent sensory nerves, with neuropeptides activating mast cells to release histamine, and, in turn, histamine inhibiting neuropeptides release.

The presence of inhibitory histamine H_3 receptors on cholinergic nerves was also demonstrated in isolated human airways by means of specific agonists and antagonists (Ichinose and Barnes, 1989b). However, a double-blind crossover study (O'Connor et al., 1993) carried out in mild atopic asthmatic subjects showed that the selective H_3 agonist (R)α-methylhistamine (10 mg) had no effect on bronchoconstriction induced by the inhaled irritant sodium metabisulphite. If this finding is confirmed by using other H_3 agonists and against other irritant stimuli, the potential therapeutic role for H_3 agonists in asthmatic patients will be minimized.

5. GENITOURINARY SYSTEM

In the guinea pig vas deferens H_3 receptors are located in the sympathetic nerves and their activation inhibits neurotransmitter release (Luo et al., 1994; Luo and Tan 1994). Contrastingly, in the isolated rat vas deferens both H_3 agonists and antagonists were found to be inactive, and the modulation of sympathetic neurotransmission was connected with prejunctional H_2 receptors (Poli et al., 1994b).

The effect of histamine on rabbit corpus cavernosum was apparently due to the activation of H_1 receptors, whereas H_3 receptors were absent. However, the effect on electrically-induced contractions was investigated only by the use of thioperamide, and the H_3 agonist was tested on basal motility and noradrenaline-induced contractions (Kim et al., 1995). We think therefore that further studies are needed to definitely exclude the presence of prejunctional H_3 receptors in this tissue.

In the mouse uterus H_3 receptors seem to be absent, since both agonists and antagonists of this receptor failed to modify noradrenaline release evoked by 100 mM K^+ or electrical stimulation (Montesino et al., 1995). It would seem that prejunctional histamine receptors which modulate noradrenergic transmission belong to the H_1 subtype.

6. IMMUNE SYSTEM

Histamine is one of the first inflammatory mediators that was considered to be important in the pathophysiology of a number of allergic diseases. Histamine is released from basophils and mast cells during hypersensitivity reactions to allergens and, -through activation of different receptor subtypes-, may provoke opposite effects on immune and inflammatory responses. While many of the inflammatory effects of histamine are mediated by H_1 receptors, H_2 receptors may mediate different immunomodulatory responses. This topic was extensively reviewed by Plaut and Lichtenstein (1982) and Falus (1994).

The third histamine receptor subtype was found to occur on mast cells and to be involved in the synthesis and release of histamine, of nitric oxide and of a series of cytokines, even if the precise immunomodulatory function of H_3 receptors is still under investigation. The bulk of data concerning H_3 receptors and immunological reactions in different species and different experimental models are summarized in Table 10.

In the rat anaphylactic histamine release from peritoneal mast cells is reduced by (R)α-methylhistamine, and this effect is reversed by thioperamide. Therefore, H_3 receptors seem to be responsible for the negative feedback regulation of histamine release from rat peritoneal mast cells. However, the high concentrations of the H_3 agonist necessary to elicit this effect, suggest that H_3 receptors on the rat peritoneal mast cells are a subtype distinct from those in the rat brain. Contrastingly, histamine release induced .by compound 48/80 was not altered by H_3 ligands (Kohno et al., 1994).

In the rat skin mast cells, histamine release induced by electrical stimulation or substance P is reduced by (R)α-methylhistamine, -an effect reversed by thioperamide (Ohkubo et al., 1994). As already mentioned for the airway system, H_3 receptors in the lung are located in sensory nerve endings, -rather than on non-neural cells-, where they mediate inhibition of tachykinins release (Dimitriadou et al., 1994). As we can see, histamine released from mast cells by peptidergic neurotransmitters exerts a negative control on C fibers *via* activation of H_3 receptors. In the rat H_3 receptors seem to regulate histamine release and also that of tumour necrosis factor α (TNF_α). This inhibition is mediated by PGE_2 and by an increase in cyclic AMP. The involvement of H_3 receptors, together with the well defined role of H_2 receptors, suggest that histamine may be an important component of the cytokine network activated during allergic reactions.

A possible role of H_3 receptors in the regulation of inflammatory reactions has been recently postulated by Rouleau et al., (1997). These authors observed that the prodrug of (R)α-methylhistamine, compound marked BP 2-94 (Krause et al., 1995), inhibited capsaicin-induced plasma protein extravasation in many rat tissues and also reduced the zymosan-induced paw swelling in mice. In our labs, preliminary experiments showed that (R)α-methylhistamine (10 mg/kg intragastrically) significantly reduced carrageenin-induced oedema, in the rat paw, with an efficacy comparable with that of the well known antiinflammatory agent, ibuprofen, 30 mg/kg intragastrically (Figure 11). The above data, although preliminary, indicate a possible therapeutic use of H_3 receptor agonists in the treatment of various inflammatory disorders.

H_3 receptor activation in the airway system of the guinea pig caused a decrease in histamine release from mast cell and/or basophils and the H_3 antagonist had the opposite effect. As far as histamine synthesis is concerned, the allergen-exposure induced histamine forming capacity could be increased by thioperamide, even though (R)α-methylhistamine was completely inactive. The explanation given for this peculiar phenomenon was that the histamine

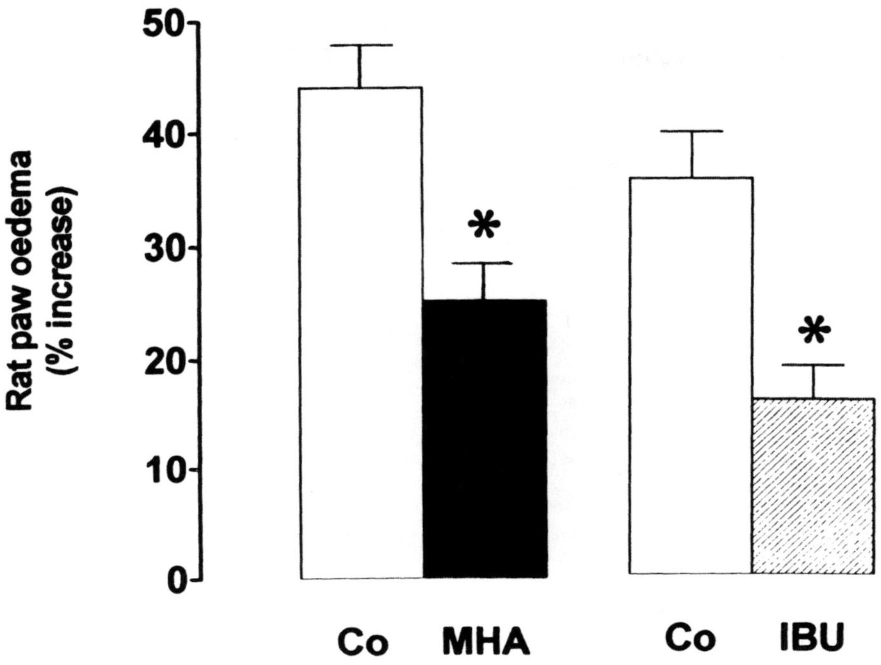

Figure 11. Antiinflammatory effect of (R)α-methylhistamine (MHA, 10 mg/Kg) and of ibuprofen (IBU, 30 mg/Kg), administered intragastrically in carboxymethylcellulose (CMC, 1%) 5 min before 1% carrageenin. The effect was measured 4 h later. On the ordinate: % increase swelling. Mean values ± SEM from 10 rats. * = $P<0.05$ *vs* control (Co, CMC alone).

H_3 receptor was already maximally stimulated by endogenous histamine during exposure to allergen (Allen et al., 1996).

Regarding human tissues, the few data available are contradictory, since according to some authors (Tedeschi et al., 1991) histamine release from basophils is not modified by H_3 receptor ligands. In other studies (Bent et al., 1991), it was found that thioperamide could increase histamine release from adenoidal mast cells, but apparently (R)α-methylhistamine was totally inactive. In any case the concentrations of thioperamide are much higher than those necessary to block H_3 receptors, thus suggesting that other mechanisms might be involved. Raible et al., (1994) hypothesized by the use of thioperamide, the presence of H_3 receptors on human eosinophils which mediate histamine-induced increase in cytosolic calcium mobilization. However, the low efficacy of the known H_3 receptor agonists, -as stimuli for eosinophils

Table 10
Histamine H_3 receptors and immmune responses

Species	Tissue	Parameter	H_3 ligand	Effect	Reference
Rat	Peritoneal mast cells	a) anaphylactic histamine release	MHA THIO	↓ ↑	Kohno et al., 1994
		b) 48/80-induced histamine release	MHA THIO	0 0	
Rat	Peritoneal mast cells	TNFα release	HA	↓*	Bissonette, 1996
Rat	Peritoneal mast cells	NO generation	MHA FUB-181[a]	↓ ↑	Mannaioni et al., 1997
Rat	Skin and peritoneal mast cells	histamine release induced by antidromic electrical stimulation or SP perfusion	MHA	↓	Ohkubo et al., 1994
Rat	Lung and abdominal skin	histamine formation	MHA	↓	Arrang et al., 1987
Guinea pig	Lung mast cells (basophils)	histamine release	MHA	↓	Barnes, 1992
Guinea pig	Bronchial mast cells	allergen-induced histamine	MHA THIO	0 ↑	Allen et al., 1996

Man	Adenoidal mast cells	histamine release	MHA	0	Bent et al., 1991
			THIO	↑	
Man	Eosinophils	chemotactic effect	HA	↑*	Raible et al., 1994
			MHA	0	
			BUR	↓	
Man	Basophils	histamine release	MHA	0	Tedeschi et al., 1991
			THIO	0	
			THIO	0	Kleine-Tebbe et al., 1990

MHA = (R)α-methylhistamine; THIO = thioperamide; TNFα = tumor necrosis factor α; HA = histamine; SP = substance P; BUR = burimamide; [a]: antagonist of histamine H_3 receptors; * antagonized by thioperamide.

calcium mobilization- plus an unexpected stimulatory effect of burimamide, suggests the existence of an unusual histamine receptor.

In conclusion, the role of histamine H_3 receptors in the immune responses has not yet been completely identified. The H_3 receptor, located on mast cells, may represent a mechanism for a negative feedback through histamine autoregulation in tissue inflammation. The presence of this receptor on both mast cells and nerve endings suggests an important role for H_3 receptors, especially when we consider that the immune system and the nervous system behave as separate arms of a unified, coordinated body defence system (McKay and Bienenstock, 1994).

7. THERAPEUTIC POTENTIAL OF HISTAMINE H_3 RECEPTOR LIGANDS IN PERIPHERAL TISSUES

The therapeutic usefulness of selective H_3 receptor ligands in peripheral tissue dysfunctions has not so far been assessed; nevertheless, support for future clinical applications of these drugs could be drawn from the experimental observations in animal or in human tissues.

The effects mediated by H_3 receptors in the gastrointestinal tract appear to be particularly attractive, in view of therapeutic applications. Although acid suppression can be better achieved by the use of H_2 blockers or proton pump inhibitors, H_3 agonists might have the advantage of combining an inhibitory effect on acid secretion with a strong gastroprotective activity. This last effect could be particularly desirable in conditions of increaseD damage to the gastric mucosa, as observed during antiinflammatory therapy. On the other hand, the inhibitory effect induced by H_3 agonists on intestinal motility seems to be less promising, since it was only evident in particular experimental conditions, and does not seem to have physiological importance. Moreover, the equivocal results obtained in animal models reproducing physiological peristalsis do not suggest any particular usefulness of H_3-receptor ligands in conditions of deranged motility of the gut.

As for the cardiovascular system, the cardioprotective effects of selective H_3-receptor agonists, demonstrated in models of protracted myocardial ischemia (Imamura et al., 1994, 1995, 1996a; Hatta et al., 1996, 1997), could be predictive of beneficial effects in coronaropatic patients. Hence, the attenuation of carrier-mediated noradrenaline release in hypoxic and/or ischemic myocardium by H_3-agonists would limit the sympathetic overactivity and the associated incidence of ventricular arrhythmias and angina, as well as the increase of metabolic demand by the myocardium, thus preventing further damage and cardiac failure.

Although α_2-adrenoceptor blockade or adenosine A_1 receptor stimulation also reduce carrier-mediated noradrenaline release (Imamura et al., 1996a), histamine H_3 receptor agonists, like imetit or (R)α-methylhistamine, may be more advantageous in correcting adrenergic hyperactivity, when compared with the other prejunctionally-acting drugs. In fact, unlike adenosine A_1-receptor stimulation (Belardinelli et al., 1989), H_3-receptor stimulation does not modify basal inotropic and chronotropic activity, thus representing a safer approach in conditions of heart failure. Moreover, H_3-receptor stimulation inhibits both exocytotic and carrier-mediated noradrenaline release, associated with acute and protracted ischemia, respectively, in contrast with α_2-adrenoceptor blockade, which actually enhances noradrenaline exocytosis (Endou et al., 1994). This difference could result in a better control of post-ischemic arrhythmias, their severity being correlated with the global amount of released noradrenaline.

The recent evidence that inhibitory histamine H_3-receptors, located on noradrenergic endings of perivascular nerves, may be activated by endogenous histamine and are operative in lowering blood pressure in spontaneously hypertensive rats (Godlewski et al., 1997a, 1997b) could suggest the investigation of specific H_3-receptor agonists also in hypertensive patients.

As for the airway system, the reduction of both cholinergic and NANC neurotransmission, together with a reduced microvascular leakage and histamine release by H_3 agonists, might suggest possible therapeutic applications in asthmatic patients. In fact, new drugs showing both bronchodilatatory and antiinflammatory activities are to be considered as elective in such kind of patients. The negative results obtained in the unique clinical study carried out so far (O'Connors et al., 1993) should not discourage new research in the field, since the unfavourable pharmacokinetics of (R)α-methylhistamine, used in that study, might be highly responsible for the lack of effect observed. Several prodrugs of (R)α-methylhistamine have been developed in order to overcome the disadvantageous pharmacokinetic profile of the histamine derivative (poor absorption from the digestive tract, rapid inactivation by histamine methyltransferase, difficult delivery into the CNS). These prodrugs, which show increased lipophilicity and, as a consequence, improved bioavailability (Krause et al., 1996), should give more relevant informations on the therapeutic potential of H_3 agonists.

The recent finding that H_3 receptor activation produce antiinflammatory effects also in tissues other than lung opens an exciting new area of clinical research with H_3 receptor agonists. The application of these drugs may attenuate the release of inflammatory mediators from mast cells and, at the same time, may affect the vascular component of neurogenic inflammation by inhibiting neuropeptide release from sensory nerves. The effects could have a major therapeutic benefit in chronic hyperrheactive disorders of the skin, like eczema, psoriasis and photodermatoses (Holzer, 1998), as well as systemic inflammations associated with infective states, rheumatoid arthritis etc.. In this regards it is important to underline that these agents should be very unusual antiinflammatory drugs, being not only devoid of gastrolesive properties, but actually displaying gastroprotective properties.

So far no H_3 ligand have been introduced into therapy, and we have to wait for results of clinical trials, which are in progress with the prodrug BP 2-94 (Rouleau et al., 1997), before the potential therapeutic usefulness of histamine H_3 receptor ligands in peripheral diseases can be definitely assessed.

8. CONCLUSIONS

Histamine H_3 receptors are well distributed in peripheral tissues, although relatively less abundant than in the CNS. As for the cellular localization, peripheral H_3 receptors are present in different cell types, like the neural, paracrine, endocrine, muscular and endothelial cells, where they subserve a predominant inhibitory role. However, the multiple location of H_3 receptors in the same tissue may lead to opposite effects on the physiologic function, thus, their role can sometimes be quite difficult to understand. As regards a possible clinical application of H_3 ligands, although many suggestions have been derived from the experimental animal data (e.g. gastric disorders, asthma, myocardial ischemia, hypertension and inflammation), to date, no clinical evidence is available for any therapeutic indication.

Therefore, H_3 ligands still represent an unusual set of "orphan drugs". All the same, a selective modulation of neurotransmission in a specific tissue by means of more and more selective ligands, could still represent an innovative possibility for therapeutic intervention.

REFERENCES

Akagi, M., Hamada, K., Nishioka, E., Fukuishi, N., Akagi, R., 1995. Role of histamine H_3 receptor on hypoxia-reoxygenation induced cardiac dysfunction in guinea pigs. Meth. Find. Exp. Clin. Pharmacol. 17 (C), 30-35.

Allen, M.C., Graham, P., Morris, G., 1996. Histamine forming capacity (HFC) and its modulation by H_3 receptor ligands in a model of bronchial hyper-responsiveness. Inflamm. Res. 45, 118-122.

Andersson, K., Mattsson, H., Larsson, H., 1990. The role of gastric mucosal in acid secretion and experimentally induced lesions in the rats. Digestion 46, 1-9.

Arakawa, T., Satoh, H., Fukuda, T., Sakuma, H., Nakamura, H., Kobayashi, K., 1988. Gastric mucosal resistance and prostanoid levels after cimetidine treatment in rats. Digestion 41, 1-8.

Arrang, J.M., Garbarg, M., Schwartz, J.C., 1983. Auto-inhibition of brain histamine release mediated by a novel class (H_3) of histamine receptors. Nature 302, 832-837.

Arrang, J.M., Garbarg, M., Lancelot, J.C., Lecomte, J.M., Pollard, M., Robba, M., Schunack, W., Schwartz. J.C., 1987. Highly potent and selective ligands for histamine H_3 receptors. Nature 327, 117-123.

Aures, D., Guth, P.H., Paulsen, G., Grossman, M.I., 1982. Effect of increased gastric mucosal histamine on alcohol-induced gastric damage in rats. Dig. Dis. Sci. 27, 347-352.

Badò, A., Dubrasquet, M., Lewin, M.J.M., 1990. Evidence of histamine H_3 receptors in the stomach: physiological role in gastric acid regulation. Gastroenterology 98 (Suppl), A17.

Badò, A., Hervatin, F., Lewin, M.J.M., 1991. Pharmacological evidence for histamine H_3 receptor in the control of gastric acid secretion in cats. Am. J. Physiol. 260, G631-G635.

Badò, A., Laigneau, J.P., Moizo, L., Cherifi, Y., Lewin, M.J.M., 1995. H_3-receptor activation inhibits cholinergic stimulation of acid secretion in isolated rabbit fundic glands. J. Pharmacol. Exp. Ther. 275, 1099-1103.

Badò, A., Moizo, L., Laigneau, J.P., Delwaide, J., Lewin, M.J.M., 1994. H_3-receptor regulation of vascular gastrin and somatostatin releases by the isolated rat stomach. Yale J. Biol. Med. 67, 113-121.

Badò, A., Moizo, L., Laigneau, J.P., Lewin, M.J.M., 1992a. Pharmacological characterization of histamine H_3 receptors in isolated rabbit gastric glands. Am. J. Physiol. 262, G56-G61.

Badò, A., Moizo, L., Laigneau, J.P., Lewin, M.J.M., 1992b. Gastrin and somatostatin secretion by isolated perfused rat stomach: regulation by the gastric histamine H_3-receptor. Gastroenterology 102, A916.

Banning, M.M., Curtis, M.J., 1994. The antiarrhythmic effects of thioperamide in isolated rat heart. Br. J. Pharmacol. 113, 179P.

Barnes, P.J., 1992. Histamine receptors in the respiratory tract. In: Schwartz, J.C., Haas, H.L. (Eds.), The histamine receptor. Wiley-Liss, New York, (NY), pp. 253-270.

Barnes, J.C., Brown, J.D., Clarke, N.P., Clapham, J., Evans, D.J., O'Shaughnessy, C.T., 1993. Pharmacological activity of VUF9153, an isothiourea histamine H_3 receptor antagonist. Eur. J. Pharmacol. 250, 147-152.

Barocelli, E., Ballabeni, V., Chiavarini, M., Impicciatore, M., 1995. R-α–methylhistamine-induced inhibition of gastric acid secretion in pylorus-ligated rats via central histamine H_3 receptors. Br. J. Pharmacol. 115, 1326-1330.

Beales, I.L.P., Calam, J., 1997. Effect of N^{α}–methyl-histamine on acid secretion in isolated cultured rabbit parietal cells: implications for *helicobacter pylori* associated gastritis and gastric physiology. Gut 40, 14-19.

Belardinelli, L., Linden, J., Berne, R.M., 1989. The cardiac effects of adenosine. Progr. Cardiovasc. Dis. 32, 73-97.

Bent, S., Fehling, U., Braam, U., Schmutzler, W., 1991. The influence of H_1, H_2- and H_3-receptors on the spontaneous and ConA induced histamine release from human adenoidal mast cells. Agents Actions 33, 67-70.

Bertaccini, G., Coruzzi, G., 1992. Histamine receptors in the digestive system. In: Schwartz, J.C., Haas, H.L. (Eds.), The histamine receptor. Wiley Liss, New York, (NY), pp. 193-230.

Bertaccini, G., Coruzzi, G., 1995. An update on histamine H_3 receptors and gastrointestinal functions. Dig. Dis. Sci. 40, 2052-2063.

Bertaccini, G., Vitali, T., 1964. Pharmacological properties of some imidazole derivatives occurring in nature. J. Pharm. Pharmacol. 16, 441-449.

Bertaccini, G., Zappia, L., 1981. Histamine receptors in the guinea-pig duodenum. J. Pharm. Pharmacol. 33, 590-593.

Bertaccini, G., Coruzzi, G., Poli, E., 1991. Review article. The histamine H_3 receptor: A novel prejunctional receptor regulating gastrointestinal functions. Aliment. Pharmacol. Ther. 5, 585-591.

Bertaccini, G., Impicciatore, M., Mossini, F., 1971. Effects of some N-substituted histamine derivatives on gastric secretion. Naunyn-Schmiedeberg's Arch. Pharmacol. 269, 418.

Bertaccini, G., Morini, G., Coruzzi, G., 1995. An update on histamine receptors and gastroprotection. Ital. J. Gastroenterol. 27, 259-264.

Bertaccini, G., Cavaggioni, A., Zappia, L., Mossini, F., 1973. Action of some N^{α}-methyl derivatives of histamine on gastric secretion of the rat. Pharmacol. Res. Commun. 5, 71-80.

Beyak, M., Vanner, S., 1993. Dual histamine vasodilator mechanisms: endothelial release of nitric oxide (NO) and presynaptic inhibition of sympathetic nerves. Gastroenterology 104 (Suppl.), A1030.

Beyak, M., Vanner, S., 1995. Histamine H_1 and H_3 vasodilator mechanisms in the guinea pig ileum. Gastroenterology 108, 712-718.

Bissonnette, E.Y., 1996. Histamine inhibits tumor necrosis factor α release by mast cells through H_2 and H_3 receptors. Am. J. Respir. Cell. Mol. Biol. 14, 620-626.

Black, J.W., Duncan, W., Durant, C., Ganellin, C.R., Parsons, M.E., 1972. Definition and antagonism of histamine H_2 receptors. Nature 236, 590-593.

Bovet, D., Staub, A.M., 1937. Action protectrice des ethers phenoliques an cours de l'intoxications histaminique. CR. Soc. Biol. (Paris) 124, 547-549.

Bull, H.A., Courtney, P.F., Rustin, M.H.A., Dowd, P.M., 1992. Characterization of histamine receptor sub-types regulating prostacyclin release from human endothelial cells. Br. J. Pharmacol. 107, 276-281.

Burgaud J.L., Oudart, N., 1993a. Bronchodilatation of guinea-pig perfused bronchioles induced by the H_3-receptor for histamine: role of epithelium. Br. J. Pharmacol. 109, 960-966.

Burgaud J.L., Oudart, N., 1993b. Effect of an histaminergic H_3 agonist on the non-adrenergic non-cholinergic contraction in guinea-pig perfused bronchioles. J. Pharm. Pharmacol. 45, 955-958.

Cardell L.O., Edvinsson, L., 1994. Characterization of the histamine receptors in the guinea-pig lung: evidence for relaxant histamine H_3 receptors in the trachea. Br. J. Pharmacol. 111, 445-454.

Cherifi, Y., Pigeon, C., Le Romancer, M., Badò, A., Reyl-Desmars, F., Lewin, M.J.M., 1992. Purification of a histamine H_3 receptor negatively coupled to phosphoinositide turnover in the human gastric cell line HGT1. J. Biol. Chem. 267, 25315-25320.

Cooke, H.J., Reddix, R.A., 1994. Neural regulation of intestinal electrolyte transport. In: Johnson, L.R. (Ed.), Physiology of the gastrointestinal tract. Raven Press, New York, (NY), pp. 2083-2132.

Cooke, H.J., Wang, Y.Z., 1994. H_3 receptors: Modulation of histamine-stimulated neural pathways influencing electrogenic ion transport in the guinea pig colon. J. Auton. Nerv. Syst. 50, 201-207.

Corbel, S., Traiffort, E., Stark, H., Schunack, W., Dy, M., 1997. Binding of histamine H_3-receptor antagonists to hematopoietic progenitor cells. FEBS Letters 404, 289-293.

Coruzzi, G., Adami, A., Bertaccini, G., 1992. Histamine H_3 receptors are not involved in the regulation of rat gastric secretion. Pharmacology 44, 190-195.

Coruzzi, G., Bertaccini, G., Schwartz, J.C., 1991a. Evidence that histamine H_3 receptors are involved in the control of gastric acid secretion in the conscious cat. Naunyn-Schmiedeberg's Arch. Pharmacol. 343, 225-227.

Coruzzi, G., Gambarelli, E., Bertaccini, G., 1996. Stimulatory effect of (R)α-methylhistamine on duodenal HCO_3^- secretion in anaesthetized rats. Ital. J. Gastroenterol. 28, 520-522.

Coruzzi, C., Poli, E., Bertaccini, G., 1991b. Histamine receptors in isolated guinea pig duodenal muscle: H_3 receptors inhibit cholinergic neurotransmission. J. Pharmacol. Exp. Ther. 258, 325-331.

Coruzzi, G., Gambarelli, E., Bertaccini, G., Timmerman, H., 1995. Cardiovascular effects of selective agonists and antagonists of histamine H_3 receptors in the anaesthetized rat. Naunyn-Schmiedeberg's Arch. Pharmacol. 351, 569-575.

Courillon-Mallet, A., Launay, J.M., Roucayrol, A.M., Callebert, J., Emond, J.P., Tabuteau, F., Cattan, D., 1995. *Helicobacter pylori* infection: physiopathologic implication of N^{α}-methylhistamine. Gastroenterology 108, 959-966.

Danko, G., Hey, J.A., Egan, R.W., Kreutner, W., Chapman, R.W., 1994. Histamine H_3 receptors inhibit sympathetic modulation of airway microvascular leakage in allergic guinea pigs. Eur. J. Pharmacol. 254, 283-286.

Dart, A.M., Du, X.J., 1993. Unexpected drug effects on autonomic function during myocardial ischemia. Cardiovasc. Res. 27, 906-914.

Delaunois, A., Gustin, P., Garbarg, M., Ansay, M., 1995. Modulation of acetylcholine, capsaicin and substance P effects by histamine H_3 receptors in isolated perfused rabbit lungs. Eur. J. Pharmacol. 277, 243-250.

Dimitriadou, V., Rouleau, A., Dam Trung Tuong, M., Newlands, G.J.F., Miller, H.R.P., Luffau, G., Schwartz, J.C., Garbarg, M., 1994. Functional relationship between mast cells and C-sensitive nerve fibers evidenced by histamine H_3-receptor modulation in rat lung and spleen. Clin. Sci. 87, 151-163.

Djuric, D.M., Andjelkovic, I.Z., 1995. The evidence for histamine H_3 receptor-mediated endothelium-dependent relaxation in isolated rat aorta. Mediators Inflammation 4, 217-221.

Djuric, D.M., Nesic, M.T., Andjelkovic, I.Z., 1996. Endothelium-dependent relaxation of rat aorta to a histamine H_3 agonist is reduced by inhibitors of nitric oxide synthase, guanylate cyclase and Na^+, K^+-ATPase. Mediators Inflammation 5, 69-74.

Ea-Kim L., Oudart, N., 1988. A highly potent and selective H_3 agonist relaxes rabbit middle cerebral artery, in vitro. Eur. J. Pharmacol. 150, 393-396.

Ea-Kim, L., Javellaud, J., Oudart, N., 1992a. In vivo cardiovascular effects of histamine H_3-agonist and H_3-antagonist. Fundam. Clin. Pharmacol. 6, 206.

Ea-Kim, L., Javellaud, J., Oudart, N., 1992b. Endothelium-dependent relaxation of rabbit middle cerebral artery to a histamine H_3-agonist is reduced by inhibitors of nitric oxide and prostacyclin synthesis. Br. J. Pharmcol. 105, 103-106.

Ea-Kim, L., Javellaud J., Oudart, N., 1993. Histamine H_3 agonist decreases arterial blood pressure in the guinea-pig. J. Pharm. Pharmacol. 45, 929-931.

Ea-Kim, L., Javellaud, J., Oudart, N., 1996. Reduction of the pressor response to nicotine in the guinea pig by a histamine (H_3) agonist is attenuated by an inhibitor of nitric oxide synthesis. J. Cardiovasc. Pharmacol. 27, 607-613.

Ea-Kim, L., Malard, N., Oudart, N., 1990. Effects of a H_3 histaminergic agonist on guinea pig basilar arteries in vitro. Fundam. Clin. Pharmacol. 4, 577.

Endou, M., Poli, E., Levi, R., 1994. Histamine H_3-receptor signaling in the heart: Possible involvement of G_1/G_0 proteins and N-type Ca^{++} channels. J. Pharmacol. Exp. Ther. 269, 221-229.

Falus, A., 1994. Inflammation: the role of histamine. In: Falus, A. (Ed.). Histamine and inflammation. R.G. Landes. Co., Georgetown, Tx. pp. 63-77.

Fargeas, M.J., Fioramonti, J., Bueno, L., 1989. Involvement of different receptors in the central and peripheral effects of histamine on intestinal motility in the rat. J. Pharm. Pharmacol. 41, 534-540.

Fuder, H., Ries, P., Schwarz, P., 1994. Histamine and serotonin released from the rat perfused heart by compound 48/80 or by allergen challenge influence noradrenaline or acetylcholine exocytotic release. Fundam. Clin. Pharmacol. 8, 477-490.

Furchgott, R.F., 1983. Role of endothelium in responses for vascular smooth muscle to drugs. Cir. Res. 53, 557-573.

Furchgott, R.F., Zawadzki, J.V., 1980. The obligatory role of endothelial cells in the relaxation of arterial smooth muscle by acetylcholine. Nature 288, 373-376.

Galli, S.J., Wershil, B.K., Bose, R., Walker, P.A., Szabo, S., 1987. Ethanol-induced acute gastric injury in mast cell-deficent and congenic normal mice. Am. J. Pathol. 128, 131-140.

Godlewski, G., Malinowska, B., Buczko, W., Schlicker, E., 1997a. Inhibitory H_3 receptors on sympathetic nerves of the pithed rat: activation by endogenous histamine and operation in spontaneously hypertensive rats. Naunyn-Schmiedeberg's Arch. Pharmacol. 355, 261-266.

Godlewski, G., Malinowska, B., Schlicker, E., Bucher, B., 1997b. Identification of histamine H_3-receptors in the tail artery from normotensive and spontaneously hypertensive rats. J. Cardiovasc. Pharmacol. 29, 801-807.

Göthert, M., Garbarg, M., Hey, J.A., Schlicker, E., Schwartz, J.C., Levi, R., 1995. New aspects of the role of histamine in cardiovascular function: identification, characterization, and potential pathophysiological importance of H_3 receptors. Can. J. Physiol. Pharmacol. 73, 558-564.

Guc, M.O., Ilhan, M., Kayaalp, S.O., 1988. Epithelium-dependent relaxation of guinea-pig tracheal smooth muscle by histamine: evidence for non-H_1- and non-H_2-histamine receptors. Arch. int. Pharmacodyn. 296, 57-65.

Gupta, S., Cragoe, E.J. JR., Deth, R.C., 1989. Influence of atrial natriuretic factor on 5-(N-ethyl-N-isopropyl)amiloride-sensitive^{22} Na^+ uptake in rabbit aorta. J. Pharmacol. Exp. Ther. 248, 991-996.

Hatta, E., Imamura, M., Yasuda, K., Levi, R., 1996. Activation of histamine H_3-receptors inhibits carrier-mediated norepinephrine release in a human model of protracted myocardial ischemia. Circulation 94, I-474.

Hatta, E., Yasuda, K., Levi, R., 1997. Activation of histamine H_3-receptors inhibits carrier-mediated norepinephrine release in a human model of protracted myocardial ischemia. J. Pharm. Expt. Thér. 283, 494-500.

Hegde, S.S., Chan, P., Eglen, R.M., 1994. Cardiovascular effects of R-α-methylhistamine, a selective histamine H_3 receptor agonist, in rats: lack of involvement of histamine H_3 receptors. Eur. J. Pharmacol. 251, 43-51.

Hew, R.W.S., Hodgkinson, C.R., Hill, S.J., 1990. Characterization of histamine H_3-receptors in guinea-pig ileum with H_3-selective ligands. Br. J. Pharmacol. 101, 621-624.

Hey, J.A., del Prado, M., Egan, R.W., Kreutner, W., Chapman, R.W., 1992a. Inhibition of sympathetic hypertensive responses in the guinea pig by prejunctional histamine H_3 receptors. Br. J. Pharmacol. 107, 347-351.

Hey, J.A., del Prado, M., Egan, R.W., Kreutner, W., Chapman, R.W., 1992b. (R)-α-methylhistamine augments neural, cholinergic bronchospasm in guinea pigs by histamine H_1-receptor activation. Eur. J. Pharmacol. 211, 421-426.

Hollande, F., Bali, J.P., Magous, R., 1993. Autoregulation of histamine synthesis through H_3 receptors in isolated fundic mucosal cells. Am. J. Physiol. 265, G1039-G1044.

Holzer, P., 1989. Ascending enteric reflex: multiple neurotransmitter systems and interactions. Am. J. Physiol. 256, G540-G545.

Holzer, P., 1998. Neurogenic vasodilatation and plasma leakage in the skin. Gen. Pharmac. 30, 5-11.

Hutchison, R.W., Hey, J.A., 1994. Pharmacological characterization of the inhibitory effect of (R)-α-methylhistamine on sympathetic cardiopressor responses in the pithed guinea-pig. J. Auton. Pharmacol. 14, 393-402.

Ichinose, M., Barnes, P.J., 1989a. Histamine H_3-receptors modulate nonadrenergic noncholinergic neural bronchoconstriction in guinea-pig in vivo. Eur. J.Pharmacol. 174, 49-55.

Ichinose, M., Barnes, P.J., 1989b. Inhibitory histamine H_3-receptors on cholinergic nerves in human aiways. Eur. J. Pharmacol. 163, 383-386.

Ichinose, M., Belvisi, M.G., Barnes, P.J., 1990. Histamine H_3-receptor inhibit neurogenic microvascular leakage in airways. J. Appl. Physiol. 68, 21-25.

Ichinose, M., Stretton, C.D., Schwartz, J.C., Barnes, P.J., 1989. Histamine H_3-receptors inhibit cholinergic neurotransmission in guinea-pig airways. Br. J. Pharmacol. 97, 13-15.

Imamura, M., Lander, H.M., Levi, R., 1996a. Activation of histamine H_3-receptors inhibits carrier-mediated norepinephrine release during protracted myocardial ischemia. Circ. Res. 78, 475-481.

Imamura, M., Poli, E., Omoniyi, A.T., Levi, R., 1994. Unmasking of activated histamine H_3-receptors in myocardial ischemia: their role as regulators of exocytotic norepinephrine release. J. Pharmacol. Exp. Ther. 271, 1259-1266.

Imamura, M., Seyedi, N., Lander, H.M., Levi, R., 1995. Functional identification of histamine H_3-receptors in the human heart. Circ. Res. 77, 206-210.

Imamura, M., Smith, N.C.E., Garbarg, M., Levi, R., 1996b. Histamine H_3-receptors-mediated inhibition of calcitonin gene-related peptide release from cardiac C-fibers. Circ. Res. 78, 863-869.

Impicciatore, M., 1978. Occurrence of H_1- and H_2 -histamine receptors in the guinea-pig gall bladder *in situ*. Br. J. Pharmacol. 64, 219-222.

Ishikawa, S., Sperelakis, N., 1987. A novel class (H_3) of histamine receptors on perivascular nerve terminals. Nature 327, 158-160.

Jansen, F.P., Wu, T.S., Voss, H.-P., Steinbusch, H.W.M., Vollinga, R.C., Rademaker, B., Bast, A., Timmerman, H., 1994. Characterization of the binding of the first selective radiolabelled histamine H_3 receptor antagonist, [^{125}I]-iodophenpropit, to rat brain. Br. J. Pharmacol. 113, 355-362.

Jennings, L.J., Salido, G.M., Pozo, M.J., Davison, J.S., Sharkey, K.A., Lea, R.W., Singh, J., 1995. The source and action of histamine in the isolated guinea-pig gallbladder. Inflamm. Res. 44, 447-453.

Jennings, L.J., Salido, G.M., Pariente, J.A., Davison, J.S., Singh, J., Sharkey, K.A., 1996. Control of exocrine secretion in the guinea-pig pancreas by histamine H_3 receptors. Can. J. Physiol. Pharmacol. 74, 744-752.

Kathmann, M., Schlicker, E., Detzner, M., Timmerman, H., 1993. Nordimaprit, homodimaprit, clobenpropit and imetit: affinities for H_3 binding sites and potencies in a functional H_3 receptor model. Naunyn-Schmiedeberg's Arch. Pharmacol. 348, 498-503.

Kidd, M., Tang, L.H., Miu, K., Lawton, G.P., Sandor, A., Modlin, I.M. (1996). Autoregulation of enterochromaffin-like cell histamine secretion via the histamine 3 receptor subtype. Yale J. Biol. Med. 69, 9-19.

Kim, Y.C., Davies, M.G., Lee, T.H., Hagen, P.O., Carson, C.C., 1995. Characterization and function of histamine receptors in corpus cavernosum. J. Urol. 153, 506-510.

Kleine-Tebbe, J., Schramm, J., Bolz, M., Lipp, R., Schunack, W., Kunkel, G., 1990. Influence of histamine H_3-antagonists on human leukocytes. Agents and Actions 30, 137-139.

Kohno, S., Nakao, S., Ogawa, K., Yamamura, H., Nabe, T., Ohata, K., 1994. Possible participation of histamine H_3-receptors in the regulation of anaphylactic histamine release from isolated rat peritoneal mast cells. Japan. J. Pharmacol.66, 173-180.

Korte, A., Myers, J., Shih, N.Y., Egan, R.W., Clark, M.A., 1990. Characterization and tissue distribution of H_3 histamine receptors in guinea pigs by N^{α}-methylhistamine. Biochemical. Biophysical. Res. Comm. 168, 979-986.

Kostic, M.M., Jakovljevic, V.L.J., 1996. Role of histamine in the regulation of coronary circulation. Physiol. Res. 45, 297-303.

Krause, M., Rouleau, A., Stark, H., Luger, P., Lipp, R., Garbarg, M. et al.,, 1995. Synthesis, X-ray cristallography, and pharmacokinetics of novel azomethine prodrugs of (R)α-methylhistamine: highly potent and selective histamine H_3 receptor agonists. J. Med. Chem. 38, 4070-4079.

Krause, M., Rouleau, A., Stark, H., Garbarg, M. Schwartz, J.-C and Schunack, W. 1996. Structure-activity relationships of novel azomethine prodrugs of the histamine H_3-receptor agonist (R)α-methylhistamine: from alkylaril to substituted diaryl derivatives. Pharmazie 51, 720-726.

Lassen, L.H., Thomsen, L.L., Olesen, J., 1995. Histamine induces migraine via the H_1-receptor. Support for the NO hypothesis of migraine. NeuroReport 6, 1475-1479.

Leurs, R., Timmerman, H., 1992. The histamine H_3-receptor: a target for developing new drugs. In: Jucker E. (Ed.), Progress in Drug Ressearch, Vol. 39, Birckhauser Verlag, Berlin, pp. 127-165.

Leurs, R., Smit, M.J., Timmerman, H., 1995a. Molecular pharmacological aspects of histamine receptors. Pharmac. Ther. 66, 413-463.

Leurs, R., Brozius, M.M., Smit, M.J., Bast, A., Timmerman, H., 1991. Effects of histamine H_1-, H_2- and H_3-receptor selective drugs on the mechanical activity of guinea-pig small and large intestine. Br. J. Pharmacol. 102, 179-185.

Leurs, R., Kathmann, M., Vollinga, R.C., Menge, W.M.P.B., Schlicker, E., Timmerman, H., 1996. Histamine homologues discriminating between two functional H_3 receptor assays. Evidence for H_3 receptor heterogeneity? J. Pharmacol. Exp. Ther. 276, 1009-1015.

Leurs, R., Tulp, M.Th.M. Menge, W.M.B.P., Adolfs, M.J.P., Zuiderveld, O.P., Timmermann, H., 1995b. Evaluation of the receptor selectivity of the H_3 receptor antagonists, iodophenpropit and thioperamide: an interaction with the 5-HT_3 receptor revealed. Br. J. Pharmacol. 116, 2315-2321.

Levi, R., Owen, D.A.A., Trzeciakowski, J., 1982. Actions of histamine on the heart and vasculature. In: Ganellin, C.R., Parsons, M.E. (Eds.), Pharmacology of histamine receptors. Wright, London, pp. 236-297.

Levi, R., Rubin, L.E., Gross, S.S 1991. Histamine in cardiovascular function and dysfunction: recent developments. In: Uvnas, B. (Ed.), Handbook Experimental Pharmacology 97, Histamine and histamine antagonists. Springer-Verlag, Berlin, FRG, pp. 347-383.

Luo, X.X., Tan, Y.H., 1994. Presynaptic histamine H_1- and H_3-receptors modulate sympathetic neurotransmission in isolated guinea pig vas deferens. Acta Pharmacol. Sin. 15, 60-64.

Luo, X.X., Tan, Y.H., Sheng, B.H., 1991. Histamine H_3-receptors inhibit sympathetic neurotransmission in guinea pig myocardium. Eur. J. Pharmacol. 204, 311-314.

Luo, X.X., Song, L., Jiang, Y.P., Tan, Y.H., 1994. Inhibition of sympathetic neurotransmission via NEM-sensitive H_3-receptors in the guinea pig vas deferens. Meth. Find. Exp. Clin. Pharmacol. 16, 185-189.

Lüscher, T.F., 1994. The endothelium in hypertension: bystander, target or mediator? J. Hypertension 12 (suppl. 10), S105-S116.

Malinowska, B., Schlicker, E., 1991. H_3 receptor-mediated inhibition of the neurogenic vasopressor response in pithed rats. Eur. J. Pharmacol. 205, 307-310.

Malinowska, B., Schlicker, E., 1993a. Effects of H_1 and H_3 agonists on the neurogenic vasopressor response in the pithed rat. Agents Actions 38, C254-C256.

Malinowska, B., Schlicker, E., 1993b. Identification of endothelial H_1, vascular H_2 and cardiac presynaptic H_3 receptors in the pithed rat. Naunyn-Schmiedeberg's Arch. Pharmacol. 347, 55-60.

Mannaioni, P.F., Bello, M.G., Di Bello, M.G., Mirabella, C., Gai, P., Schunack, W., Masini, E., 1997. Interaction between histamine and nitric oxide in rat mast cells and in isolated guinea pig hearts. Int. Arch. Allergy Immunol. 113, 297-299.

Mansfield, L.E., 1990. The role of antihistamine therapy in vascular headaches. J. Allergy. Clin. Immunol. 86, 673-676.

Martinez. A.C., Novella, S., Raposo, R., Recio, P., Labadia, A., Costa, G., 1997. Histamine receptors in isolated bovine oviductal arteries. Eur. J. Pharmacol. 326, 163-173.

Matsubara, T., Moskowitz, M.A., Huang, Z., 1992. UK-14,304, R(-)-a-methyl-histamine and SMS 201-995 block plasma protein leakage within dura mater by prejunctional mechanisms. Eur. J. Pharmacol. 224, 145-150.

McGowan, C.C., Cover, T.L., Blaser, M.J., 1996. *Helicobacter pylori* and gastric acid: biological and therapeutic implications. Gastroenterology 110, 926-938.

McKay, D.M., Bienenstock, J., 1994. The interaction between mast cells and nerves in the gastrointestinal tract. Immunol. Today 15, 533-538.

McLeod, R.L., Gertner, S.B., Hey, J.A., 1991. Modulation of cardiovascular function by central histamine H_3 receptors in conscious guinea pigs. Eur. J. Pharmacol. 206, 141-142.

McLeod, R.L., Gertner, S.B., Hey, J.A., 1993. Production by R-α-methylhistamine of a histamine H_3 receptor-mediated decrease in basal vascular resistance in guinea-pigs. Br. J. Pharmacol. 110, 553-558.

McLeod, R.L., Gertner, S.B., Hey, J.A., 1994. Species differences in the cardiovascular responses to histamine H_3 receptor activation. Eur. J. Pharmacol. 259, 211-214.

McNeill, J.H., 1984. Histamine and the heart. Can. J. Physiol. Pharmacol. 62, 271-280.

Menkveld, G.J., Timmerman, H., 1990. Inhibition of electrically-evoked contractions of guinea-pig ileum preparations mediated by the histamine H_3 receptor. Eur. J. Pharmacol. 186, 343-347.

Modlin, I.M., Lawton, G.P., Tang, L.H., Miu, K., Schwartz, J.C., 1995. Histamine-3 receptor modulation of ECL cell histamine secretion and DNA synthesis. Gastroenterology 108 (Suppl), A992.

Molderings, G.J., Weißenborn, G., Schlicker, E., Likungu, J., Gothert, M., 1992. Inhibition of noradrenaline release from the sympathetic nerves of the human saphenous vein by presynaptic histamine H_3 receptors. Naunyn-Schmiedeberg's Arch. Pharmacol. 346, 46-50.

Montesino, H., Villar, M., Vega, E., Rudolph, M.I., 1995. Histamine, a neuromodulator of noradrenergic transmission in uterine horns from mice in diestrus. Biochem. Pharmacol. 50, 407-411.

Morini, G., Grandi, D., Bertaccini, G., 1995. (R)-alpha-methylhistamine inhibits ethanol-induced gastric lesions in the rat: involvement of histamine H_3 receptors? Digestion 56, 145-152.

Morini, G., Grandi, D., Bertaccini, G., 1996a. Comparison of the protective effect of (R)-α-methylhistamine and its prodrugs on gastric mucosal damage induced by ethanol in the rat. Gen. Pharmac. 27, 1391-1394.

Morini, G., Grandi, D., Gentili, S., Bertaccini, G., 1996b. Prevention by (R)-α-methylhistamine of ethanol-induced gastric mucosal lesions in rats: importance of adherent mucus gel layer. Acta Bio-Med. (Ateneo Parmense) 67, 71-78.

Morini, G., Grandi, D., Krause, M., Schunack, W., 1997. Gastric mucosal injury by nonsteroidal anti-inflammatory drugs is reduced by (R)-α-methylhistamine and its pro-drugs in the rat. Inflamm. Res. 46 (Suppl. 1), S101-S102.

Moskowitz, M.A., 1991. The visceral organ brain: implications for the pathophysiology of vascular headache. Neurology 41, 182-186.

Muller, M.J., Padol, I., Hunt, R.H., 1993. Acid secretion in isolated murine gastric glands: classification of histaminergic and cholinergic receptors. Gastroenterology 104 (Suppl), A151.

Nishiwaki, H., Takeuchi, K., Okada, M., Tanaka, H., Okabe, S., 1989. Stimulation of gastric alkaline secretion by histamine in rats: possible involvement of histamine H_2-receptors and endogenous prostaglandins. J. Pharmacol. Exp. Ther. 248, 793-798.

O'Connor, B.J., Lecomte, J.M., Barnes, P.J., 1993. Effect of an inhaled histamine H_3-receptor agonist on airway responses to sodium metabisulphite in asthma. Br. J. Clin. Pharmacol. 35, 55-57.

Ohkubo, T., Shibata, M., Inoue, M., Kaya, H., Takahashi, H., 1994. Autoregulation of histamine release via the histamine H_3 receptor on mast cells in the rat skin. Arch. int. Pharmacodyn. 328, 307-314.

Ohkubo, T., Shibata, M., Inoue, M., Kaya, H., Takahashi, H., 1995. Regulation of substance P release mediated via prejunctional histamine H_3 receptors. Eur. J. Pharmacol. 273, 83-88.

Oike, M., Kitamura, K., Kuriyama, H., 1992. Histamine H_3-receptor activation augments voltage-dependent Ca^{2+} current via GTP hydrolysis in rabbit saphenous artery. J. Physiol. 448, 133-152.

Oishi, R., Adachi, N., Saeki, K., 1993. N^{α}-methylhistamine inhibits intestinal transit in mice by central histamine H_1 receptor activation. Eur. J. Pharmacol. 237, 155-159.

Oudart, N., Javellaud. J., Ea Kim, L., 1995. A histamine H_3-agonist attenuates the cardiovascular response to nicotine in rats. Pharmacol. Res. 31 (Suppl), 250.

Owen, D.A.A., Poy, E., Woodward, D.F., Daniel, D., 1980. Evaluation of the role of histamine H_1- and H_2-receptors in cutaneous inflammation in the guinea-pig produced by histamine and mast cell degranulation. Br. J. Pharmacol. 69, 615-623.

Palitzsch, K.D., Horales, R.E., Kronauge, J.F., Szabo, S., 1989. Biphasic effect of histamine on hemorrhagic mucosal lesions is related to vascular permeability: studies with histamine, H_1-, H_2- and H_3-agonists and bradykinin. Gastroenterology 96, (Suppl.) A381.

Pile, A.J., Smaje, L.H., 1992. The H_3 antagonist thioperamide inhibits the permeability effect of histamine in rat microvessels via an H_1 receptor effect. Br. J. Pharmacol. 105, 210P.

Plaut, M., Lichtenstein, L.M., 1982. Histamine and immune responses. In: Pharmacology of histamine receptors. Ganellin, C.R. Parsons M.E. (Eds.), Wright-PSG, London, UK. pp. 392-435.

Poli, E., Pozzoli, C., 1997. Histamine H_3 receptors do not modulate reflex-evoked peristaltic motility in the isolated guinea-pig ileum. Eur. J. Pharmacol. 327, 49-56.

Poli, E., Coruzzi, G., Bertaccini, G., 1991. Histamine H_3 receptors regulate acetylcholine release from the guinea pig ileum myenteric plexus. Life Sci. 48, PL63-PL68.

Poli, E., Pozzoli, C., Bertaccini, G., 1997. Is there a role for histamine H_3-receptors in the control of intestinal peristalsis? Inflamm. Res. 46 (Suppl.1), S99-S100.

Poli, E., C. Pozzoli, C., Coruzzi, G., Bertaccini, G., 1993. Histamine H_3 receptor-mediated inhibition of duodenal cholinergic transmission is independent of intracellular cyclic AMP and GMP. Gen. Pharmacol. 24, 1273-1278.

Poli, E., Pozzoli, C., Coruzzi, G., Bertaccini., G., 1994a. Signal transducing mechanisms coupled to histamine H_3 receptors and *alpha*-2 adrenoceptors in the guinea pig duodenum: possible involvement of N-type Ca^{++} channels. J. Pharmacol. Exp. Ther. 270, 788-794.

Poli, E., Todorov, S., Pozzoli, C., Bertaccini, G., 1994b. Presynaptic histamine H_2 receptors modulate the sympathetic nerve transmission in the isolated rat vas deferens; no role for H_3-receptors. Agents Actions 42, 95-100.

Pozzoli, C., Poli, E., Costa, A., De Ponti, F., 1997. Absence of histamine H_3 receptors in the rabbit colon: species difference. Gen. Pharmacol. 28, 217-221.

Prinz, C., Kajimura, M., Scott, D.R., Mercier, F., Helander, H.F., Sachs, G., 1993. Histamine secretion from rat enterochromaffin-like cells. Gastroenterology 105, 449-461.

Raible, D.G., Lenahan, T., Fayvilevich, Y., Kosinski, R., Schulman, E.S., 1994. Pharmacological characterization of a novel histamine receptor on human eosinophils. Am. J. Respir. Crit. Care Med. 149, 1506-1511.

Rangachari, P.K., Prior, T., 1994. Functional subtyping of histamine receptors on the canine proximal colonic mucosa. J. Pharmacol. Exp. Ther. 271, 1016-1026.

Reeder, L., DeFilippi, V.J., Ferguson, M.K., 1996. Characterization of the effects of histamine on porcine tracheobronchial lymph vessels. Am. J. Physiol. 271, H2501-H2507.

Rizzo, C.A., Tozzi, S., Monahan, M.E., Hey., J.A., 1995. Pharmacological characterization of histamine H_3 receptors in isolated guinea pig pulmonary artery and ileum. Eur. J. Pharmacol. 294, 329-335.

Rosic, M., Collis, C.S., Andjelkovic, I.Z., Segal, M.B., Diuric, D., Zlokovic, B.V., 1991. The effects of (R)α-methylhistamine on the isolated guinea pig aorta. In: Timmerman, H., Van Der Goot, H. (Eds.). New perspectives in histamine research. Birkauser, Basel, pp. 283-287.

Rouleau, A., Garbarg, M., Ligneau, X., Mantion, C., Lavie, P., Advenier, C., Lecomte, J.M., Krause, M., Stark, H., Schunack, W., Schwartz, J.C., 1997. Bioavailability, antinociceptive and antiinflammatory properties of BP 2-94, a histamine H-3 receptor agonist prodrug. J. Pharmacol. Exp. Ther. 281, 1085-1094.

Sandvik, A.K., Lewin, M.J.M., Waldum, H.L., 1989. Histamine release in the isolated vascularly perfused stomach of the rat: regulation by autoreceptors. Br. J. Pharmacol. 96, 557-562.

Schlicker, E., Behling, A., Lummen, G., Gothert. M., 1992. Histamine H_{3A} receptor-mediated inhibition of noradrenaline release in the mouse brain cortex. Naunyn-Schmiedeberg's Arch. Pharmacol. 345, 489-493.

Schlicker, E., Malinowska, B., Kathmann, M., Gothert, M., 1994a. Modulation of neurotransmitter release via histamine H_3 heteroreceptors. Fundam. Clin. Pharmacol. 8, 128-137.

Schlicker, E., Kathmann, M., Reidemeister, S. Stark, H., Schunack. W., 1994b. Novel histamine H_3 receptor antagonists: affinities in an H_3 receptor binding assay and potencies in two functional H_3 receptor models. Br. J. Pharmacol. 112, 1043-1048.

Schlicker, E., Kathmann, M., Bitschnau, S., Marr, I., Reidemeister, S., Stark, H., Schunack, W., 1996. Potencies of antagonists chemically related to iodoproxyfan at histamine H_3 receptors in mouse brain cortex and guinea-pig ileum: evidence for H_3 receptor heterogeneity? Naunyn-Schmiedeberg's Arch. Pharmacol. 353, 482-488.

Schneider, D., Schlicker, E., Malinowska, B., Molderings, G., 1991. Noradrenaline release in the rat vena cava is inhibited by γ-aminobutyric acid via $GABA_B$ receptors but not effected by histamine. Br. J. Pharmacol. 104, 478-482.

Schomig, A., 1990. Catecholamines in myocardial ischemia: systematic and cardiac release. Circulation 82, 13-22.

Schubert, M.L., Harrington, L., Makhlouf,. G.M., 1993. Histamine H_3-receptors are coupled to inhibition of somatostatin secretion in the fundus and antrum of the stomach. Gastroenterology 104 (Suppl), A854.

Schwörer, H., Katsoulis, S., Racke', K., 1992. Histamine inhibits 5-hydroxytryptamine release from the porcine small intestine: involvement of H_3 receptors. Gastroenterology 102, 1906-1912.

Schwörer, H., Reimann, A., Ramadori, G., Racke', K., 1994. Characterization of histamine H_3 receptors inhibiting 5-HT release from porcine enterochromaffin cells: further evidence for H_3 receptor heterogeneity. Naunyn Schmiedeberg's Arch. Paharmacol. 350, 375-379.

Smit, J., Coppes, R.P., van Tintelen, E.J.J., Roffel, A.F., Zaagsma, J., 1997. Prejunctional histamine H_3-receptors inhibit electrically evoked endogenous noradrenaline overflow in the portal vein of freely moving rats. Naunyn-Schmiedeberg's Arch. Pharmacol. 355, 256-260.

Soldani, G., Garbarg, M., Intorre, L., Bertini, S., Rouleau, A., Schwartz, J.C., 1996. Modulation of pentagastrin-induced histamine release by histamine H_3 receptors in the dog. Sc. J. Gastroenterol. 31, 631-638.

Soldani, G., Intorre, L., Bertini, S., Luchetti, E., Coruzzi, G., Bertaccini, G., 1994. Regulation of gastric acid secretion by histamine H_3 receptors in the dogs: an investigation into the site of action. Naunyn-Schmiedeberg's Arch. Pharmacol. 350, 218-223.

Soldani, G., Mengozzi, G., Intorre, L., De Giorgi, G., Coruzzi, G., Bertaccini, G., 1993. Histamine H_3 receptor-mediated inhibition of gastric acid secretion in conscious dogs. Naunyn-Schmiedeberg's Arch. Pharmacol. 347, 61-65.

Tamura, K., Palmer, J.M., Wood, J.D., 1988. Presynaptic inhibition produced by histamine at nicotinic synapses in enteric ganglia. Neuroscience 25, 171-179.

Takeuchi, K., Nishiwaki, H., Furukawa, O., Okabe, S., 1987. Cytoprotective action of histamine against 0.6 N HCL-induced gastric mucosal injury in rats: comparative study with adaptive cytoprotection induced by exogenous acid. Japan. J. Pharmacol. 44, 335-344.

Taylor, S.J., Kilpatrick, G.J., 1992. Characterization of histamine-H_3 receptors controlling non-adrenergic non-cholinergic contractions of the guinea pig isolated ileum. Br. J. Pharmacol. 105, 667-674.

Tedeschi, A., Lorini, M., Arquati, M., Miadonna, A., 1991. Regulation of histamine release from human basophil leucocytes: role of H_1, H_2 and H_3 receptors. Allergy 46, 26-31.

Trzeciakowski, J.P., 1987. Inhibition of guinea pig ileum contractions mediated by a class of histamine receptors resembling the H_3 subtype. J. Pharmacol. Exp. Ther. 243, 874-880.

Velasquez, R.D., Brunner, G., Varrentrapp, M., Tsikas, D., Frolich, J.C., 1996. Helicobacter pylori produces histamine and spermidine. Z Gastroenterology 34, 116

Vollinga, R.C., Menge, W.M.P.B., Leurs, R., Timmerman, H., 1995. Homologs of histamine as histamine H_3 receptor antagonists: a new potent and selective H_3 antagonist, 4(5)-(5-aminopentyl)-1*H*-imidazole. J. Med. Chem. 38, 266-271.

Vuyyuru, L., Schubert, M.L., 1993. Participation on histamine H_3-receptors of the fundus in the regulation of gastric acid secretion. Gastroenterology 104 (Suppl), A220.

Vuyyuru, L., Harrington, L., Arimura, A., Schubert, M.L. (1997). Reciprocal inhibitory paracrine pathways link histamine and somatostatin secretion in the fundus of the stomach. Am. J. Physiol. 273, G106-G111.

Vuyyuru, L., Schubert, M.L., Harrington, L., Arimura, A., Makhlouf, G.M., 1995. Dual inhibitory pathways link antral somatostatin and histamine secretion in human, dog, and rat stomach. Gastroenterology 109, 1566-1574.

Wang, Y.Z., Cooke, H.J., 1990. Presynaptic H_3 receptors regulate cyclical Cl^- secretion in guinea pig distal colon. Gastroenterology 98 (Suppl), A400.

West, R.E., Zweig, A., Shih, N.Y., Siegel, M.I., Egan, R.W., Clark, M.A., 1990. Identification of two H_3 histamine receptor subtypes. Mol. Pharmacol. 38, 610-613.

Yau, W.M., Youther, M.L., 1993. Selective inhibition of acetylcholine release from myenteric plexus by $histamine_3$ receptor activation. Gastroenterology 104 (Suppl), A605.

Acknowledgements

Research made by the authors in the field of histamine H_3 receptors have been carried out by the help of Drs. G. Morini, M. Adami, C. Pozzoli and E. Gambarelli (Inst. of Pharmacology, University of Parma). Thanks are due to Mrs. S. Spaggiari for her skilful technical assistance.

R. Leurs and H. Timmerman (Editors)
The Histamine H_3 Receptor

Biochemical properties of the histamine H_3 receptor.

Marcel Hoffmann, Henk Timmermann and Rob Leurs.

Leiden/Amsterdam Centre for Drug Research, Division of Medicinal Chemistry, Faculty of Chemistry, Vrije Universiteit De Boelelaan 1083, 1081 HV Amsterdam the Netherlands

1. INTRODUCTION.

During the last decade many genes encoding G-protein coupled receptors (GPCRs) have been cloned, including the histamine H_1 and H_2 receptor [1-11]. The cloning of these genes has offered new possibilities in the field of pharmacological research. Some ten years ago GPCRs could only be studied in tissue preparations and cell lines, endogenously expressing low levels of receptors. With cloned receptor genes it is now possible to express these receptors heterologously in cultured cells and study a single receptor type in an isolated system. This technique offers the possibility not only to express and study the receptor protein itself but also to look in more detail to signal transduction routes that are activated by receptors. Furthermore, it is now possible to introduce changes in the amino acid composition of the receptor protein in order to test receptor-ligand interaction models or to mimic polymorphic receptor gene variants. With the ongoing synthesis of new compounds, cloned receptor cDNAs also offer the possibility to design high-throughput screening assay's to test drugs at the same rate as they become available. Although the cloning of receptor protein genes has opened new insights in receptor pharmacology and fundamental routes of signal transduction one has to be aware of the fact that the transfected cells are just models. Therefore a combination of modern and classical pharmacology remains necessary to fully understand receptor function and action in vivo.
Since the discovery of the histamine H_3 receptor by Arrang and co-workers in 1983 [12] a large amount of pharmacological data on this receptor has been obtained by many research groups (recently reviewed by Leurs et al. [13]. All this research has led to a fairly good understanding of the H_3 receptor although some questions remain unanswered. Among these there are questions involving the exact signal transduction pathway the receptor activates, and the existence of receptor subtypes.

Despite all the research done on the H_3 receptor, the gene encoding the H_3 receptor has not been cloned yet, although considerable efforts probably have been made in the past. In this chapter we will discuss in detail the biochemical properties of the H_3 receptor and potential strategies that may end in the cloning of the H_3 receptor cDNA.

2. SIGNAL TRANSDUCTION

At present, little is known about the signal transduction pathways that are stimulated via the H_3 receptor. From radioligand binding studies in which guanyl-nucleotids have been shown to inhibit the binding of H_3 agonists to the H_3 receptor, the interaction with a G-protein is suggested [14-18]. In 1994 Vollinga et al. [19] showed results similar to the results obtained with tritiated H_3 agonist using an iodinated H_3 antagonist. In this study they showed that displacement of the H_3 antagonist [^{125}I]iodophenpropit ([^{125}I]IPP) [17, 20] binding to rat cortical membranes resulted in shallow displacement curves for the H_3 specific agonist immepip (figure 1.). Computer analysis of the curve revealed a high and a low affinity binding site for immepip. Displacement experiments in the presence of GTPγS showed only the low affinity binding site (figure 1).

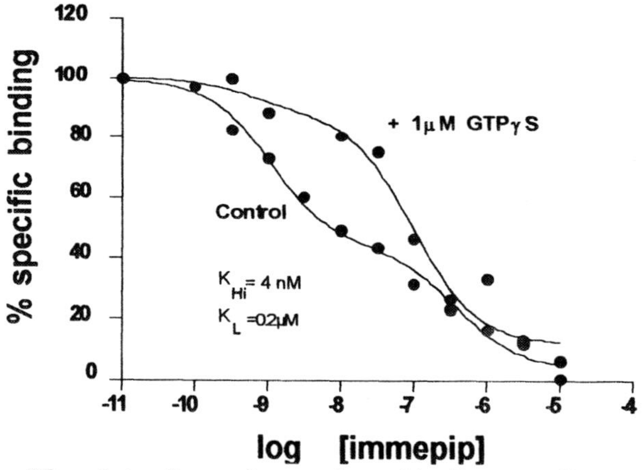

Figure 1. **Inhibition of [^{125}I]IPP (0.4 nM) binding to rat cortical membranes by the histamine H_3 receptor agonist immepip.** Competition binding curves in the absence and presence of 1 μM GTPγS were compared. Data are expressed as percentage of the specific binding. A clear shift from biphasic to monophasic curves can be observed by the addition of 1μM GTPγS.
Figure extracted from the data of Vollinga et al [19].

The data from the various binding studies are a strong indication for the involvement of G-proteins interacting with the H_3 receptor. More evidence for the involvement of G-proteins was recently provided by Clark and Hill [21]. In their study they showed the sensitivity of H_3 receptor agonist induced [^{35}S]GTPγS binding to rat cortical membranes for pertussis toxin (figure 2). Figure 2A shows the specific [^{35}S]GTPγS binding in relation to H_3 receptor stimulation by N^{α}methylhistamine (EC_{50} =0.25 nM) as well as by (R)α-methylhistamine (EC_{50} =0.42 nM). The stimulation by the H_3 receptor agonists was attenuated in the presence of 1 μM thioperamide (figure 2A). The specific [^{35}S]GTPγS binding could completely be abolished by the ADP-ribosylation of $G_{i/o}$-subunits by pertussis toxin (figure 2B). The technique of [^{35}S]GTPγS binding allows measurement of agonist stimulated G-protein activation independently of the second messenger system involved. This technique is applied successfully to study other GPCRs (recently reviewed by Lazareno [22]).

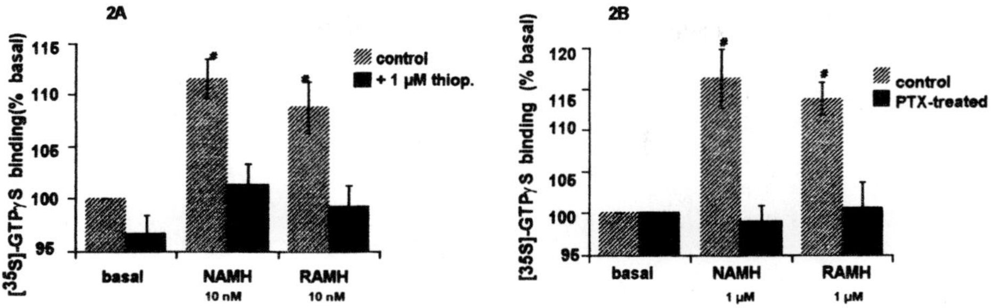

Figure 2. **Effect of thioperamide (2A) and pertussis toxin (2B) on histamine H_3 receptor agonist stimulated [^{35}S]GTPγS binding in rat cerebral cortical membranes.** Hatched bars represent control values and solid bars represent specific binding in the presence of (A) 1 μM thioperamide and (B) membranes pre-treated with pertussis toxin. # $p<0.05$ to basal or $p<0.05$ compared to control.
Figure adapted from Clark and Hill [21].

The reported H_3 receptor sensitivity to pertussis toxin is consistent with the electrophysiological responses to histamine H_3 receptor agonists in the guinea-pig myocardium [23]. In this study the attenuation of adrenergic response by (R)α-methylhistamine were prevented by thioperamide, a selective H_3 -receptor antagonist and attenuated by pertussis-toxin pre-treatment. Taken these data together it suggests that the histamine H_3 receptor belongs to the family of G_i or G_o coupled receptors. Since several authors failed to show a negative coupling of the H_3 receptor to intracellular cAMP levels [16, 24, 25], it is likely that the H_3 receptor has a preference to couple to G_o class of G-proteins. In addition, Cherifi et al. [16] showed negative coupling to phospholipase C in the human gastric tumour cell line HGT-1. Both the basal and the muscarinic receptor-induced IP_3 production were inhibited within a few seconds, and this response could be inhibited with cholera- as well as pertussis toxin. It is not known whether the inhibitory effects on the IP_3 production are due to a direct effect on phospholipase C or must be explained by an inhibition of the influx of extracellular Ca^{2+} which in some cases can activate phospholipase C [26]. Such a regulation of Ca^{2+} influx generally is thought to be responsible for the H_3 receptor effect on neurotransmitter release. It is known that the release of [^{3}H] histamine from neurons is dependent upon the presence of extracellular Ca^{2+} [12, 27]. Recently Garcia reported on the dependency for Ca^{2+} of histamine and immepip induced inhibition of [^{3}H]GABA release in rat striatal slices [28].

The modulation of the N-type Ca^{2+} channels has been shown for some presynaptic receptors to be the mechanistic basis for the inhibition of Ca^{2+} influx [29]. In 1989 Takemura et al. [30] reported on the effective inhibition of histamine release from rat hypothalamic slices by the N-type Ca^{2+} -channel blocker ω-conotoxin. In addition Endou et al. [23] showed that ω-conotoxin greatly potentiated the modulatory effect of (R)α-methylhistamine on cardiac adrenergic responses. Yang and Hatton [31] provided direct evidence for an H_3 receptor-mediated modulation of ion permeability of neurons. They showed that in magnocellular histaminergic neurons from the rat posterior hypothalamus, H_3

receptor activation can depress both spontaneous and current-evoked spikes and leads to hyperpolarisation of these cells via activation of a K^+ channel [31]. These results suggest that histamine can reduce neuronal Ca^{2+} entry and subsequent histamine release by inhibition of N-type Ca^{2+} channels via hyperpolarisation of the neurons. Yet it is still unclear how the H_3 receptor can be linked to this K^+ channel. K^+ channels can either be modulated directly via G-proteins or indirectly by diffusible second messengers.

In conclusion, no firm statements can be made with regard to the signal transduction pathway(s) of the H_3 receptor. Many findings indicate the involvement of a G-protein(s), probably from the G_o class, The recent identification of cell-lines expressing the H_3 receptor [16, 32] can be of great value in the elucidation of this issue. Moreover, the cloning of the gene encoding the H_3 receptor will be of major importance for the investigations in the histamine H_3 receptor field.

3. CLONING OF HISTAMINE RECEPTORS GENES.

In 1991 Sugama and co-workers showed that H_1 agonists can cause an increase in calcium-dependent, inward chloride currents in Xenopus oocytes injected with poly(A)$^+$ mRNA from bovine adrenal medulla [33]. Based upon these findings Yamashita and co-workers [11] succeeded in expression cloning the bovine histamine H_1 receptor cDNA. To do so, poly(A)$^+$ mRNA was isolated from bovine adrenal medulla, size fractionated, injected into Xenopus oocytes and assayed for H_1 agonist induced calcium-depend inward chloride currents. Two pools of RNA were found to be positive and after progressive rounds of panning of one of the positive pools a single clone was isolated. This cloned cDNA showed after expression in Cos7 cells specific [^{3}H]mepyramine binding clearly indicating that the cloned cDNA actually was the bovine H_1 receptor.

The cloned receptor cDNA encodes for a protein of 491 amino acids with all the characteristics of a typical GPCR, as there are 7 putative transmembrane domains, N-terminal glycosylation sites and phosphorylation sites for protein kinase A and C (resp PKA and PKC) [11].

The cloning of the bovine H_1 receptor resulted soon thereafter in the cloning of the genes encoding the rat [2], guinea-pig [6, 9], mouse [34] and the human H_1 receptor [1-3]. All the cloned genes are intronless and the predicted amino acid sequences show a high degree of similarity (90%) indicating that these clones can be considered as true species homologues.

Shortly before the H_1 receptor gene was cloned Gantz et al cloned the canine H_2 receptor [4], using a PCR based strategy with degenerated primers directed against regions conserved among other GPCRs. With this technique a partial clone was obtained from a canine gastric parietal cDNA library. This partial clone was used to screen a canine genomic library to obtain a full length H_2 receptor gene. The gene encodes a protein of 359 amino acids and shows the typical characteristics of a GPCR as well. At this moment many species homologues of the H_2 receptor gene have been cloned by various groups (rat: [8], human: [4], guinea-pig: [10], mouse: [7]).

With all the knowledge gained on the GPCRs in general and the H_1 and H_2 receptors in specific it is surprising to see that the gene encoding the H_3 receptor has not been cloned yet. On the other hand, comparing the amino acid sequence of the H_1- and H_2 receptor it is clear that these two receptors are not that closely related to each other as expected. The overall homology is very low (21 % between the human amino acid sequences) and the highest homology can be found in transmembrane domains three and five. Figure 2 shows a phylogenetic tree (adapted from the GPCR database at http://www.swift.embl-heidelberg.de/7tm/phylo/phylo.html) of all human aminergic GPCRs. From this phylogenetic tree it is clear that the human H_1 receptor is more closely related to the family of human muscarinic receptors than to the human H_2 receptor. The human H_2 receptor is more related to the human dopamine D_1 and serotonin $5HT_1$ receptor families. This relationship might explain the lack of success of cloning the H_3 receptor gene on the basis of expected homology with the H_1 and H_2 receptor.

4. SUBTYPES OF THE H_3 RECEPTOR.

In 1990 West and co-workers reported on the identification of two histamine H_3 receptor subtypes [15]. They identified two different H_3 receptor binding sides in rat brain on the basis of inhibition of $[^3H]N^\alpha$-methylhistamine binding by thioperamide and burimamide. Saturation curves for $[^3H]N^\alpha$-methylhistamine were consistent with a single class of binding sites (K_D =0.37 nM, B_{max} = 73 fmol/mg of protein). For thioperamide, West and co-workers defined two classes of binding sites with K_i and B_{max} values of 5 nM, 30 fmol/mg of protein (H_{3a}) and 68nM, 48 fmol/mg of protein (H_{3b}) respectively. Besides differences in affinity for thioperamide the two subtypes also differed in their sensitivity towards guanyl-nucleotides. Whereas the affinity of the H_{3a} receptor for $[^3H]N^\alpha$-methylhistamine is only reduced twofold, binding of $[^3H]N^\alpha$-methylhistamine to the H_{3b} receptor is completely abolished. However, the results with radioligand binding studies must be examined carefully since the use of different radioligands may lead to different results. In studies carried out with the iodinated antagonist $[^{125}I]$iodophenpropit biphasic displacement curves were obtained for burimamide and dimaprit [17, 20, 35] but, in contrast to the results of West et al [15], monophasic displacement was observed for thioperamide [17, 35]. Recently Leurs et al [36] investigated several histamine homologues as potential ligands for the H_3 receptor. In this studies they used both $[^{125}I]$IPP and $[^3H]N^\alpha$-methylhistamine binding to rat cortical membrane preparations. Besides the binding assays they also used two functional assays: inhibition of neurogenic contraction of the guinea pig jejunum and $[^3H]$noradrenaline in mouse brain cortex slices. Since in this study four different assays were used one would expect to see a clear difference between H_3 receptor subtypes. However, only the binding data obtained with compound VUF4732 constitute evidence for the existence of H_3 receptor heterogeneity. The discrepancies between the functional data obtained with homohistamine, VUF4701 and impentamine can be explained by the existence of H_3 receptor subtypes but could also indicate species difference [36].

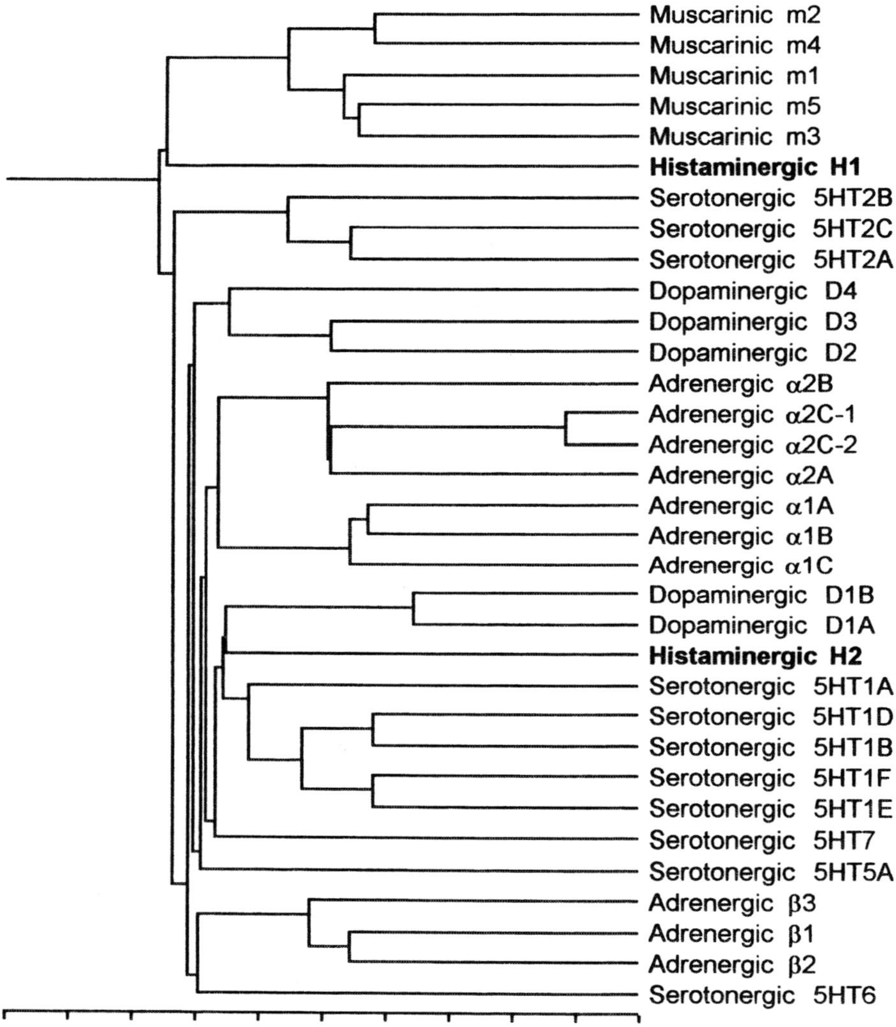

Figure 2. **Phylogenetic tree of cloned human aminergic GPCRs aminoacid sequences.** Evolutionary alignment of the deduced amino acid sequences of cloned human GPCR genes. Clearly from this tree is the distant relation between the human H_1 and H_2 receptor both indicated in bold. (Figure adapted from GPCR Database, http://www.swift.embl-heidelberg.de/7tm/phylo/phylo.html.)

Still additional studies are necessary to solve the nature of H_3 receptor heterogeneity. Cloning of the H_3 receptor and potential subtype will be of great value, especially from a therapeutic point of view.

5. PURIFICATION OF H_3 RECEPTOR PROTEIN.

Until now only limited information on the structural properties of the histamine H_3 receptor is available. This is mainly due to the lack of success in cloning the cDNA that encodes the H_3 receptor. Two papers report the purification of H_3 receptor protein.

In 1992 Zweig and co-workers reported on the solubilisation of the histamine H_3 receptor from total bovine brain using [^{3}H]histamine [18]. [^{3}H]histamine bound with high affinity to an apparently single class of bovine brain membrane sites with the characteristics of an histamine H_3 receptor ($K_D = 4.6 \pm 1.3$ nM; $B_{max} = 78 \pm 20$ fmol/mg of protein). Experiments with various detergents such as digitonin, octylglucoside, CHAPS, Triton X-100 and deoxycholate showed an incompatibility between Triton X-100 and deoxycholate with the binding of [^{3}H]histamine. Not only interfered Triton X-100 and deoxycholate with the binding of [^{3}H]histamine to solubilised receptors but also with the binding to membrane fractions. Digitonin turned out to be the best detergent used for solubilisation of the histamine H_3 receptor from bovine brain [18].

Solubilisation had little effect on the K_D of [^{3}H]histamine binding. Saturation binding to the solubilised receptor protein yielded a K_D of 6.1 nM and a B_{max} of 92 fmol/mg of protein. The differences between the inhibitory potencies of H_3 specific agonists and antagonists between the particulate and the solubilised protein were generally minor. (table 1). The rank order of inhibitor potencies confirms that the binding was to the H_3 receptor.

Table 1. K_i values of several H_3 receptor agonists and antagonists. [^{3}H]histamine displacement was done on bovine whole brain membrane preparations of total bovine brain (particulate) and digitonin treated material (solubilised)

	K_i (mean ± range, nM)	
Compound	Particulate	Solubilised
N^{α}-methylhistamine	0.67 ± 0	2.8 ± 0.3
(R)α-methylhistamine	1.2 ± 0.3	1.8 ± 0.3
Histamine	8.2 ± 5.7	11 ± 5
Thioperamide	15 ± 6	32 ± 7
Burimamide	140 ± 100	78 ± 19
Dimaprit	260 ± 110	700 ± 60

table adapted from Zweig et al ([18])

Size exclusion chromatography revealed a protein with an average mass of 220 kDa [18]. Compared to the H_1 and H_2 receptor proteins which are 67 and 45 kDa respectively, the mass of the H_3 receptor seems to be overestimated. Especially considering the guanine nucleotide sensitivity of the solubilised protein it is likely that the molecular mass of 220 kDa represents a complex of receptor, G-protein and digitonin [18].

In 1992 Cherifi et al. [16] published a Triton X-100 solubilisation and purification of the histamine H_3 receptor from the human gastric tumour cell line HGT-1 (clone 6). These HGT-1 cells were shown to bind [^{3}H]N$^\alpha$-methylhistamine in a specific manner. Scatchard analysis was consistent with the presence of one major population of high affinity sites with an apparent K_D of 0.61 nM and a B_{max} of 54 fmol/mg of protein. In the presence of GTP[γ]S, the K_D value increased up to 2.2 nM but the B_{max} values remained unchanged. The Triton X-100 solubilisation did in their hands not interfere with [^{3}H]N$^\alpha$-methylhistamine binding which is in contrast to the results of Zweig et al [18]. With [^{3}H]histamine solubilised sites had a K_D value of 2 nM which is close to the K_D value of the membrane binding sites in the presence of GTP[γ]S and solubilisation must have unmasked additional binding sites, since the B_{max} value increased almost 2.5 fold. To remove potential H_2 binding sites from the [^{3}H]N$^\alpha$-methylhistamine binding sites of the solubilised material it was treated with a Sepharose-famotidine affinity gel [37]. No [^{3}H]N$^\alpha$-methylhistamine binding sites were retained on the gel. However [^{3}H]N$^\alpha$-methylhistamine binding sites were retained on a sepharose-thioperamide gel. After elution, this material was analysed on SDS-polyacrylamide gel electrophoresis and revealed a single band of 70 kDa. Analysis of the binding of [^{3}H]N$^\alpha$-methylhistamine to the purified preparation, indicated the presence of a single binding site with a K_D of 1.6 nM and a B_{max} of 12,000 pmol/mg of protein. [^{3}H]N$^\alpha$-methylhistamine binding was totally displaced by N$^\alpha$-methylhistamine, (R)α-methylhistamine and thioperamide (in order of potency). The typical H_2 and H_1 antagonists famotidine and mepyramine did not displaced [^{3}H]N$^\alpha$-methylhistamine at 1 μM concentrations indicating that the purified receptor is not the H_1 or H_2 receptor. The K_D value of the purified sites was close to that of the native sites in the presence of GTP[γ]S suggesting that the purified sites are not associated with the G-protein anymore. This would lead to the conclusion that the H_3 receptor is a protein of approximately 70 kDa.

In 1991 Reyl-Desmars et al. [37] purified the histamine H_2 receptor from the same HGT-1 cell line. The histamine H_2 receptor had, after affinity chromatography and HPLC purification, a size of 70 kDa as well. Surprisingly both receptors show the same size after analysis. Cloning of the human histamine H_2 receptor [4] revealed a coding region of 1080 bp encoding a seven transmembrane domain protein of 359 amino acids. Roughly calculated this would lead to a protein of approximately 45 kDa. Conflicting results concerning the molecular weight of the H_2 receptor have been obtained in various systems. In 1990 Ruat et al [38] identified proteins in guinea-pig brain by use of the photoaffinity probe [^{125}I]-iodoazidopotentidine with molecular weights of 87, 59, 51, 32 kDa. Only the labelling of the 59 and 32 kDa bands could be prevented upon incubations with H_2 antagonists. In contrast, western blot analysis of histidine tagged rat H_2 receptors expressed in Sf9 cells revealed an apparent molecular weight of 36 kDa [39]. Since the calculated protein size derived from the cloned human histamine H_2 gene and the results from western blot and photoaffinity labelling analysis differ significantly from the size determined from the protein solubilisation and purification experiments, it seems likely that the purified protein is still associated with the G-protein and/or digitonin present in the final HPLC purification step. This might be explained by aggregate formation or association with a G-protein. The same may then hold true for the estimated size of the histamine H_3 receptor.

6. STRATEGIES FOR CLONING THE H_3 RECEPTOR GENE.

To obtain detailed insights on the molecular architecture, signal transduction, and putative receptor subtypes, there is a clear need to clone the gene encoding the H_3 receptor.

The most obvious way to start would be to use a suspected homology between the H_1, H_2 and the H_3 receptor. Yet as one can see from the phylogenetic tree in figure 3 the homology between the human H_1 and H_2 receptor is not as close as to other members of the aminergic family of GPCRs. For that reason an approach that is based upon the homology between the H_1 and H_2 receptor is not very likely to succeed. However, a more general approach on the basis of the overall homology between members of the aminergic GPCRs family could be more successful; a degenerated PCR strategy, using the most conserved transmembrane domains (TM3, TM6 and TM7) could be used. (figure 4). In this case degenerated primers should be designed in the indicated transmembrane domains on the basis of the consensus sequence shown on the last line of the alignment. In addition the alignment clearly shows that the receptors can be divided into subgroups on the basis of the homology they show in the indicated transmembrane domains. By choosing different degenerated primers it is possible to bias the outcome of such a degenerated PCR towards sequences that will fit in one of the subgroups. By choosing a high degree of degeneration the outcome of the PCR reaction will be targeted towards GPCR sequences in general. Various examples of this approach can be found in literature [40-53].

A more complex, but also more specific cloning strategy based on the pharmacology of the H_3 receptor instead of on a suspected homology, might be more successful. To do so, either one uses a radioligand to detect specific binding in with cDNAs transfected cells or one uses a reporter-gene assay that will be activated by the signal transduction route the receptor uses. The use of radioligands for expression cloning has recently be reviewed by Simonsen and Lodish [54]. In recent years two iodinated H_3 radioligands have been synthesised; $[^{125}I]$iodoproxyfan ($[^{125}I]$IPF) [55] and $[^{125}I]$iodophenpropit ($[^{125}I]$IPP) [17, 35].Until now, the availability of these ligands has not resulted in the cloning of the receptor. Nevertheless several GPCRs have been cloned in this manner [56-64].

The other method of expression cloning makes use of the signal transduction route of a specific receptor. The basis of this method is that triggering of a signal transduction route will activate the transcription of a so-called reporter-gene, that can easily be detected. The products of such a reporter-genes are usually enzymes such as β-galactosidase, luciferase, or alkaline phosphatase which are detectable in a simple manner. In order to use the reporter-gene method for

	TM3		TM6		TM7
H_1	**----FWLSMDYVASTASIFSVFILCIDRY**	A1A	--TL-AIVVGVFVLCWFPFFFVLPLGSL-	M5	---------LGYWLCYVNSTVNPICYALC--
$5H_2B$	----AWLFLDVLFSTASIMHLCAISVDRY	A1AB	---L-GIVVGMFILCWLPFFIALPLGSL-	M3	---------LGYWLCYINSTVNPVCYALC--
$5H_2A$	--CAVWIYLDVLFSTASIMHLCAISLDRY	A1AC	---L-GIVVGCFVLCWLPFFLVMPIGSF-	M1	---------LGYWLCYVNSTINPMCYALC--
$5H_2C$	---PVWISLDVLFSTASIMHLCAISLDRY	A2AB	---L-AVVIGVFVLCWFPFFFSYSLGAI-	M4	---------IGYWLCYVNSTINPACYALC--
H_2	**----IYTSLDVMLCTASILNLFMISLDRY**	A2AA	--VL-AVVIGVFVVCWFPFFFTYTLTAV-	M2	---------IGYWLCYINSTINPACYALC--
D3	----VFVTLDVMMCTASILNLCAISIDRY	A2AD	---L-AVVMGVFVLCWFPFFFSYSLYGI-	B2A	EVY----ILLNWIGYVNSGFNPLIY-CRS-
D2	----IFVTLDVMMCTASILNLCAISIDRY	A2AC	---L-AVVMGVFVLCWFPFFFIYSLYGI-	B1A	RLF----VFFNWLGYANSAFNPIIY-CRS-
D4	----ALMAMDVMLCTASIFNLCAISVDRY	B3A	--TL-GLIMGTFTLCWLPFFLANVL----	B3A	-AF---LALNWLGYANSAFNPLIYC-----
A2AB	E---VYLALDVLFCTSSIVHLCAISLDRY	B2A	---L-GIIMGTFTLCWLPFFIVNIVHVI-	5H1A	TLL---GAIINWLGYSNSLLNPVIYAYF--
A2AA	E---IYLALDVLFCTSSIVHLCAISLDRY	B1A	---L-GIIMGVFTLCWLPFFLANVVKAF-	D3	---------ATTWLGYVNSALNPVIYTTFNI
A2AD	G---VYLALDVLFCTSSIVHLCAISLDRY	5H7	---L-GIIVGAFTVCWLPFFLLSTARPF-	D2	----VLYSAFTWLGYVNSAVNPIIYTTF--
A2AC	G---VYLALDVLFCTSSIVHLCAISLDRY	D3	--MV-AIVLGAFIVCWLPFFLTHVL----	D4	---PRLVSAVTWLGYVNSALNPVIYT----
$5H_1A$	D---LFIALDVLCCTSSILHLCAIALDRY	D2	--ML-AIVLGVFIICWLPFFITHIL-NI-	5H6	CISPGLFDVLTWLGYCNSTMNPIIY-----
A1A	---DVWAAVDVLCCTASILSLCTISVDRY	5H1F	ATTL-GLILGAFVICWLPFFVKEL-----	5H1E	-----VADFLTWLGYVNSLINPLLYTSF--
A1AB	IFCDIWAAVDVLCCTASILSLCAISIDRY	5H1E	--IL-GLILGAFILSWLPFFIKELIV---	5H1F	-----MSNFLAWLGYLNSLINPLIYTIFN-
A1AC	---NIWAAVDVLCCTASIMGLCIISIDRY	5HT1B	---L-GIILGAFIVCWLPFFIISLVMPIC	**H2**	**VLE----AIVLWLGYANSALNPILYA----**
5H7	----VFIAMDVMCCTASIMTLCVISIDRY	5H1D	---L-GIILGAFIICWLPFFVVSLVLPI-	**H1**	**MF-------TIWLGYINSTLNPLIYPL---**
5H6	----LWTAFDVMCCSASILNLCLISLDRY	5H1A	T--L-GIIMGTFILCWLPFFIVALV----	5H7	V-E----RTFLWLGYANSLINPFIYAFF--
B2A	---EFWTSIDVLCVTASIETLCVIAVDRY	5H5A	AALMVGILIGVFVLCWIPFFL--------	5H5A	--------SIFLWLGYSNSFFNPLIYTAF--
B1A	---ELWTSVDVLCVTASIETLCVIALDRY	D4	--VL-PVVVGAFLLCWTPFFVVHITQ---	5H2C	--LL---NVFVWIGYVCSGINPLVYTLF--
B3A	----LWTSVDVLCVTASIETLCALAVDRY	5H6	ASLTLGILLGMFFVTWLPFFVANIV----	5H2B	--LL---EIFVWIGYVSSGVNPLVYTLF--
$5H_1D$	ILCDIWLSSDITCCTASILHLCVIALDRY	5H2C	--VL-GIVFFVFLIMWCPFFITNIL----	5H2A	--------NVFVWIGYLSSAVNPLVYTLFN-
$5H_1B$	VVCDFWLSSDITCCTASILHLCVIALDRY	5H2B	--VL-GIVFFLFLLMWCPFFITNI-----	5HT1B	LAIF---DFFTWLGYLNSLINPIIYTMSN-
$5H_1F$	-VCDIWLSVDITCCTCSILHLSAIALDRY	5H2A	---L-GIVFFLFVVMWCPFFITNIMA---	5H1D	-ALF---DFFTWLGYLNSLINPIIYTVFN-
$5H_1E$	----VWLSVDMTCCTCSILHLCVIALDRY	**H2**	**T--L-AAVMGAFIICWFPYFTAFVY--RG**	A2AD	---------FFFWIGYCNSSLNPVIYTVFN-
5H5A	----LWIACDVLCCTASIWNVTAIALDRY	**H1**	**---F---IMAAFILCWIPYFIFFMVI---**	A2AC	---------FFFWIGYCNSSLNPVIYTVFN-
M4	----LWLALDYVVSNASVMNLLIISFDRY	M4	---I-FAILLAFILTWTPYNVMVLV----	A2AB	GL-F---QFFFWIGYCNSSLNPVIYTIF--
M2	----LWLALDYVVSNASVMNLLIISFDRY	M2	---I-LAILLAFIITWAPYNVMVLI----	A2AA	-L-F---KFFFWFGYCNSSLNPVIYTIFN-
M5	----LWLALDYVASNASVMNLLVISFDRY	M5	---L-SAILLAFIITWTPYNIMVLV----	A1A	EGVF---KVIFWLGYFNSCVNPLIYP-CS-
M1	----LWLALDYVASNASVMNLLLISFDRY	M3	---L-SAILLAFIITWTPYNIMVLV----	A1AC	-TVF---KIVFWLGYLNSCINPIIYP-CS-
M3	----LWLAIDYVASNASVMNLLVISFDRY	M1	---L-SAILLAFILTWTPYNIMVLV----	A1AB	DAVF---KVVFWLGYFN-------------
Cons.	**----.W.A.DV..CTASI..LC.IS.DRY**	**Cons**	**---L-GI..G.F.LCW.PFF.....----**	**Cons**	**--------....WLGY.NS..NP.IY...--**

Figure 4. Alignment of the transmembrane domains 3 (TM 3), 6 (TM 6) an d 7 (TM 7) of human aminergic GPCRs. Alignment of the most homologue regions between GPCRs. Last line of every alignment shows the consensus. On the basis of this consensus degenerated primers could be designed.

cloning the H_3 receptor, specific knowledge about the signal transduction route used is necessary. Since this is lacking for the H_3 receptor, it is difficult to design such an reporter-gene assay. However, recently, Stables and co-workers [65] developed a bioluminescent assay that will work with virtually any GPCR. The basis of this assay is $G_{\alpha 16}$. This specific G protein is promiscuous in its nature and will couple almost any G-protein to a phospholipase C (measured as aequorin luminescence).

7. Conclusion

Since the discovery of the histamine H_3 receptor in 1983 Arrang and co-workers [12] many research groups have been active in this field and the pharmacology of this receptor has been well studied by now. Important information needed for the cloning of the H_3 receptor can be extracted from these pharmacological studies. The observed guanine-nucleotide sensitivity of agonist binding to the receptor as observed by different authors suggest the receptor to be a GPCR. The pertussis toxin sensitivity of the agonist stimulated binding of [^{35}S]GTPγS as reported by Clark and Hill [21] suggests coupling of the H_3 receptor to G_i or G_o classes of G-proteins. Since many authors fail to show a negative coupling to adenylyl cyclase, coupling to G_o class of G-proteins seems more likely. More detailed information on the signal transduction pathways triggered by the H_3 receptor can be obtained from expression studies using the cloned receptor cDNA.

Many GPCRs are cloned on the basis of their homology with other members of their specific family or on the basis of general characteristics of GPCRs. The H_3 receptor has not been cloned sofar although probably several attempts have been made. Cloning the H_3 receptor on basis of the homology between the H_1 and H_2 receptor is not a strategy likely to succeed, whereas the homology between the H_2 and H_1 receptor is less then the homology these receptors share with other GPCRs (dopamine and muscarinic receptors respectively). Assuming that the H_3 receptor belongs to the family of aminergic seven transmembrane receptors an approach using degenerated PCR primers in the most conserved regions of these receptors may lead to the cloning of the receptor. Even more successful could be an expression cloning strategy using one of the available iodinated radioligands. Finally, as a result of the many ongoing sequencing projects nowadays, the sequence encoding the histamine H_3 receptor will be present in one of these sequence databanks at some point. However, the question remains if the H_3 sequence will be recognised as such.

REFERENCES

1. M.D. De Backer, W. Gommeren, H. Moereels, G. Nobels, P. Van Gompel, J.E. Leysen, and W.H. Luyten, Biochem Biophys Res Commun, 197(3) (1993) 1601-8.
2. K. Fujimoto, Y. Horio, K. Sugama, S. Ito, Y.Q. Liu, and H. Fukui, Biochem Biophys Res Commun, 190(1) (1993) 294-301.
3. H. Fukui, K. Fujimoto, H. Mizuguchi, K. Sakamoto, Y. Horio, S. Takai, K. Yamada, and S. Ito, Biochem Biophys Res Commun, 201(2) (1994) 894-901.
4. I. Gantz, G. Munzert, T. Tashiro, M. Schaffer, L. Wang, J. DelValle, and T. Yamada, Biochem Biophys Res Commun, 178(3) (1991) 1386-92.
5. I. Gantz, M. Schaffer, J. DelValle, C. Logsdon, V. Campbell, M. Uhler, and T. Yamada, Proc. Natl. Acad. Sci. USA., 88(13) (1991) 5937.
6. Y. Horio, Y. Mori, I. Higuchi, K. Fujimoto, S. Ito, and H. Fukui, J Biochem (Tokyo), 114(3) (1993) 408-14.
7. T. Kobayashi, I. Inoue, N.A. Jenkins, D.J. Gilbert, N.G. Copeland, and T. Watanabe, Genomics, 37(3) (1996) 390-4.
8. M. Ruat, E. Traiffort, J.M. Arrang, R. Leurs, and J.C. Schwartz, Biochem Biophys Res Commun, 179(3) (1991) 1470-8.
9. E. Traiffort, R. Leurs, J.M. Arrang, J. Tardivel-Lacombe, J. Diaz, J.C. Schwartz, and M. Ruat, Journal of Neurochemistry, 62(2) (1994) 507-18.
10. E. Traiffort, M.L. Vizuete, J. Tardivellacombe, E. Souil, J.C. Schwartz, and M. Ruat, Biochem Biophys Res Commun, 211(2) (1995) 570-7.
11. M. Yamashita, H. Fukui, K. Sugama, Y. Horio, S. Ito, H. Mizuguchi, and H. Wada, Proc. Natl. Acad. Sci. USA. 88(24) (1991) 11515-9.
12. J.M. Arrang, M. Garbarg, and J.C. Schwartz, Nature, 302(5911) (1983) 832-7.
13. R. Leurs, R.C. Vollinga, and H. Timmerman, Progress in Drug Research, 45(1995) 107-65.
14. J.M. Arrang, J. Roy, J.L. Morgat, W. Schunack, and J.C. Schwartz, European Journal of Pharmacology, 188(4-5) (1990) 219-27.
15. R.E. West, Jr., A. Zweig, N.Y. Shih, M.I. Siegel, R.W. Egan, and M.A. Clark, Mol Pharmacol, 38(5) (1990) 610-3.
16. Y. Cherifi, C. Pigeon, M. Le Romancer, A. Bado, F. Reyl-Desmars, and M.J. Lewin, J Biol Chem, 267(35) (1992) 25315-20.
17. F.P. Jansen, B. Rademaker, A. Bast, and H. Timmerman, European Journal of Pharmacology, 217(2-3) (1992) 203-5.
18. A. Zweig, M.I. Siegel, R.W. Egan, M.A. Clark, R.G. Shorr, and R.E. West, Jr., J Neurochem, 59(5) (1992) 1661-6.
19. R.C. Vollinga, J.P. de Koning, F.P. Jansen, R. Leurs, W.M. Menge, and H. Timmerman, Journal of Medicinal Chemistry, 37(3) (1994) 332-3.
20. W.M.P.B. Menge, H. Van der Goot, H. Timmerman, J.L.H. Eersels, and J.D.M. Herscheid, J. Labelled Comp. Radiopharm., 31(1992) 781-6.
21. E.A. Clark and S.J. Hill, European Journal of Pharmacology, 296(2) (1996) 223-5.
22. S. Lazareno, Methods in Molecular Biology, 83(1997) 107-16.

23. M. Endou, E. Poli, and R. Levi, Journal of Pharmacology & Experimental Therapeutics, 269(1) (1994) 221-9.
24. M. Garbarg, M.D. Tuong, C. Gros, and J.C. Schwartz, European Journal of Pharmacology, 164(1) (1989) 1-11.
25. E. Schlicker, K. Fink, G. Molderings, and M. Göthert, J. Neurochem., 57(1991) S37.
26. S. Cockcroft and G.M.H. Thomas, Biochem. J., 288(1992) 1-14.
27. J.F. van der Werf, G.J. Bijloo, A. van der Vliet, A. Bast, and H. Timmerman, Agents & Actions, 20(3-4) (1987) 239-43.
28. M. Garcia, B. Floran, J.M. Young, J. Aceves, and J.A. Arias-Montano. in *Society for neuroscience, 27th annual meeting*. 1997. New Orleans, USA.
29. N.I. Chernevskaya, A.G. Obukhov, and O.A. Krishtal, Nature, 349(6308) (1991) 418-20.
30. M. Takemura, J. Kishino, A. Yamatodani, and H. Wada, Brain Res, 496(1989) 351-6.
31. O.Z. Yang and G.I. Hatton, Soc. Neurosci. Abstr., 17(1991) 409.
32. M.A. Clark, A. Korte, J. Myers, and R.W. Egan, European Journal of Pharmacology, 210(1992) 31-5.
33. K. Sugama, M. Yamashita, H. Fukui, S. Ito, and H. Wada, Japanese Journal of Pharmacology, 55(2) (1991) 287-90.
34. I. Inoue, I. Taniuchi, D. Kitamura, N.A. Jenkins, D.J. Gilbert, N.G. Copeland, and T. Watanabe, Genomics, 36(1) (1996) 178-81.
35. F.P. Jansen, T.S. Wu, H.-P. Voss, H.W.M. Steinbusch, R.C. Vollinga, B. Rademaker, A. Bast, and H. Timmerman, Br. J. Pharmacol., 113(1994) 355-62.
36. R. Leurs, M. Kathmann, R.C. Vollinga, W.M. Menge, E. Schlicker, and H. Timmerman, Journal of Pharmacology & Experimental Therapeutics, 276(3) (1996) 1009-15.
37. F. Reyl-Desmars, Y. Cherifi, M. Le Romancer, C. Pigeon, S. Le Roux, and M.J.M. Lewin, C. R. Acad. Sci. Paris, III(1991) 221-4.
38. M. Ruat, E. Traiffort, M.L. Bouthenet, J.-C. Schwartz, J. Hirschfeld, A. Buschauer, and W. Schunack, Proc. Natl. Acad. Sci. USA, 87(1990) 1658-62.
39. M.W. Beukers, C.H.W. Klaassen, W.J. de Grip, D. Verzijl, H. Timmerman, and R. Leurs, Br. J. Pharmacol., 122(1997) 867-74.
40. N. Adham, H.T. Kao, L.E. Schecter, J. Bard, M. Olsen, D. Urquhart, M. Durkin, P.R. Hartig, R.L. Weinshank, and T.A. Branchek, Proc. Natl. Acad. Sci. USA, 90(2) (1993) 408-12.
41. K.A. Eidne, J. Zabavnik, T. Peters, S. Yoshida, L. Anderson, and P.L. Taylor, FEBS Lett, 292(1-2) (1991) 243-8.
42. C.M. Gerard, C. Mollereau, G. Vassart, and M. Parmentier, Biochem J, 279(Pt 1) (1991) 129-34.
43. R. Janssens, D. Communi, S. Pirotton, M. Samson, M. Parmentier, and J.M. Boeynaems, Biochem Biophys Res Commun, 221(3) (1996) 588-93.
44. J.D. Kursar, D.L. Nelson, D.B. Wainscott, M.L. Cohen, and M. Baez, Mol Pharmacol, 42(4) (1992) 549-57.

45. T.W. Lovenberg, M.G. Erlander, B.M. Baron, M. Racke, A.L. Slone, B.W. Siegel, C.M. Craft, J.E. Burns, P.E. Danielson, and J.G. Sutcliffe, Proc. Natl. Acad. Sci. USA, 90(6) (1993) 2184-8.
46. F. Libert, M. Parmentier, A. Lefort, C. Dinsart, J. Van Sande, C. Maenhaut, M.J. Simons, J.E. Dumont, and G. Vassart, Science, 244(4904) (1989) 569-72.
47. L.C. Mahan, L.D. McVittie, E.M. Smyk-Randall, H. Nakata, F.J. Monsma, Jr., C.R. Gerfen, and D.R. Sibley, Mol Pharmacol, 40(1) (1991) 1-7.
48. M.I. Masana, R.C. Brown, H. Pu, M.E. Gurney, and M.L. Dubocovich, Recept Channel, 3(4) (1995) 255-62.
49. G. McAllister, A. Charlesworth, C. Snodin, M.S. Beer, A.J. Noble, D.N. Middlemiss, L.L. Iversen, and P. Whiting, Proc. Natl. Acad. Sci. USA, 89(12) (1992) 5517-21.
50. F.J. Monsma, Jr., L.C. Mahan, L.D. McVittie, C.R. Gerfen, and D.R. Sibley, Proc. Natl. Acad. Sci. USA, 87(17) (1990) 6723-7.
51. C.J. Raport, V.L. Schweickart, D. Chantry, R.L. Eddy, T.B. Shows, R. Godiska, and P.W. Gray, J Leukocyte Biol, 59(1) (1996) 18-23.
52. F. Yamaguchi, A.D. Macrae, and S. Brenner, Genomics, 35(3) (1996) 603-5.
53. A. Sarkar and I.M. Dickerson, Journal of Neurochemistry, 69(2) (1997) 455-64.
54. H. Simonsen and H.F. Lodish, Trends Pharmacol Sci, 15(12) (1994) 437-41.
55. X. Ligneau, M. Garbarg, M.L. Vizuete, J. Diaz, K. Purand, H. Stark, W. Schunack, and J.C. Schwartz, Journal of Pharmacology & Experimental Therapeutics, 271(1) (1994) 452-9.
56. T.J. Murphy, K. Takeuchi, and R.W. Alexander, Am J Hypertens, 5(12 Pt 2) (1992) 236S-42S.
57. T. Macneil, K.K. Bierilo, J.G. Menke, and J.F. Hess, Bba-Gene Struct Express, 1264(2) (1995) 223-8.
58. R. Chen, K.A. Lewis, M.H. Perrin, and W.W. Vale, Proc. Natl. Acad. Sci. USA, 90(19) (1993) 8967-71.
59. F.W. Kluxen, C. Bruns, and H. Lubbert, Proc. Natl. Acad. Sci. USA, 89(10) (1992) 4618-22.
60. A.S. Kopin, Y.M. Lee, E.W. McBride, L.J. Miller, M. Lu, H.Y. Lin, L.F. Kolakowski, Jr., and M. Beinborn, Proc. Natl. Acad. Sci. USA, 89(8) (1992) 3605-9.
61. T. Ebisawa, S. Karne, M.R. Lerner, and S.M. Reppert Proc. Natl. Acad. Sci. USA, 91(13) (1994) 6133-7.
62. D.R. Gehlert, L.S. Beavers, D. Johnson, S.L. Gackenheimer, D.A. Schober, and R.A. Gadski, Mol Pharmacol, 49(2) (1996) 224-8.
63. C. Gerald, M.W. Walker, P.J. Vaysse, C. He, T.A. Branchek, and R.L. Weinshank, J Biol Chem, 270(45) (1995) 26758-61.
64. T.J. Murphy, R.W. Alexander, K.K. Griendling, M.S. Runge, and K.E. Bernstein, Nature, 351(6323) (1991) 233-6.
65. J. Stables, A. Green, F. Marshall, N. Fraser, E. Knight, M. Sautel, G. Milligan, M. Lee, and S. Rees, Analytical Biochemistry, 252(1) (1997) 115-26.

R. Leurs and H. Timmerman (Editors)
The Histamine H_3 Receptor

Radioligands for the histamine H_3 receptor and their use in pharmacology.

F.P. Jansen, R. Leurs and H. Timmerman

Leiden/Amsterdam Center for Drug Research, Department of Pharmacochemistry, Faculty of Chemistry, Vrije Universiteit, De Boelelaan 1083, 1081 HV Amsterdam, the Netherlands.

1. PRELUDE

Receptor radioligands play a crucial role in receptor research, covering many aspects of receptor pharmacology, e.g. receptor-ligand binding dynamics, tissue distribution of the receptor, G-protein coupling, receptor heterogeneity and structural identification of the receptor protein. From the discovery of the histamine H_3-receptor in 1983 until now, eight different radioligands have been used in H_3-receptor binding studies. The first three ligands described are tritiated agonists. More recently, the availability of different selective H_3-antagonists led to the introduction of three tritiated and of two radioiodinated antagonists. The pharmacology of the radiolabelled agonists and antagonists will be reviewed in the present chapter.

2. RADIOLABELLED H_3-AGONISTS

The agonists [^{3}H](R)α-methylhistamine, [^{3}H]N$^{\alpha}$-methylhistamine and [^{3}H]histamine have been used to characterize histamine H_3-receptors in various tissues and species.

(R)α-methylhistamine

N$^{\alpha}$-methylhistamine

histamine

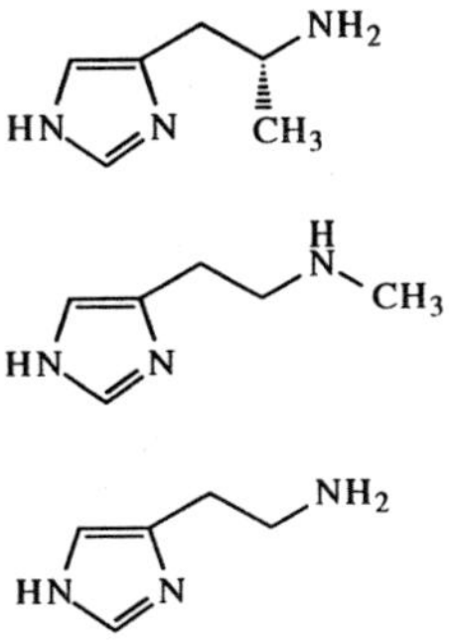

Figure 1. The chemical stuctures of H_3-agonists used a radioligands.

In Table 1 the main characteristics of the radiolabelled agonists are summarized, based on results obtained in cerebral tissues. All three ligands were shown to bind reversibly, saturably and with high affinity to H_3-receptors in rat brain membrane preparations. $[^3H](R)\alpha$-Methylhistamine was the first radioligand described (Arrang *et al.*, 1987). The compound has been used at different laboratories to study H_3-receptor characteristics and distribution from rodents to primates (Arrang *et al.*, 1987; Martinez-Mir *et al.*, 1990; West *et al.*, 1990a; Fujimoto *et al.*, 1991; Yanai *et al.*, 1992; Pollard *et al.*, 1993). N^{α}-Methylhistamine has a lower H_3-receptor selectivity as compared to $(R)\alpha$-methylhistamine (Schwartz *et al.*, 1990; Oishi *et al.*, 1993). Nevertheless, $[^3H]N^{\alpha}$-methylhistamine has proven to be a very suitable radioligand for H_3-receptor binding studies and it has widely been used as well (Korte *et al.*, 1990; Cumming *et al.*, 1991; Cherifi *et al.*, 1992; Clark *et al.*, 1993; Kathmann *et al.*, 1993; West *et al.*, 1994). $[^3H]N^{\alpha}$-Methylhistamine has the advantage of a higher specific activity and a lower nonspecific binding as compared to $[^3H](R)\alpha$-methylhistamine (Table 1). $[^3H]$Histamine has a relatively low selectivity towards H_3-receptors. A few reports have described its use as a radioligand to study histamine H_3-receptors however (Cumming *et al.*, 1991; Zweig *et al.*, 1992) and the ligand has been successfully used to partially purify the H_3-receptor protein from bovine brain (Zweig *et al.*, 1992). In 1980, Barbin *et al.* already found that the binding profile of tritiated histamine to rat brain membranes did not correspond to H_1- and H_2-receptors known at that time (Barbin *et al.*, 1980). Hence, $[^3H]$histamine may be the first radioligand used to 'unintentionally' label H_3-receptors at the time the H_3-receptor had not yet been identified. Nevertheless, $[^3H]$histamine cannot be regarded as a radioligand of first choice. At nanomolar concentrations $[^3H]$histamine binds to multiple sites in guinea-pig cerebral cortex membranes (Sinkins *et al.*, 1993, Sinkins and Wells, 1993). Its binding was displaced in a complex manner by a variety of H_2- and H_3-agonists and antagonists; only a model of co-operative interactions could explain the observed displacement curves (Sinkins *et al.*, 1993, Sinkins and Wells, 1993). The authors suggested on the basis of their findings the involvement of receptor dimers in the binding of $[^3H]$histamine. In view of convincing data on receptor dimerization obtained recently for the β_2-adrenoceptor, the dopamine D_2 and the $5HT_{1B}$-receptor (Herbert *et al.*, 1996; Ng *et al.*, 1996), the findings of Sinkins and collegues with $[^3H]$histamine could indicate the existence of H_3-receptor dimers.

Binding of the three tritiated agonists to brain tissue appeared to be sensitive to guanine nucleotides (Table 1) and to pertussis toxin ($[^3H]N^{\alpha}$-methylhistamine; Clark *et al.*, 1993), indicating that histamine H_3-receptors, like H_1- and H_2-receptors, belong to the family of G-protein coupled receptors. Guanine nucleotides were shown to inhibit agonist receptor binding (Arrang *et al.*, 1990; West *et al.*, 1990b; Cumming *et al.*, 1991; Kilpatrick & Michel, 1991; Zweig *et al.*, 1992; Clark *et al.*, 1993; Clark & Hill, 1995). This phenomenon is generally observed for G-protein coupled receptors and is usually explained by the abolishment of the

Table 1. Characteristics of radiolabelled agonists used to study histamine H_3-receptors.

	[^{3}H](R)α-MeHA	[^{3}H]N^α-MeHA	[^{3}H]histamine
Dissociation constant (K_D, nM)^a	0.4 - 2.5	0.4 - 2	5 - 8
Receptor density (B_{max}, fmol/mg)^a	30 - 190	40 - 150	20 - 80
Specific binding (% of total binding)^a	60 - 90	> 90	~ 75
Specific activity (Ci/mmol)	20 - 40	80	6 - 43
Sensitivity to guanine nucleotides	+	+	+
Functional potency (EC_{50}-value, nM)	4	15	62
Species characterized	rat guinea-pig primate	rat guinea-pig mouse	rat bovine
References used	[1-5]	[6-14]	[8,9]

[1] Arrang *et al.*, 1987, [2] Arrang *et al.*, 1990, [3] West *et al.*, 1990a, [4] Kilpatrick & Michel, 1991, [5] Yanai *et al.*, 1994, [6] Korte *et al.*, 1990, [7] West *et al.*, 1990b, [8] Cumming *et al.*, 1991, [9] Zweig *et al.*, 1992, [10] Clark *et al.*, 1993, [11] Kathmann *et al.*, 1993, [12] West *et al.*, 1994, [13] Clark & Hill, 1995, [14] Brown *et al.*, 1996. Data are derived from receptor binding studies in the species indicated, using brain membranes preparations (except for Cumming, 1991; autoradiographic study). ^{a}Values correspond to experiments using rat whole brain or rat cerebral cortex. The specific binding is related to radioligand concentrations around or below the K_D-value. EC_{50}-Values correspond to the inhibition of the potassium induced [^{3}H]histamine release from rat cerebral cortex slices, as taken from Leurs et al. (1992). Abbreviations: [^{3}H](R)α-MeHA, [^{3}H](R)α-methylhistamine; [^{3}H]N^α-MeHA, [^{3}H]N^α-methylhistamine.

high affinity binding state of the receptor. However, for the H_3-receptor the underlying mechanism for the effect of guanine nucleotides on radioligand receptor binding seems to be more complex. The reduction of [^{3}H]N^α-methylhistamine binding induced by GTPγS might result from both a reduced B_{max} or from a reduced affinity of the ligand (West *et al.*, 1990b; Clark *et al.*, 1993). It has also been suggested that guanine nucleotides affect a subpopulation of [^{3}H]N^α-methylhistamine binding sites (West *et al.*, 1990b). In one report, using [^{3}H](R)α-methylhistamine, the guanine nucleotide induced reduction of agonist binding has been described to be dependent on the presence of calcium in the incubation buffer (Arrang *et al.*, 1990). It is obvious that the different radiolabelled H_3-agonists do not display a straightforward binding profile with respect to the G-protein coupling which may partly explain the inconsistencies found in literature.

A remarkable discrepancy exists between the receptor binding affinities and the functional potencies of H_3-agonists (see also Table 1). In general, the affinity of agonists observed in binding studies exceeds their functional potency by about 10-fold (Arrang *et al.*, 1990;

Schwartz *et al.*, 1990; Leurs *et al.*, 1995b). In contrast, for H_3-antagonists, a good correlation between receptor binding affinity and functional potencies is generally obtained (Kathmann *et al.*, 1993; Jansen *et al.*, 1994; Schlicker *et al.*, 1994; Leurs *et al.*, 1995b; see also next section). The discrepancy between binding affinities and functional potencies as observed with agonists could result from the involvement of G-protein coupling in agonist binding, i.e. the radiolabelled agonists may bind to the high affinity receptor state predominantly. This phenomenon might also explain the relatively low H_3-receptor densities observed using tritiated agonists (30 to 190 fmol/mg of protein, Table 1) as compared to the densities observed for most radiolabelled antagonists (70 to 400 fmol/mg of protein, Table 2). The involvement of G-protein coupling in agonist binding can be regarded as a general drawback of the use of radiolabelled agonists as tools for receptor binding, complicating the interpretation of the binding data. Nevertheless, H_3-receptor agonists proved to be valuable H_3-receptor binding tools.

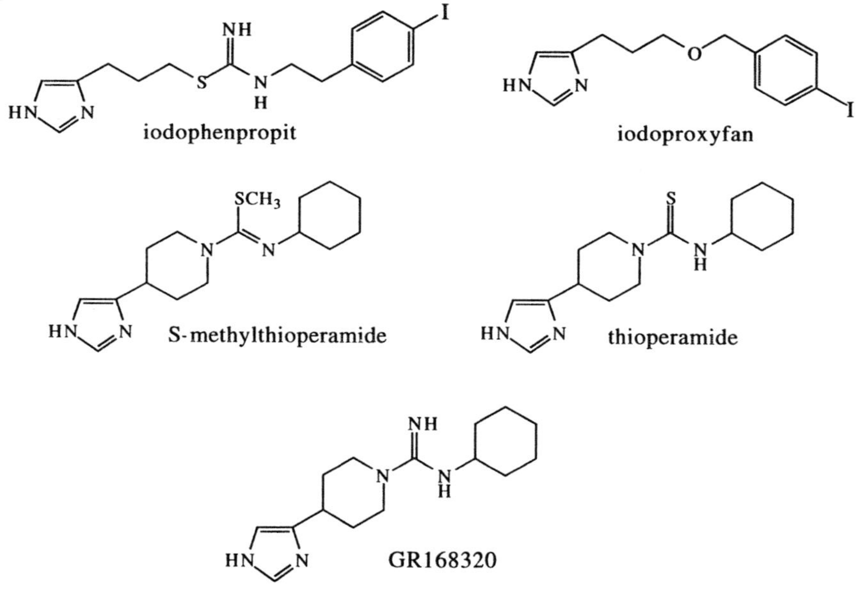

Figure 2. The chemical stuctures of H_3-antagonists used a radioligands.

3. RADIOLABELLED H_3-ANTAGONISTS

The complexity of the radiolabelled agonists binding profiles stressed the need to develop radiolabelled antagonists for H_3-receptor binding studies. In the last five years two radioiodinated and three tritiated antagonists have been reported. The main characteristics of these ligands are summarized in Table 2. Binding of the antagonists to rat brain show a high affinity, and is saturable and reversible. Like the agonists, also the radiolabelled antagonists described so far are imidazoles, differing in their side chain (see Figure 2).
The introduction of $[^{125}I]$-labelled antagonists has led to probes with a high specific activity, yielding a higher sensitivity as compared to tritiated compounds. The high sensitivity of the radioiodinated probes has been especially beneficial with respect to the exposure time required to obtain receptor binding autoradiograms (i.e. several hours versus several months, using conventional Hyperfilm). It may be noted though that the exposure times of tritiated radioligands could be significantly reduced to several days using the recently developed storage

Table 2. Characteristics of radiolabelled antagonists used to study histamine H_3-receptors.

	$[^{125}I]$IPP[a]	$[^{125}I]$IPF[b]	$[^3H]$SMT[c]	$[^3H]$thioperamide[a]	$[^3H]$GR168320[a]
Dissociation constant (K_D, nM)	0.3 - 0.6	0.1	2.1	0.8	0.1
Receptor density (B_{max}, fmol/mg)	250 - 350	78	550	73	412
Specific binding (% of total binding)	45 - 55[d]	50 - 60	70 - 80	50 - 60	>90
Specific activity (Ci/mmol)	2000	2000	50	6	5
Sensitivity to guanine nucleotides	-	+	N.D.	N.D.	N.D.
Functional potency (K_B-value, nM)	0.3	1 - 5	N.D.	4	0.2
Species characterized	rat mouse	rat	rat	rat	rat
References used	[1,2]	[3]	[4]	[5]	[6]

[1] Jansen *et al.*, 1992, [2] Jansen *et al.*, 1994, [3] Ligneau *et al.*, 1994, [4] Yanai *et al.*, 1994, [5] Alves-Rodrigues *et al.*, 1996, [6] Brown *et al.*, 1996. Data are derived from receptor binding studies using brain membranes preparations, and values correspond to experiments using [a]rat cerebral cortex, [b]rat striatum or [c]rat forebrain. [d] ~75% as determined in mouse whole brain. The specific binding is related to radioligand concentrations around or below the K_D-value. Functional potencies correspond to the inhibition of the potassium induced $[^3H]$histamine release from rat cerebral cortex slices or to H_3-receptor mediated inhibition of the electrically contracted guinea-pig jejunum. Abbreviations: $[^{125}I]$IPP, $[^{125}I]$iodophenpropit; $[^{125}I]$IPF, $[^{125}I]$iodoproxyfan; $[^3H]$SMT, $[^3H]$S-methylthioperamide; N.D., not described.

phosphor imaging method, which has been applied to study the autoradiographic distribution of $[^3H](R)\alpha$-methylhistamine and of $[^3H]$S-methylthioperamide (Yanai *et al.*, 1992; Yanai *et al.*, 1994).

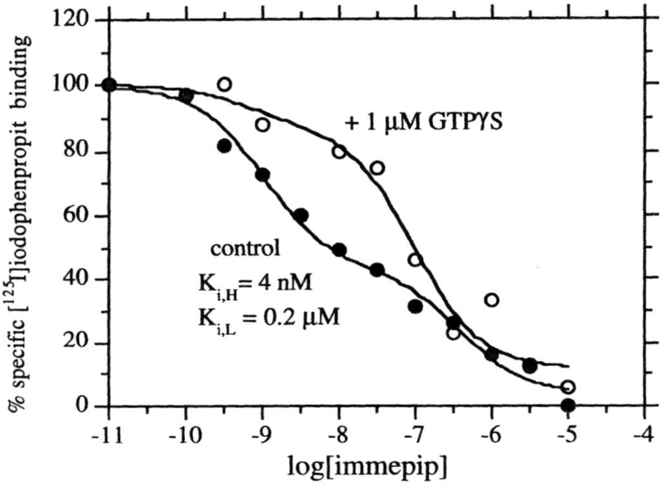

Figure 3. Effect GTPγS on the displacement of $[^{125}I]$iodophenpropit from rat cerebral cortex membranes by immepip.

Consistent with conventional models of antagonist receptor binding, saturation binding curves of $[^{125}I]$iodophenpropit were not affected by guanine nucleotides. Displacement of $[^{125}I]$iodophenpropit by H_3-agonists was usually biphasic in nature and displacement curves were shifted to the right (towards a monophasic curve) after inclusion of GTPγS in the incubation medium, consistent with an involvement of G-proteins in the binding of the agonists (Jansen *et al.*, 1994; Leurs *et al.*, 1996: see also Figure 3). In contrast to $[^{125}I]$iodophenpropit, the binding of $[^{125}I]$iodoproxyfan to rat striatal membranes was shown to be partly sensitive to guanine nucleotides (Ligneau *et al.*, 1994). This observation might be related to the partial agonistic nature of iodoproxyfan recently described (Schlicker *et al.*, 1996). Also for $[^{125}I]$iodoproxyfan biphasic agonist displacement binding curves were reported. Histamine competition binding curves were shifted to the right by guanine nucleotides (Ligneau *et al.*, 1994).

The effect of guanine nucleotides on the binding curves of the tritiated antagonists has not been described so far. Using tritiated antagonists, agonist competition binding curves were found to be biphasic. Competition binding curves of (R)α-methylhistamine were not affected by GTPγS, using [^{3}H]thioperamide as a radioligand (Alves-Rodrigues *et al.*, 1996). This rather unexpected finding might result from a changed G-protein coupling at the relatively low temperature (i.e. 4°C) used in the study (Alves-Rodrigues *et al.*, 1996). Altogether, the radiolabelled antagonists provided further evidence for an involvement of G-proteins in H_3-receptor mediated signal transduction. Not all the antagonists display a 'classical' behaviour regarding the sensitivity towards guanine nucleotides however.

Using radiolabelled antagonists, the affinities of different unlabelled antagonists generally correlate well with their functional potencies (pA_2-values). For [^{3}H]S-methylthioperamide a detailed binding analysis including competition binding curves of different ligands has not been presented however. Considering competition binding curves of agonists, with [^{125}I]iodophen-propit, [^{3}H]thioperamide and [^{3}H]GR168320, a good correlation has been described between the high affinity binding site of agonists and their EC_{50}-values found in functional studies (Jansen *et al.*, 1994; Alves-Rodrigues *et al.*, 1996; Brown *et al.*, 1996). For [^{125}I]iodoproxyfan, agonist affinities obtained from competition binding experiments were 3 to 10-fold higher as compared to their functional potencies. In this respect, [^{125}I]iodoproxyfan displays characteristics comparable to radiolabelled agonists (see previous section).

A remarkable feature of the radiolabelled H_3-antagonists is that their nonspecific binding generally appeared to be high, except for [^{3}H]GR168320. The definition of the nonspecific binding of the radiolabelled antagonists needs critical consideration. For several antagonists, radioligand binding displaced by antagonists was found to exceed radioligand binding displaced by agonists. An obvious interpretation of these observations is that the radiolabelled antagonists bind to 'non-H_3-receptor components' from which they are readily displaced by H_3-antagonists, and not by H_3-agonists. The phenomenon has been observed for [^{125}I]iodophenpropit, [^{125}I]iodoproxyfan, [^{3}H]thioperamide and for [^{3}H]S-methylthioperamide (Jansen, 1997; Ligneau *et al.*, 1994; Alves-Rodrigues *et al.*, 1996; Yanai *et al.*, 1994). The magnitude of the different displacement seems to be different for each ligand and is largely dependent on the tissue and species used. For [^{125}I]iodophenpropit, the difference between agonist and antagonist displacement largely varies between different rat brain regions. In cortical brain areas and in the basal ganglia a difference of 10 to 20% was found between [^{125}I]iodophenpropit binding displaced by (R)α-methylhistamine and by the H_3-receptor antagonist thioperamide (Jansen, 1997). Yet, the non-H_3-receptor component could amount 30-40% in regions with lower H_3-receptor densities, like the thalamus and the hippocampus (Alves-Rodrigues, 1996). In contrast, (R)α-methylhistamine and thioperamide displaced the same fraction of [^{125}I]iodophenpropit binding in mouse brain (Jansen, 1997). A difference of about 40% between binding displaced between agonists and antagonists has been observed for [^{125}I]iodoproxyfan, in rat striatal tissue (Ligneau *et al.*, 1994). Comparable differences were

found for [^{3}H]thioperamide binding to rat cerebral cortex membranes (Alves-Rodrigues *et al.*, 1996). From these results it may be concluded that, in general, H_3-agonists can be regarded as more reliable tools to define the nonspecific binding of the radiolabelled antagonists.
At present, the origin of the non-H_3-receptor binding component(s) of H_3-antagonists is largely unknown. Recently, iodophenpropit and thioperamide have been screened on about forty different receptor assays (Leurs *et al.*, 1995a). The screening did not yield a possible candidate for the nonspecific binding of both radioligands. Both iodophenpropit and thioperamide display a relatively high affinity for $5HT_3$-receptors (K_i-values of 11 nM and 120 nM, respectively). At the experimental conditions used, $5HT_3$-receptor binding does not interfere with the the assay for both radioligands however. For [^{3}H]thioperamide, binding to cytochrome P_{450} isoenzymes may be involved (Alves-Rodrigues *et al.*, 1996). The same has been suggested for [^{3}H]S-methylthioperamide (Yanai *et al.*, 1994). Interestingly, [^{125}I]iodoproxyfan was recently reported to bind with high affinity to a histamine transporter present in murine hematopoietic progenitor cells (Corbel *et al.*, 1997). Also thioperamide displayed a high affinity to these binding sites, in contrast to H_3-agonists. Whether this interaction could contribute to high nonspecific binding of [^{125}I]iodoproxyfan in the striatum remains to be determined.
The origin of the nonspecific binding may of course differ between the radiolabelled antagonists. Illustrative for this rationale is the observation that the thioperamide related compound [^{3}H]GR168320 does not seem to exhibit the phenomenon of a differential displacement by agonists and antagonists, at least in rat cerebral cortex membranes (Brown *et al.*, 1996). The relatively high nonspecific binding of most radiolabelled antagonists can be regarded as a less favourable feature. In this respect, [^{3}H]GR168320 may be the most appropriate radioligand. However, due to its low specific activity of 4.8 Ci/mmol, a relatively high amount of protein is required in the receptor binding assay, especially when tissues containing lower H_3-receptor densities are assessed. The low specific activity of [^{3}H]GR168320 seems to be inadequate for the generation of autoradiographic images. Therefore, a [^{125}I]-labelled radioligand with properties similar to [^{3}H]GR168320 would be desirable.

4. LOCALIZATION OF H_3-RECEPTOR BINDING SITES IN THE CNS

The autoradiographic distribution of H_3-receptor binding sites has been studied in rat brain using different radioligands, i.e. [^{3}H](R)α-methylhistamine (Arrang *et al.*, 1987; Yanai *et al.*, 1992; Pollard *et al.*, 1993; Ryu *et al.*, 1995), [^{3}H]N$^{\alpha}$-methylhistamine (Cumming *et al.*, 1991; Cumming *et al.*, 1994) and more recently with [^{125}I]iodophenpropit (Jansen *et al.*, 1994), [^{125}I]-iodoproxyfan (Ligneau *et al.*, 1994) and [^{3}H]S-methylthioperamide (Yanai *et al.*, 1994). In contrast to the differences in binding characteristics between the H_3-receptor

Table 3. Comparison of [^{125}I]iodophenpropit and [^{3}H](R)α-methylhistamine binding sites in rat brain.

	radioligand binding (% of total cerebral cortex)	
	[^{125}I]iodophenpropit[1)] *(autoradiography)*	[^{3}H](R)α-methylhistamine[2)] *(membrane preparations)*
anterior cerebral cortex	97 ± 8	108 ± 2
medial cerebral cortex	96 ± 11	100 ± 2
posterior cerebral cortex	110 ± 13	85 ± 6
olfactory tubercle	121 ± 13	103 ± 8
hippocampus	53 ± 8	48 ± 4
caudate putamen	127 ± 10	108 ± 11
nucleus accumbens	133 ± 7	126 ± 12
septum	77 ± 4	N.D.
hypothalamus (anterior)	67 ± 5	70 ± 2
hypothalamus (posterior)	61 ± 7	54 ± 8
hypothalamus (lateral)	68 ± 4	N.D.
hypothalamus (VMH)	78 ± 13	N.D.
thalamus	54 ± 16	N.D.
anterior amygdaloid area	98 ± 6	N.D.
amygdala (posterior)*	69 ± 10	N.D.
substantia nigra	141 ± 21	97 ± 7
pons	33 ± 14	28 ± 5
cerebellum	8 ± 5	7 ± 3

[1)]Specific binding was determined using 1 μM (R)α-methylhistamine. [2)]Values reported by Pollard *et al.*, 1993. *Including: amygdalohippocampal area (AHi) and posteromedial cortical amygdaloid nucleus (PMCo). N.D.: not described. The density of [^{125}I]-iodophenpropit binding sites in the cortex is 268 fmol/mg of protein (Jansen et al., 1994). Randomized brain sections of three to five rats were used. Values are expressed as mean ± SD of three to five separate determinations of which each was performed at least in triplicate.

radioligands found using brain membrane preparations, a consistent overlap is observed so far with respect to the autoradiographic distribution of the radioligand binding sites. In Table 3 a comparison of the distribution of [^{3}H](R)α-methylhistamine and [^{125}I]iodophenpropit binding sites in rat brain is given.

A comprehensive description of the distribution of [^{3}H](R)α-methylhistamine binding sites in the rat CNS has been given by Pollard *et al.* (Pollard *et al.*, 1993). In brief, highest densities are observed in the cerebral cortex, the olfactory tubercles, the caudate putamen, the nucleus accumbens and the substantia nigra. Moderate densities are found in the hippocampus, the globus pallidus, the thalamus and the hypothalamus, including the histaminergic perikarya in the posterior area. (For a more detailed overview of the H_3-receptor binding sites in the CNS see Chapter 1).

H_3-Receptor binding sites in the CNS display a distribution pattern distinct from the localization of the histaminergic varicosities, which may in part be explained by the existence of H_3-heteroreceptors. A presynaptic localization of H_3-receptors on noradrenergic (Schlicker *et al.*, 1989), and on serotonergic (Fink *et al.*, 1990; Alves-Rodrigues *et al.*, 1995) nerve terminals in the cerebral cortex has been indicated from functional studies. Receptor binding studies provided evidence for a presynaptic localization of H_3-receptors on GABA neurons in the substantia nigra (Cumming *et al.*, 1991; Ryu *et al.*, 1994). The presynaptic localization of H_3-receptors in the substantia nigra has recently been confirmed with superfused rat brain slices, demonstrating an inhibition by H_3-agonists of dopamine D_1-receptor stimulated GABA release (Garcia *et al.*, 1997).

In addition to presynaptic receptors, autoradiographic studies have indicated the presence of postsynaptic H_3-receptor binding sites as well. Chemical destruction of postsynaptic structures in the striatum using quinolinic and kainic acid resulted in a marked decrease of striatal H_3-receptor binding sites (Cumming *et al.*, 1991; Pollard *et al.*, 1993; Ryu *et al.*, 1994). Consequently, a major part of the striatal H_3-receptors may be located on striatal GABA neurons, representing more than 85% of the striatal efferents (Kita & Kitai, 1988).

5. H_3-RECEPTOR BINDING STUDIES IN PERIPHERAL TISSUES

The densities of H_3-receptors in the periphery appeared to be much lower as compared to densities in the CNS. This makes peripheral H_3-receptors less accessible for receptor binding studies, and consequently explains that only a limited number of studies on H_3-receptor binding in peripheral tissues has been described so far.

H_3-Receptors in the periphery of the guinea-pig have been characterized with $[^3H]N^{\alpha}$-methylhistamine (Korte *et al.*, 1990). In most tissues H_3-receptor densities were below 1 fmol/mg of protein. Highest densities (between 4 and 8 fmol/mg protein) were found in the large intestine, the ileum, the pancreas and the pituitary. A full pharmacological characterization of the $[^3H]N^{\alpha}$-methylhistamine binding sites in the peripheral tissues was not presented (Korte *et al.*, 1990).

H_3-Receptors were also detected in the human gastric mucosa (Courillon-Mallet *et al.*, 1995). $[^3H]N^{\alpha}$-Methylhistamine saturation binding to mucosal H_3-receptors yielded a receptor density of 10 fmol/mg of protein. H_3-Receptor binding was reduced in Heliobacter *pylori* infected patients (Courillon-Mallet *et al.*, 1995). Gastric H_3-receptors have also been characterized using a human fundic tumor cell line (HGT-1). Binding of $[^3H]N^{\alpha}$-methylhistamine to these cells was sensitive to GTPγS and to both cholera and pertussis toxin, again indicating the coupling of the gastric H_3-receptors to G-proteins (Cherifi *et al.*, 1992). Similar results have been obtained in the murine pituitary tumor cell line AtT-20 (Clark *et al.*, 1993; West *et al.*, 1994).

In guinea-pig lung the distribution of H_3-receptors has been visualized by receptor autoradiography (Schwartz *et al.*, 1990). [^{3}H](R)α-Methylhistamine binding was scattered in the parenchyma. A more dense labelling was observed in the bronchioles (Schwartz *et al.*, 1990).

Except for [^{3}H]S-methylthioperamide, receptor binding studies to peripheral tissues have not been described for radiolabelled antagonists. [^{3}H]S-Methylthioperamide showed a considerably high amount of nonspecific binding, which interfered with the accurate determination of H_3-receptors in peripheral tissues (Yanai *et al.*, 1994). Based on the relatively high amount of nonspecific binding observed with most radiolabelled H_3-antagonists, similar limitations may evolve for other radiolabelled antagonists.

6. HETEROGENEITY OF RADIOLIGANG BINDING SITES

6.1. Radiolabelled H_3-agonist binding sites

In 1990, West *et al.* reported that thioperamide and burimamide discriminated [^{3}H]N$^\alpha$-methylhistamine binding to rat brain membranes into high and low affinity binding sites (West *et al.*, 1990b). [^{3}H]N$^\alpha$-Methylhistamine binding was partly decreased by the GTP analogue GTPγS. In the presence of GTPγS, thioperamide and burimamide yielded monophasic competition binding curves, with affinities corresponding to their high affinity binding sites. From these results the existence of subtypes of H_3-receptors i.e. H_{3A}- and H_{3B}-receptors was proposed, the latter being sensitive towards guanine nucleotides (West *et al.*, 1990b). [^{3}H]N$^\alpha$-Methylhistamine itself did not discriminate between the proposed H_{3A}- and H_{3B}-receptors.

In a study by Arrang and co-workers, using the agonist [^{3}H](R)α-methylhistamine as the radioligand, biphasic competition binding curves in rat cerebral cortex membranes were obtained for burimamide, but not for thioperamide (Arrang *et al.*, 1990). A guanine nucleotide sensitivity of the burimamide binding sites was not reported in this study. In contrast to [^{3}H]N$^\alpha$-methylhistamine (West *et al.*, 1990b), in the standard incubation medium, binding of [^{3}H](R)α-methylhistamine was not sensitive to the GTP analogue Gpp(NH)p. However, when calcium was added to the incubation buffer, two binding sites were found for [^{3}H](R)α-methylhistamine, the low affinity site being abolished by Gpp(NH)p (Arrang *et al.*, 1990). From these observations it may be suggested that the possible heterogeneity of burimamide binding sites and of [^{3}H](R)α-methylhistamine binding sites are unrelated phenomena. The heterogeneity of [^{3}H](R)α-methylhistamine binding sites was suggested to result from the conversion of a subpopulation of the receptors into low-affinity binding sites, triggered by

calcium (Arrang *et al.,* 1990).
A heterogeneity of [^{3}H](R)α-methylhistamine binding sites has also been found in kinetic studies, using buffer without calcium (West *et al.,* 1990a). In this study a homogeneous population of [^{3}H](R)α-methylhistamine binding sites was observed at equilibrium conditions (i.e. saturation binding analysis) however. Thioperamide and burimamide yielded monophasic competition binding curves in this report (West *et al.,* 1990a).
The three reports cited illustrate the complexity of the receptor binding data obtained with the radiolabelled agonists and the controversies in literature with respect to a heterogeneity of H_3-receptor binding sites. Biphasic competition binding curves for burimamide have been described in several reports using [^{3}H](R)α-methylhistamine (Arrang *et al.,* 1990) and [^{3}H]N$^\alpha$-methylhistamine (West *et al.,* 1990b; Kathmann *et al.,* 1993; Cumming & Gjedde, 1994; Brown *et al.,* 1996). Accordingly, different studies reported a heterogeneous displacement of [^{3}H]N$^\alpha$-methylhistamine by thioperamide (West *et al.,* 1990b; Cumming & Gjedde, 1994; Clark & Hill, 1995; Brown *et al.,* 1996). Controversially, other studies did not confirm the presence of two distinct binding sites for burimamide (West *et al.,* 1990a; Kilpatrick & Michel, 1991; Clark & Hill, 1995) and for thioperamide (Arrang *et al.,* 1990; West *et al.,* 1990a; Kilpatrick & Michel, 1991; Kathmann *et al.,* 1993).
One explanation for the different observations concerning heterogeneity of thioperamide and burimamide binding sites is the relatively small difference in affinity between the two separate binding sites, making it difficult to discriminate them statistically. In addition, the controversies concerning the heterogeneity of binding sites may arise from different experimental conditions used, like the choice of buffer (Tris-HCl, phosphate, HEPES), the ionic composition of the buffer (mono- and divalent cations) and the tissue preparation used (cerebral cortex versus whole brain). For example, it has been reported that the affinity of thioperamide for [^{3}H](R)α-methylhistamine binding sites was 10-fold higher in phosphate buffer as compared to Tris-HCl buffer (West *et al.,* 1990a). In contrast the affinity of the agonists histamine, (R)α-methylhistamine and N$^\alpha$-methylhistamine were not substantially different when phosphate and Tris-HCl buffer are compared (Arrang *et al.,* 1987; West *et al.,* 1990a).
The ionic composition of the buffer has been indicated to differentially affect binding characteristics of ligands. As previously cited, guanine nucleotide sensitivity of [^{3}H](R)α-methylhistamine (but not of [^{3}H]N$^\alpha$-methylhistamine; West *et al.,* 1990b) was dependent on the presence of calcium in the buffer (Arrang *et al.,* 1990). Sodium ions were shown to abolish the low affinity binding site of thioperamide, whereas the binding affinities of clobenpropit and N$^\alpha$-methylhistamine were not affected (Clark & Hill, 1995). From these results, it was suggested that the H_3-receptor exists in different conformations, for each of which thioperamide has a different affinity (Clark & Hill, 1995). Hence, contribution of differential allosteric effects dependent of the buffer composition may relate to the observed heterogeneity of binding sites and to the controversy in literature in this respect. A differential allosteric action of sodium

has also been reported for other receptor systems including the binding of H_1-receptor antagonists (Treherne *et al.*, 1991; Gibson *et al.*, 1994). The allosteric effect may also be related to the involvement of G-proteins in the binding of agonists to the receptor, further complicating the interpretation of the binding data. Altogether, the complexity of the binding profile of radiolabelled agonists does not provide a sound basis for the definition of H_3-receptor subtypes.

An important criterion for the identification of receptor subtypes is that they are related to distinct functional responses. Based on the functional potencies of thioperamide and of tiotidine, H_{3A}- and H_{3B}-receptors were suggested to be linked to H_3-receptor mediated inhibition of histamine release and synthesis, respectively (West *et al.*, 1990b). At present, not much additional evidence for this suggestion has been presented. Histamine H_3-receptors inhibiting noradrenaline release in mouse brain cortex slices have been suggested to represent the H_{3A}-receptor subtype (Schlicker et al., 1992; Schlicker et al., 1994). To our knowledge, functional responses in brain tissue related to the H_{3B}-receptor have never been observed however.

6.2. Radiolabelled H_3-antagonist binding sites

[^{125}I]Iodophenpropit was biphasically displaced from rat cortex membranes by the antagonists burimamide and dimaprit (Jansen *et al.*, 1994). In contrast to agonist binding, antagonist binding was not affected by GTPγS. Hence, biphasic competition binding curves of burimamide and dimaprit were likely not related to the G-protein coupling of the [^{125}I]iodophenpropit binding sites. For the other radiolabelled antagonists, biphasic competition binding curves of antagonists have so far not been demonstrated. Remarkably, thioperamide and burimamide yielded steep competition binding curves (Hill-coefficients of 1.7 and 1.9, respectively) in rat striatal membranes using the [^{125}I]-iodoproxyfan assay (Ligneau *et al.*, 1994).

A heterogeneous distribution of putative H_3-receptor subtypes has not been demonstrated so far. Using a receptor autoradiographic approach, we have recently found that [^{125}I]iodophenpropit binding to ten different rat brain areas was not discriminated by a chemically heterogeneous group of H_3-receptor antagonists (Jansen, 1997).

As mentioned before, for all radiolabelled antagonists, biphasic competition binding curves were reported for H_3-agonists (Jansen *et al.*, 1994; Yanai *et al.*, 1994; Ligneau *et al.*, 1994; Alves-Rodrigues *et al.*, 1996; Brown *et al.*, 1996). In the [^{125}I]iodophenpropit and [^{125}I]-iodoproxyfan binding assays, agonist competition binding curves were sensitive to guanine nucleotides (Jansen *et al.*, 1994; Ligneau *et al.*, 1994). The apparent heterogeneity of agonist binding may therefore be attributed to the involvement of G-proteins in the agonist receptor binding rather than to a receptor heterogeneity. For the tritiated antagonists, the sensitivity to guanine nucleotides was not studied or could not be demonstrated at the experimental conditions used (Alves-Rodrigues *et al.*, 1996). Yet, it can be concluded that receptor binding studies with both, radiolabelled agonists and with radiolabelled antagonists did not reveal exclusive evidence for H_3-receptor heterogeneity. In general, the exploration of H_3-receptor subtypes requires the availability of ligands with a higher selectivity towards one of these putative subtypes.

7. RESUMPTION AND CONCLUDING REMARKS

Studies performed with H_3-receptor radioligands have substantially contributed to the current knowledge of the characteristics, distribution and function of the histamine H_3-receptor. Tritiated agonists were successfully used to study H_3-receptors in rodent and primate CNS. Binding studies with radiolabelled agonists provided evidence for a role of G-proteins in H_3-receptor mediated signal transduction. The apparent involvement of G-protein coupling in the binding of the radiolabelled agonists may underlie two less favourable features of the radioligands however. At first, an overestimation of H_3-agonists potencies is obtained in competition binding studies. Secondly, the complexity of the radiolabelled agonists binding dynamics makes it difficult to distinguish binding phenomena related to G-protein coupling, allosteric interactions, and receptor heterogeneity in terms of H_3-receptor subtypes. Radiolabelled agonists are advantageous with respect to their low nonspecific binding in the rat CNS. [^{3}H](R)α-Methylhistamine and [^{3}H]N$^\alpha$-methylhistamine were both shown to be very useful studying the distribution of H_3-receptors by autoradiography.

The introduction of radiolabelled H_3-receptor antagonists yielded improved tools for H_3-receptor binding studies. With the use of these ligands additional evidence was provided for the interaction of H_3-receptors with G-proteins. As compared to radiolabelled agonists, [^{125}I]iodophenpropit, [^{3}H]GR168320 and [^{3}H]thioperamide exhibit the advantage of a good correlation between agonist binding affinities and their functional potencies.

So far, studies with radiolabelled H_3-antagonists did not provide significant progress in the search for H_3-receptor heterogeneity. Ligands which more clearly discriminate between putative subtypes are still awaited, and a link between binding heterogeneity and functional receptor responses will be indispensable. Not all radiolabelled antagonists display a straightforward binding profile, which may in part be due to the relatively high amount of nonspecific binding, to be considered as a disadvantage. In this respect [^{3}H]GR168320 is a promising ligand, displaying a negligible amount of nonspecific binding, allowing a unambiguous interpretation of receptor binding data.

REFERENCES

Alves-Rodrigues, A. (1996). Doctoral thesis: The histamine H3-receptor in the rat brain: pharmacological and (patho)physiological aspects. Vrije Universiteit, Amsterdam, The Netherlands.

Alves-Rodrigues, A., Jansen, F.P., Leurs, R., Timmerman, H. and Prell, G.D. (1995). Interaction of clozapine with the histamine H3-receptor in rat brain. *Br. J. Pharmacol.* **114**, 1523-1524.

Alves-Rodrigues, A., Leurs, R., Wu, T.S.W., Prell, G.D., Foged, C. and Timmerman, H. (1996). [^{3}H]-Thioperamide as a radioligand for the histamine H3-receptor in rat cerebral cortex. *Br. J. Pharmacol.* **118**, 2045-2052.

Arrang, J.M., Garbarg, M., Lancelot, J.-C., Lecomte, J.-M., Pollard, H., Robba, M., Schunack, W. and Schwartz, J.C. (1987). Highly potent and selective ligands for histamine H3-receptors. *Nature* **327**,117-125.

Arrang, J.M., Roy, J., Morgat, J.-L., Schunack, W. and Schwartz, J.C. (1990). Histamine H3-receptor binding sites in rat brain membranes: modulations by guanine nucleotides and divalent cations. *Eur. J. Pharmacol.* **188**, 219-227.

Barbin, G., Palacios, J.M., Rodergas, E., Schwartz, J.C. and Garbarg, M. (1980). Characterization of the high-affinity binding sites of [^{3}H]histamine in rat brain. *Mol. Pharmacol.* **18**, 1-10.

Brown, J.D., Oshaughnessy, C.T., Kilpatrick, G.J., Scopes, D.I.C., Beswick, P., Clitherow, J.W. and Barnes, J.C. (1996). Characterisation of the specific binding of the histamine H3-receptor antagonist radioligand [^{3}H]GR168320. *Eur J Pharmacol* **311**, 305-310.

Cherifi, Y., Pigeon, C., Leromancer, M., Bado, A., Reyldesmars, F. and Lewin, M.J.M. (1992). Purification of a histamine H3-receptor negatively coupled to phosphoinositide turnover in the human gastric cell line Hgt1. *J. Biol. Chem.* **267**, 25315-25320.

Clark, E.A. and Hill, S.J. (1995). Differential effect of sodium ions and guanine nucleotides on the binding of thioperamide and clobenpropit to histamine H3-receptors in rat cerebral cortical membranes. *Br. J. Pharmacol.* **114**, 357-362.

Clark, M.A., Korte, A. and Egan, R.W. (1993). Guanine nucleotides and pertussis toxin reduce the affinity of histamine H3-receptors on AtT-20 cells. *Agents and Actions* **40**, 129-134.

Corbel, S., Traiffort, E., Stark, H., Schunack, W. and Dy, M. (1997). Binding of histamine H3-receptor antagonists to hematopoetic progenitor cells. *Febs Lett.* **404**, 289-293.

Courillon-Mallet, A., Launay, J.M., Roucayrol, A.M., Callebert, J., Emond, J.P., Tabuteau, F. and Cattan, D. (1995). Helicobacter pylori infection: Physiopathologic implication of Nα-methylhistamine. *Gastroenterology* **108**, 959-966.

Cumming, P. and Gjedde, A. (1994). Subclasses of histamine H3-antagonist binding sites in rat brain. *Brain Res.* **641**, 203-207.

Cumming, P., Gjedde, A. and Vincent, S. (1994). Histamine H_3 binding sites in rat brain - localization in the nucleus of the solitary tract. *Brain Res.* **641**, 198-202.

Cumming, P., Shaw, C. and Vincent, S.R. (1991). High affinity histamine binding site is the H_3-receptor - characterization and autoradiographic localization in rat brain. *Synapse* **8**, 144-151.

Fink, K., Schlicker, E., Neise, A. and Göthert, M. (1990). Involvement of presynaptic H_3-receptors in the inhibitory effect of histamine on serotonin release in the rat brain cortex. *Naunyn-Schmied. Arch. Pharmacol.* **342**, 513-519.

Fujimoto, K., Mizuguchi, H., Fukui, H. and Wada, H. (1991). Presynaptic localization of histamine H_3-receptors in rat brain. *Biochem. Biophys. Res. Commun.* **177**, 907-912.

Garcia, M., Floran, B., Arias-Montano, J.A., Young, J.M. and Aceves, J. (1997). Histamine H_3-receptor activation selectively inhibits dopamine D_1-dependent [^{3}H]-γ-aminobutyric acid release from depolarisation-stimulated slices of rat substantia nigra pars reticulata. *Neuroscience,* in press.

Gibson, W.J., Roques, T.W. and Young, J.M. (1994). Modulation of antagonist binding to histamine H_1-receptors by sodium ions and by 2-amino-2-hydroxymethyl-propan-1,3-diol HCl. *Br. J. Pharmacol.* **111**, 1262-1268.

Herbert T.E., Moffett, S., Morello, J.P., Loisel, T.P., Bichet, D.G., Barret, C. and Bouvier, M. (1996). A peptide derived from a beta$_2$-adrenergic receptor transmembrane domain inhibits both receptor dimerization and activation. *J. Biol. Chem.* **271**, 16384-16392.

Jansen, F.P. (1997). Doctoral thesis: The histamine H_3-receptor in the rat brain: pharmacological and (patho)physiological aspects. Vrije Universiteit, Amsterdam, The Netherlands.

Jansen, F.P., Rademaker, B., Bast, A. and Timmerman, H. (1992). The first radiolabelled histamine H_3-receptor antagonist, [^{125}I]iodophenpropit: saturable and reversible binding to rat cortex membranes. *Eur. J. Pharmacol.* **217**, 203-205.

Jansen, F.P., Wu, T.S., Voss, H.P., Steinbusch, H.W.M., Vollinga, R.C., Rademaker, B., Bast, A. and Timmerman, H. (1994). Characterization of the binding of the first selective radiolabelled histamine H_3-receptor antagonist, [^{125}I]-iodophenpropit, to rat brain. *Br. J. Pharmacol.* **113**, 355-362.

Kathmann, M., Schlicker, E., Detzner, M. and Timmerman, H. (1993). Nordimaprit, homodimaprit, clobenpropit and imetit - affinities for H_3 binding sites and potencies in a functional H_3-receptor model. *Naunyn Schmied. Arch. Pharmacol.* **348**, 498-503.

Kilpatrick, G.J. and Michel, A.D. (1991) Characterisation of the binding of the histamine H_3-receptor agonist [^{3}H](R)α-methylhistamine to homogenates of rat and guinea-pig cortex. In: *New perspectives in histamine research* (Timmerman, H. & Van der Goot, H., Eds) pp 69-75, Birkhäuser Verlag, Basel.

Kita, H. and Kitai, S.T. (1988). Glutamate decarboxylase immunoreactive neurons in rat neostriatum: their morphological types and populations. *Brain Res.* **447**, 346-352.

Korte, A., Myers, J., Shih, N.-Y., Egan, R.W. and Clarck, M.A. (1990). Characterization and tissue distribution of H3-histamine receptors in guinea-pigs by N^{α}-methylhistamine. *Biochem. Biophys. Res. Commun.* **168**, 979-986.

Leurs, R., Kathmann, M., Vollinga, R.C., Menge, W.M.P.B., Schlicker, E. and Timmerman, H. (1996). Histamine homologues discriminating between two functional H3-receptor assays. Evidence for H3-receptor heterogeneity? *J. Pharmacol. Exp. Ther.* **276**, 1009-1015.

Leurs, R. and Timmerman, H. (1992). The histamine H3-receptor: a target for developing new drugs. *Prog. Drug Res.* **39**, 127-165.

Leurs, R., Tulp, M.T.M., Menge, W.M.B.P., Adolfs, M.J.P., Zuiderveld, O.P. and Timmerman, H. (1995a). Evaluation of the receptor selectivity of the H3-receptor antagonists, iodophenpropit and thioperamide: an interaction with the 5-HT3- receptor revealed. *Br. J. Pharmacol.* **116**, 2315-2321.

Leurs, R., Vollinga, R.C. and Timmerman, H. (1995b). The medicinal chemistry and therapeutic potentials of ligands of the histamine H3-receptor. *Prog. Drug Res.* **45**, 107-165.

Ligneau, X., Garbarg, M., Vizuete, M.L., Diaz, J., Purand, K., Stark, H., Schunack, W. and Schwartz, J.C. (1994). [^{125}I]Iodoproxyfan, a new antagonist to label and visualize cerebral histamine H3-receptors. *J. Pharmacol. Exp. Ther.* **271**, 452-459.

Martinez-Mir, M.I., Pollard, H., Moreau, J., Arrang, J.M., Ruat, M., Traiffort, E., Schwartz, J.C. and Palacios, J.M. (1990). Three histamine receptors (H_1, H_2 and H_3) visualized in the brain of human and non-human primates. *Brain Res.* **526**, 322-327.

Ng, G.Y.K., O'Dowd, B.F., Lee, S.P., Chung, H.T., Brann, M.R., Seeman, P. and George, S.R. (1996). Dopamine D_2 receptor dimers and receptor blocking peptides, Biochem. Biophys Res, Commun. 227, 200-204.

Oishi, R., Adachi, N. and Saeki, K. (1993). N^{α}-Methylhistamine inhibits intestinal transit in mice by central histamine H_1-receptor activation. *Eur. J. Pharmacol.* **237**, 155-159.

Pollard, H., Moreau, J., Arrang, J.M. and Schwartz, J.C. (1993). A detailed autoradiographic mapping of histamine H3-receptors in rat brain areas. *Neuroscience* **52**, 169-189.

Ryu, J.H., Yanai, K., Sakurai, E., Kim, C.Y. and Watanabe, T. (1995). Ontogenetic development of histamine receptor subtypes in rat brain demonstrated by quantitative autoradiography. *Developmental Brain Res.* **87**, 101-110.

Ryu, J.H., Yanai, K. and Watanabe, T. (1994). Marked increase in histamine H3-receptors in the striatum and substantia nigra after 6-hydroxydopamine-induced denervation of dopaminergic neurons: an autoradiographic study. *Neurosci. Lett.* **178**, 19-22.

Schlicker, E., Behling, A., Lümmen, G. and Göthert, M. (1992). Histamine H_{3A} receptor-mediated inhibition of noradrenaline release in the mouse brain cortex. *Naunyn-Schmied Arch Pharmacol* **345**, 489-493.

Schlicker, E., Kathmann, M., Bitschnau, H., Marr, I., Reidemeister, S., Stark, H. and Schunack, W. (1996). Potencies of antagonists chemically related to iodoproxyfan at histamine H_3-receptors in mouse brain cortex and guinea-pig ileum: evidence for H_3-receptor heterogeneity? *Naunyn-Schmied Arch Pharmacol* **353**, 482-488.

Schlicker, E., Kathmann, M., Reidemeister, S., Stark, H. and Schunack, W. (1994). Novel histamine H_3-receptor antagonists: affinities in an H_3-receptor binding assay and potencies in two functional H_3-receptor models. *Br. J. Pharmacol.* **112**, 1043-1048.

Schwartz, J.C., Arrang, J.M., Garbarg, M. and Pollard, H. (1990). A third histamine receptor subtype: characterisation, localisation and functions of the H_3-receptor. *Agents and Actions* **30**, 13-23.

Sinkins, W.G., Kandel, M., Kandel, S.I., Schunack, W. and Wells, J.W. (1993). Protein-linked receptors labeled by [^{3}H]histamine in guinea-pig cerebral cortex. 1. Pharmacological characterization. *Mol. Pharmacol.* **43**, 569-582.

Sinkins, W.G. and Wells, J.W. (1993). Protein-linked receptors labeled by [^{3}H]histamine in guinea-pig cerebral cortex. 2. Mechanistic basis for multiple states of affinity. *Mol. Pharmacol.* **43**, 583-594.

Treherne, J.M., Stern, J.S., Flack, W.J. and Young, J.M. (1991). Inhibition by cations of antagonist binding to histamine H_1-receptors: differential effect of sodium ions on the binding of two radioligands. *Br. J. Pharmacol.* **103**, 1745-1751.

West, R.E., Myers, J., Zweig, A., Siegel, M.I., Egan, R.W. and Clark, M.A. (1994). Steroid-sensitivity of agonist binding to pituitary cell line histamine H_3-receptors. *Eur. J. Pharmacol.* **267**, 343-348.

West, R.E., Zweig, A., Granzow, R.T., Siegel, M.I. and Egan, R.W. (1990a). Biexponential kinetics of (R)α-[^{3}H]methylhistamine binding to the rat brain histamine H_3-receptor. *J. Neurochem.* **55**, 1612-1616.

West, R.E.J., Zweig, A., Shih, N.-Y., Siegel, M.I., Egan, R.W. and Clarck, M.A. (1990b). Identification of two H_3-histamine receptor subtypes. *Mol. Pharmacol.* **38**, 610-613.

Yanai, K., Ryu, J.H., Sakai, N., Takahashi, T., Iwata, R., Ido, T., Murakami, K. and Watanabe, T. (1994). Binding characteristics of a histamine H_3-receptor antagonist, [^{3}H]S-methylthioperamide: Comparison with [^{3}H](R)α-methylhistamine binding to rat tissues.*Jap. J. Pharmacol.* **65**, 107-112.

Yanai, K., Ryu, J.H., Watanabe, T., Iwata, R. and Ido, T. (1992). Receptor autoradiography with [^{11}C] and [^{3}H]-labelled ligands visualized by imaging plates. *Neuroreport* **3**, 961-964.

Zweig, A., Siegel, M.I., Egan, R.W., Clark, M.A., Shorr, R.G.L. and West, R.E. (1992). Characterization of a digitonin-solubilized bovine brain H_3-histamine receptor coupled to a guanine nucleotide-binding protein. *J. Neurochem.* **59**, 1661-1666.

R. Leurs and H. Timmerman (Editors)
The Histamine H_3 Receptor

Substituted imidazoles, the key to histaminergic receptors

W. M. P. B. Menge, H. Timmerman

Division of Medicinal Chemistry, Leiden/Amsterdam Center for Drug Research, Vrije Universiteit, De Boelelaan 1083, 1081 HV, Amsterdam, The Netherlands

1. INTRODUCTION

Substituted imidazoles are an example of a pharmaceutically important class of heterocyclic compounds, several of which have been incorporated in drugs that have reached the market, e.g. cimetidine, ondansetron and losartan (figure 1).

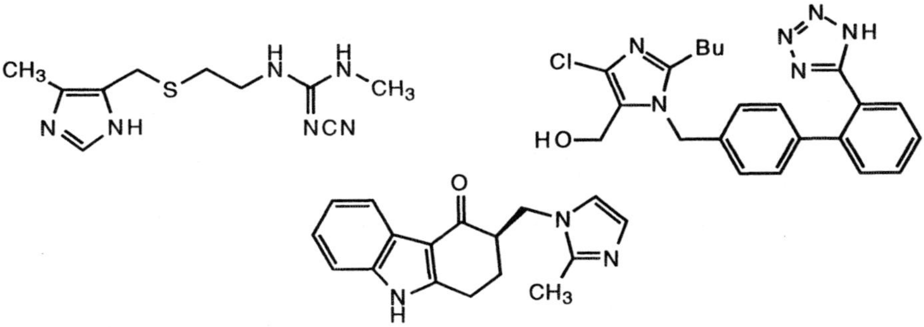

Figure 1: Drugs containing an imidazole ring, cimetidine, ondansetron and losartan

Also in the development of specific ligands for the histamine receptors numerous substituted imidazoles have been synthesised during the last decades. This is not surprising since the natural ligand, histamine, is also a substituted imidazole. Via a change in the substitution pattern of the imidazole nucleus or a modification of the existing substituent many histamine analogues and imidazole derivatives, with potent and selective agonistic or antagonistic activity, have been prepared (figure 2). These compounds and their structure-activity relationships have played an important role in the classification of the histamine receptors into the three, currently known, subtypes[1].

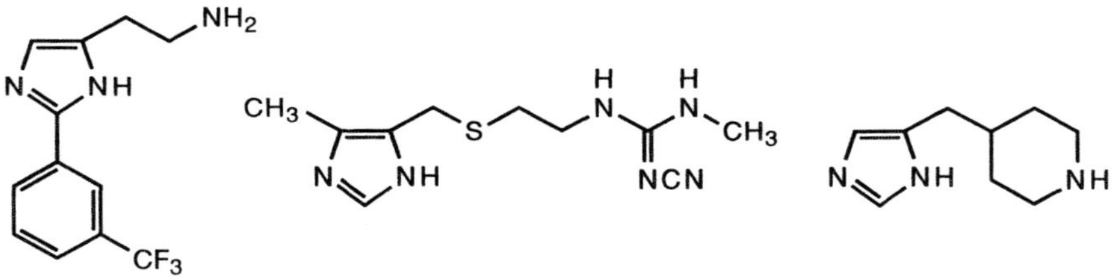

Figure 2: Examples of potent and selective histaminergic ligands

Continued investigations of the structure-activity relationships of ligands for these receptors have also led to the development of potent non-imidazolic ligands for the histamine H_1- and H_2-receptor subtypes. However, the state of the art on the medicinal chemistry of the histamine H_3-receptor indicates that this most recently discovered receptor subtype has a remarkable preference for substituted imidazoles[2] as ligands. Replacement of the imidazole ring in known potent H_3-ligands by other heterocycles leads consistently to a complete loss or, at the very least, a drastic decrease in affinity[3]. Therefore, the research on the structure-activity relationships for the histamine H_3-receptor still depends on the availability of new substituted imidazoles as ligands. Also, for the two other histamine receptor subtypes there is a continuing interest in the synthesis of new substituted imidazoles and their pharmacological evaluation as histaminergic ligands.

For example; the recent discovery of the inverse agonism of several compounds formerly classified as histamine H_2-antagonists[4], the systematic studies on the binding interactions of the agonistic and antagonistic ligands with specific amino acid residues of the histamine H_1-[5] and H_2-receptor[6] or the study on the molecular mechanism of activation[6,7] of these receptors. And last but not least, the continuous refinement of pharmacological and molecular modelling techniques demand a (re-) investigation of old and new compounds with the imidazole pharmacophore.

The synthesis of imidazoles and the evaluation of their pharmacological activities can therefore be considered to be the key to a better understanding of the histamine receptors. Further new developments such as combinatorial technologies and the uprise of molecular biology will certainly play an important role in the need for and the synthesis and evaluation of new substituted imidazoles in the coming years.

2. SYNTHESES OF IMIDAZOLES

A comprehensive review on imidazole chemistry by Grimmett has appeared recently[8]. Therefore, the next paragraphs will focus on the most recent developments in the synthesis of substituted imidazoles used as ligands for the histamine receptors. Special attention will be pay to methods with perspective for the synthesis of histaminergic ligands.

2.1. Modification of imidazole precursors

A simple and direct approach to new imidazole containing ligands is the modification of commercially available imidazoles. Unfortunately, only a limited number of substituted imidazoles is available and most of them are not ideally substituted for further modification because they have only plain alkyl or phenyl substituents. An example of this approach is the synthesis of 3-aminopropylimidazole[9] and 3-hydroxy-propylimidazole[10] from urocanic acid (figure 3).

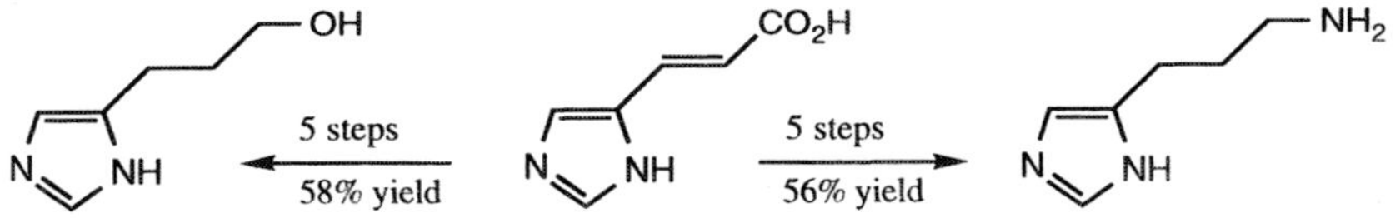

Figure 3: Urocanic acid as a source for fragments of histaminergic ligands

These compounds are valuable intermediates for the synthesis of histamine H_3- and H_2-ligands.

2.1. Condensation approaches

Most substituted imidazoles are obtained via the condensation of non-cyclic fragments to form the desired imidazole. The majority of these condensation reactions has recently been reviewed by Grimmett[8]. The largest number of imidazoles obtained via condensation approaches is only of limited use for the development of histaminergic ligands as they are mainly polysubstituted imidazoles. For the histamine receptors proper substitution of the imidazole ring is usually limited to the 4(5)-position, with the exception of the histamine H_1-receptor, where a broader range of substituents on the 2-position is tolerated.

The most frequently applied condensation approach for histaminergic ligands is the condensation of an α-substituted carbonyl compound, C4-C5 backbone, with an amidine or activated amide, C2-precursor (figure 4). The Bredereck[11] method, in which an α-bromo carbonyl compound is condensed with an amidine or another activated precursor to form the heterocyclic ring[12,13,14], is a popular example of such an approach.

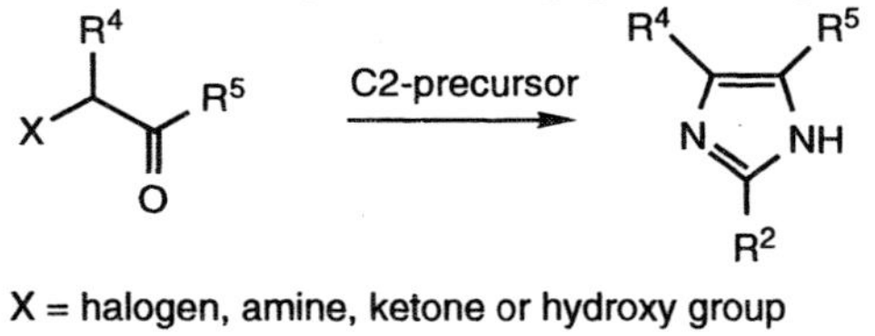

Figure 4. The synthesis of imidazoles using conventional condensation aproaches.

The isolated yields of the Bredereck method are satifactory for the synthesis of di-substituted compounds (figure 5) but are generally lower for 4(5)-mono-substituted imidazoles.

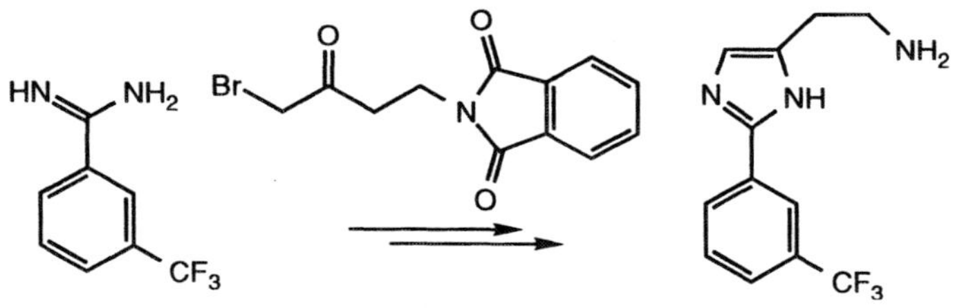

Figure 5. Synthesis of histamine H_1-agonists via the Bredereck method

Other important methods involve the condensation of an aminoketone[15,16] (figure 6) or a diketone[17] with an imidate or other activated amide. In general, the methods are of a broad scope and have been used to synthesize a variety of imidazole derivatives.

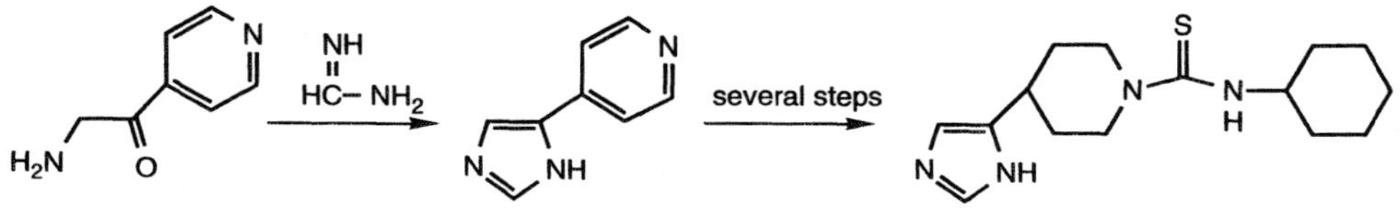

Figure 6: The synthesis of thioperamide

Major drawbacks of the condensation approaches[8] are the difficult syntheses of the starting materials, the low yields of the end products and the lack of flexibility in the approach. Each new target compound requires a different precursur and thus a completely new synthesis route including all the difficulties associated with the isolation of the product.

In conclusion, although satisfactory condensation reactions are available to prepare most of the desired substituted imidazoles, there is a continuing interest in new, less elaborate and more flexible, synthesis routes.

2.2. The synthon approach

An alternative way to synthesize imidazoles in a more flexible manner is to use a synthon. An example of such a synthon approach, is the synthesis of imidazoles using tosylmethyl isocyanide (TOSMIC). The original procedure using this synthon gave only moderate overall yields of the substituted imidazoles[18], mainly due to the poor yield in the first step (figure 7).

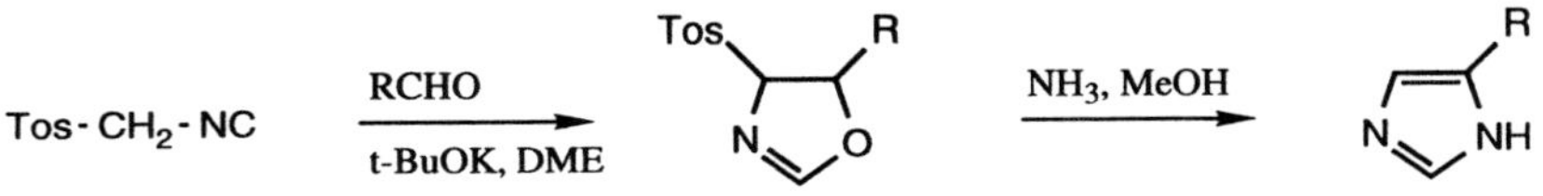

Figure 7. The synthesis of imidazoles using TosMIC.

However, recently the yield of the first step of this imidazole synthesis was improved dramatically by replacement of the basic catalyst (tBuOK) by a milder basic catalyst, sodium cyanide[19]. Due to these milder reaction conditions a greater variety of aldehydes, one of the starting materials for the synthesis of imidazoles, can be used increasing the flexibility of the method[20] even more. This synthon approach can also be used to prepare the bioiscsteric substituted thiazole analogs[21].

Other recent improvements of the TOSMIC synthon approach include the use of different precursors for the C5-N1 part of the imidazole ring. For example by the use of silyl imines[22].

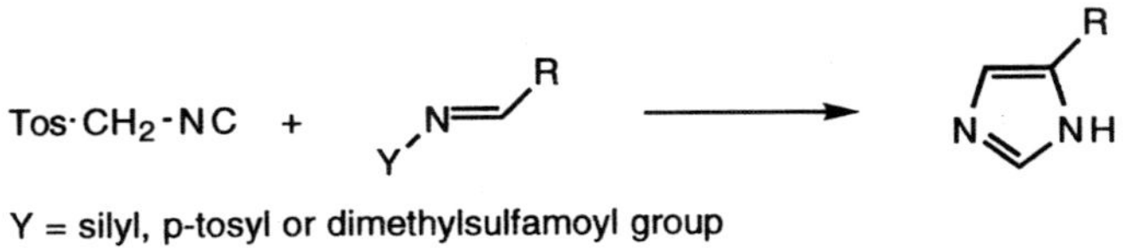

Figure 8. Improved synthesis of imidazoles using TosMIC.

These silyl imines are generated *in situ* from the corresponding aldehydes resulting in a reduction of the number of reaction steps and simplifies the work-up procedures (figure 8). Another option is to convert the aldehyde into a N-tosyl or a N,N-dimethylsulfamoyl imine[23].

In this latter approach the product of the reaction is the NH protected imidazole. The isolation the imidazole in the protected form can be advantageous in the further derivatisation of the compound.

2.3. Solid Phase Synthesis approaches

Conventional liquid phase synthesis suffers from the limitation that each product or intermediate has to be separated from the other components of the reaction mixture. An elegant answer to this problem is to use a solid phase synthesis (SPS) approach. In such an approach the compounds are synthesized on a solid support and simple washing steps replace the laborious work up and isolation procedures. At the end of the synthesis the product is released from the solid support. The SPS of oligomers of amino acids or nucleotides is well estabilished and task chemists are facing now is the development of SPS routes for small organic molecules.

In the field of imidazole chemistry the first example of a SPS approach to synthesize imidazoles has allready appeared . The group of Mjalli[24] reported a SPS of imidazoles on the basis of the Ugi, four component condensation reaction[25] (figure 9).

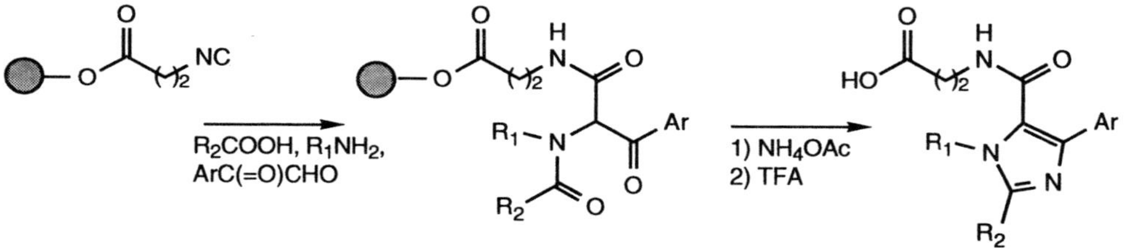

Figure 9. A Solid Phase Synthesis route to substituted imidazoles.

Allthough this SPS route averted some of the problems inherent to the synthesis of substituted imidazoles via condensation approaches, the value of the synthesized libraries of compounds is of a limited interest for the histamine receptor research field. First, there is only a limited number of glyoxals, one of the four reaction components, available as precursor. Secondly, only tri- and tetra-substituted imidazoles can be prepared via this method. And finally, the linker (HO-C(=O)-$(CH_2)_2$-), a pharmacophore not common to histaminergic ligands, remains present in the final product.

These drawbacks ask for further development of new solid phase synthesis methods to prepare imidazole libraries for the discovery of compounds active at histaminergic receptors.

3. SYNTHESES OF IMIDAZOLES VIA DIRECT FUNCTIONALISATION OF THE IMIDAZOLE RING

A method to avoid the lack of flexibility inherent to the condensation approach is to functionalise an imidazole ring in a direct manner. Although organo-lithium chemistry is widely used in organic chemistry its use in the substitution of the heterocyclic ring remains limited to a few examples[8]. Indeed for imidazoles, the most popular method of preparation seems still to be the condensation approach. This condensation approach works well when only one or a few specific target structures are aimed for. However, if a series of imidazoles with a range of substituents is desired the organometallic approach has clear advantages as far as flexibility and diversity is concerned. The synthesis of imidazoles with the aid of organometallic reagents has evolved into three different approaches;

- the deprotonation approach, in combination with use of protective groups,
- a metal-halogen exchange approach, also making use of protective groups,
- and recently the scope of the former methods has been broadened even more through the use of transition metal catalysed transformations.

In the following paragraphs the general strategies used in organometallic transformations of imidazoles to prepared histaminergic ligands will be reviewed.

3.1. The deprotonation-functionalisation approach

Since the imidazole nucleus is prone to react with various reagents under all kinds of reaction conditions, quite early strategies were developed to tame this unruly heterocyclic ring. This led to the development of protective groups for the imidazole nitrogen. Among the first are acyl and urethane based protective groups[26], which are commonly used in peptide chemistry. However, these protective groups are labile under the more drastic deprotonation reaction conditions and were replaced by the more stable benzyl-[27,28], trityl-[29] and methoxymethyl-protective groups[27].

The use of these specific NH-protective groups allows deprotonation of the C2 position and eventually also deprotonation of the C5 position of the imidazole ring.

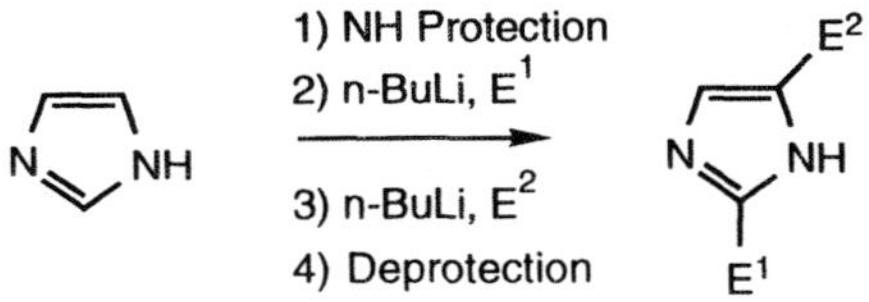

Figure 10. Functionalisation of imidazoles via a deprotonation approach.

The reaction of these anions with electrophiles presents a general synthesis route to 2-, 5-monosubstituted or 2,5-disubstituted imidazoles (figure 10).

These and other protective groups have been reviewed by the groups of Iddon[30] and Chadwick[31] and evaluated for their suitability in the C2 deprotonation of imidazoles (see table 1). The benzyl-group, for example, proved to be a stable and easily removable protective group but led to side reactions in the deprotonation approach in a number of cases[32,33]. The more stable alkoxymethyl-groups work well in the lithiation step but the removal of this protective group can not always be accomplised[27] easily. Another candidate, the tosyl-group is effective in the protection of the NH function but the intermediate anion is not reactive enough towards most commonly used electrophiles. The trityl-group proved to be the first reliable NH protective group for use in the C2 deprotonation of imidazoles[29], although deprotonation was slow.

Table 1

Protective groups for the NH function

Protective group	lithiation at C2, temperature,time	deprotection conditions	reference
methyl-	-60, 1 hr	none reported	34
benzyl-	-78, 30 min	Na/NH_3 or $H_2/Pd(C)$	28,31,35
trityl-	RT, 2 hrs	HCl conc, RT, 17 hrs	29,36
(m)ethoxymethyl-	-60, 45 min	EtOH/HCl conc, reflux, 8 hrs	27,37
1-(ethoxy)ethyl-	-30, 30 min	0.1N HCl/MeOH, reflux, 4 hrs	38
diethoxymethyl-	-40, 15 min	dil HCl, RT, few min	33
dimethylamino-methyl-	-70, 1 hr	2N HCl, RT, few minutes	39
2-(trimethylsilyl)-ethoxymethyl-	-78, 30 min	3N HCl/EtOH, reflux, 1 hr $n\text{-}Bu_4NF/THF$, reflux, 2 hrs	40
benzenesulfonyl-	-20, 1 hr	1N NaOH, RT, 1 hr	31,41
dimethylsulfamoyl-	-65, 15 min	30% HBr, reflux, 7 hrs 2% KOH, reflux, 12 hrs	31,42,43,44

To facilitate the deprotonation conditions the diethoxymethyl- group, an acid labile protective group, was introduced. This group not only protected the NH function but also stabilised the organolithium intermediate.

Currently, a wide range of protective groups is available for protection of the NH function during the functionalisation of the C2 position. Yet, large differences in protection, deprotonation and deprotection conditions exist, leaving the task to the chemist to evaluate those prior to the introduction of the protective group. In general, the diethoxymethyl-, dimethylaminomethyl- and trityl-groups are preferred because of their ease of introduction and deprotection.

For the functionalisation of the C5 position of the imidazole nucleus much stronger basic conditions are needed than for the functionalisaation of the C2 position. The hydrogen atom on the C5 position has a much weaker acidity than the hydrogen atom on the C2 position. Therefore, functionalisation of the C5 proved to be much more difficult and requires the use of an additional protective group for the C2 position if 4(5) mono-substituted imidazoles are desired (table 2).

Table 2

Protective groups for the C2-position

C2 protective.group	NH protective group	lithiation conditions temperature, time	deprotection	reference
Phenylthio-	(m)ethoxy-methyl-	-70, 2hrs	Al(Hg) water, RT, 13 hrs	30,37
trimethylsilyl-	trimethylsilyl-ethoxyethyl-	-78, 30 min	water, RT, 1 hr	40,45
triethylsilyl-	dimethyl-sulfamoyl-	-78, 30 min	2N HCl, RT, 30 min	43
t-butyldimethyl-silyl-	dimethyl-sulfamoyl-	-78, 30 min	2N HCl, RT, 30 min	44,46

Functionalisation of the C5 position also puts a larger strain on the stability of the NH protective group[46]. A large number of protective groups cannot cope with this demand and deprotonation at other sites is found, for example in case of the benzyl group[47]. Besides deprotonation at other sides, also nucleophilic cleavage of the protective group occurs, as is observed for the diethoxymethyl-, 1-ethoxyethyl- and benzenesulfonyl-groups. Another aspect, which is unfortunately neglected by some authors, is the last step of the reaction sequence, the deprotection step. It is not obvious that removal of the protective groups is equally easy for the (poly-)substituted compounds as for the unsubstituted imidazoles[37,43,46]. Some cases have been reported in which removal of the protective group could only be achieved under very harsh conditions[31].

The trityl-, dimethylsulfamoyl- and the SEM-groups work best for NH protection in the C5 functionalisation of the imidazoles. Protective groups for the C2 position such as the triethylsilyl and t-butyl-dimethylsilyl groups have proven their effectiveness both in ease of use and in their stability during the lithiation step.

This sequential functionalisation of the 2- and 5-position can be performed in an one-pot procedure[48] (figure 11).

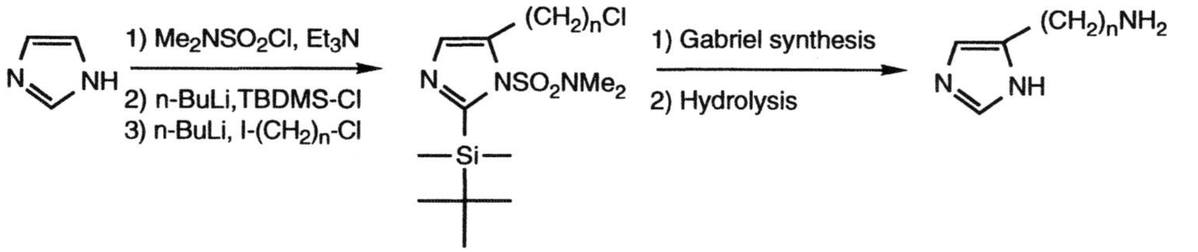

Figure 11. Synthesis of homologs of histamine

Since the electrophile is introduced adjacent to the NH protective group, substantial steric hindrance may be encountered in the following reaction steps. In case of the dimethylsulfamoyl protected imidazole-5-carboxaldehyde, a rapid isomerisation to the 4-substituted product can be induced catalytically by traces of triethylamine or by mere standing at RT for several days[42]. The effect of steric hindrance by the protective group was also observed in the reduction of ethyl dimethylsulfamoyl-imidazolecarboxylate with DIBAH. The 5-isomer could not be reduced, whereas the 4-isomer is reduced easily to the imidazole carboxaldehyde under the standard conditions[49].

In conclusion, the deprotonation approach to functionalise imidazoles has proved to be feasable and constitutes a new flexible method for the preparation of especially 4(5)-mono-substituted imidazoles in a straight forward manner (figure 12).

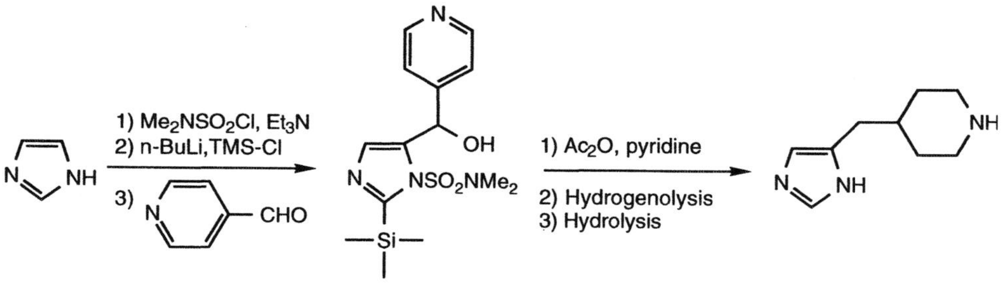

Figure 12: Synthesis of Immepip

3.2. The metal-halogen exchange approach

The metal-halogen approach emerged in 1981 when brominated imidazoles were treated with t-butyl lithium to give the corresponding lithium anions[8]. The reaction works best when a halogen was exchanged on the 2-position. If exchange was attempted at the 4(5)-position a rapid equilibration of the intermediate anion on the 4(5)-position to the more stable anion on the 2-position[50] occured. However, in a proper reaction sequence (C2, C5 and C4) all three positions on the imidazole ring can be substituted via a metal-halogen exchange. Alternatively, the C2 position can be protected followed by another metal-halogen exchange step to give the 5-mono-substituted product[51].

An recent improvement of the metal-halogen exchange is the use of a Grignard reagent instead of a alkyl-lithium reagent. The magnesium anion on the C4 atom is stable and reacts with a series of electrophiles[52,53] without isomerisation. This approach has been applied succesfully in the synthesis of new histamine H_3 receptor agonists[54] (figure 13) and in an improved synthesis of thioperamide[55].

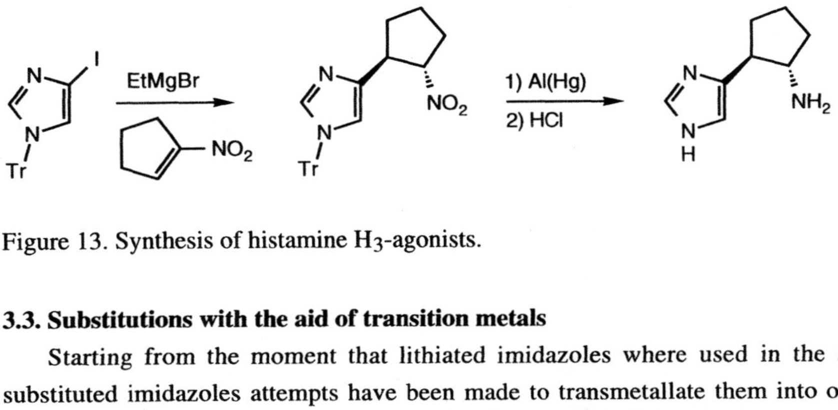

Figure 13. Synthesis of histamine H_3-agonists.

3.3. Substitutions with the aid of transition metals

Starting from the moment that lithiated imidazoles where used in the synthesis of substituted imidazoles attempts have been made to transmetallate them into organozinc[53], organocopper and organopalladium[56] species. These transmetallations are a further extension of the scope of the organometallic methods (figure 14).

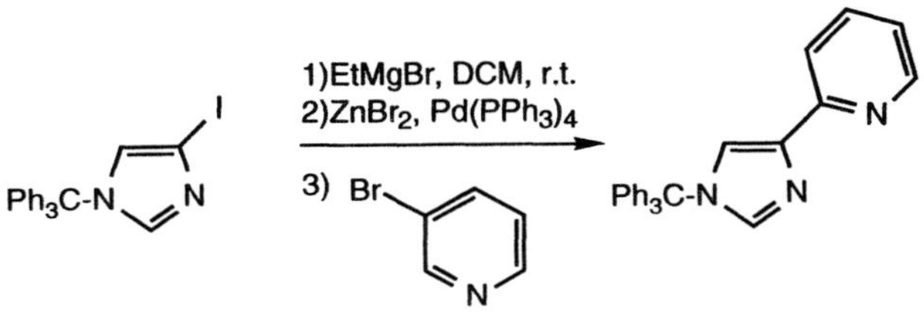

Figure 14: Arylation of imidazoles in a direct fashion.

For example, via cross-coupling reactions arylated imidazoles, which are otherwise difficult to prepare from imidazoles, can be prepared in a direct reaction.

However, only a few examples of this approach have been reported sofar.

4. CONCLUSIONS

The field of imidazole chemistry is still full of new and exciting developments. Both in the field of condensation approaches as in the direct functionalisation with the aid of organometallic reagents many new methods and approaches have emerged in the last decade. The use of a solid phase synthesis approach proved to be possible and it is exciting see what the future might bring in this respect.

These developments will help the medicinal chemists in their search for new and selective ligands for the characterisation of histamergic receptors and in the development of ligands for other biological targets.

REFERENCES

1. R. Leurs, M.J. Smit and H. Timmerman, Pharmac. Ther., 66 (1995) 413
2. R. Leurs, R.C. Vollinga and H. Timmerman, In Progress in Drug Research., J. Jucker (Ed.), Birkhäuser Verlag, Basel, 45 (1995) 107
3. K. Kiec-Kononowics, X. Ligneau, J-C. Schwartz, W. Schunack. Arch. Pharm. 328 (1995) 445, 469
4. M.J. Smit, R. Leurs, A.E. Alewijnse, J. Blauw, G.P. Van Nieuw Amerongen, Y. Van de Vrede, E. Roovers and H. Timmerman, Proc. Natl. Acad. Sci. USA, 93 (1996) 6802
5. A.M. Ter Laak, J. Venhorst, G.M. Donné-op den Kelder and H. Timmerman, J. Med. Chem., 38 (1996) 3351
6. P.H.J. Nederkoorn, J. Van Lenthe, H. Van der Goot, G.M. Donné-Op den Kelder and H. Timmerman, J. Comp.-Aided Mol. Design, 10 (1996) 461
7. P.H.J. Nederkoorn, E.M. Van Gelder, G.M. Donné-Op den Kelder and H. Timmerman, J. Comp.-Aided Mol. Design, 10 (1996) 479
8. M.R. Grimmett, Imidazole and Benzimidazole Synthesis, Academic Press, San Diego, (1997)
9. H. Stark, K. Piunand, X. Ligneau, A. Rouleau, J-M. Arrang, J-C. Schwartz, W. Schunack. J. Med. Chem. 39 (1996) 1220
10. C. Sellier, A. Buschauer, S. Elz, W. Schunack. Liebigs Ann. Chem. (1992) 317
11. H. Bredereck, R. Gompper, H.G.V. Shuh and W. Theilig, Angew. Chem., 71 (1959) 753

H. Bredereck, R. Gompper and D. Hayer, Chem. Ber., 92 (1959) 338
12. P. Dziuron and W. Schunack, Arch. Pharm., 307 (1974) 89
13. S. Elz and W. Schunack, Z. Naturforsch., 42 (1987) 238
14. H. Detert, A. Hagelüken, R. Seifert and W. Schunack, Eur. J. Med. Chem., 30 (1995) 271
15. W. Markwald, Ber. Dtsch. Chem. Ges., 52 (1892) 2354
16. W. Schunack, Arch. Pharm., 306 (1973) 934
17. B. Radziszewski, Chem. Ber., 15 (1882) 2706
18. A.M. van Leusen, F.J. Schaart and D. van Leusen, Recl. Trav. Chim. Pays-Bas, 98 (1979) 258
A.M. van Leusen, J. Wildeman and O.H. Oldenziel, J. Org. Chem., 42 (1977) 1153
19. D.A. Horne, K. Yakushijin and G. Büchi, Heterocycles, 39 (1994) 139
20. I.J.P. de Esch, The synthesis of cyclopropylhistamine, a new rigid and selective Histamine H3-agonist. Submitted for publication.
21. D.E. Bergstrom, P. Zhang and J. Zhou, J. Chem. Soc. Perkin Trans., (1994) 3029
D.E. Bergstrom and P. Zhang, Tetrahedron Lett., 32 (1991) 6485
22. N.-Y. Shih, Tetrahedron Lett., 34 (1993) 595
23. R. ten Have, M. Huisman, A. Meetsma and A.M. van Leusen, Tetrahedron, 53 (1997) 11355
24. C. Zhang, E.J. Moran, T.F. Woiwode, K.M. Short and A.M.M. Mjalli, Tetrahedron Lett., 37 (1996) 751
25. S. Sarshar, D. Siev and A.M.M. Mjalli, Tetrahedron Lett., 37 (1996) 835
26. E. Muller (Ed), Houben-Weyls Methoden der organischen Chemie, Peptides I. Georg Thieme Verlag, Stuttgart, 1980
27. A.M. Roe, J. Chem. Soc., (1963) 2195
28. D.A. Shirley and P.W. Alley, J. Am. Chem. Soc., (1957) 4922
29. K.L. Kirk, J. Org. Chem., 43 (1978) 4381
30. B. Iddon, Heterocycles, 23 (1985) 417
31. D.J. Chadwick and R.I. Ngochindo, J. Chem. Soc. Perkin Trans. I, (1984) 481
32. R. Breslow, J.T. Hunt, R. Smiley and T. Tarnowski, J. Am. Chem. Soc., 105 (1983) 5337
33. N.J. Curtis and R.S. Brown, J. Org. Chem., 45 (1980) 4038
34. M. Begtrup and P. Larsen, Acta. Chem. Scand., 44 (1990) 1050
35. D.J. Chadwick and S.T. Hodgson, J. Chem. Soc. Perkin Trans. I, (1982) 1833
36. D.P. Davis, K.L. Kirk and L.A. Cohen, J. Heterocyclic Chem., 19 (1982) 253
37. C.C. Tang, D. Davalian,P. Huang and R. Breslow, J. Am. Chem. Soc., 100 (1978) 3918
38. T.S. Manoharan and R.S. Brown, J. Org. Chem., 53 (1988) 1107
39. A.R. Katritzki. G.W. Newcastle and W.-Q. Fan, J. Org. Chem., 53 (1988) 5685

40. B.H. Lipshutz, B. Huff and W. Hagen, Tetrahedron Lett., 29 (1988) 3411
41 R.J. Sundberg, J. Heterocyclic Chem., 14 (1977) 517
42. J.-W. Kim, S.M. Abdelaal, L. Bauer and N.E. Heimer, J. Heterocyclic Chem., 32 (1995) 611
43. A.J. Carpenter and D.J. Chadwick, Tetrahedron, 42 (1986) 2351
44. R.C. Vollinga, W.M.P.B. Menge, R. Leurs and H. Timmerman, J. Med. Chem., 38 (1995) 266
45. J.P. Whitten, D.P. Matthews and J.R. McCarthy, J. Org. Chem., 51 (1986) 1891
46. R.I. Ngochindo, J. Chem. Soc. Perkin Trans. I, (1990) 1645
47. H. Ogura and H. Takahashi, J. Org. Chem., 39 (1974) 1374
48. R.C. Vollinga, W.M.P.B. Menge and H. Timmerman, Recl. Trav. Chim. Pays-Bas. 112 (1993) 123
49. R.C. Vollinga, PhD thesis Amsterdam (1995)
50. M.P. Groziak and L. Wei, J. Org. Chem., 56 (1991) 4296
51. M.P. Groziak and L. Wei, J Org. Chem., 57 (1992) 3776
52. D.S . Carver, S.D. Lindell and E.A. Savillestones, Tetrahedron, 63 (1997) 1448
53. R.M. Turner, S.V. Lewy and S.D. Lindell, Synlett., (1993) 748
54. N.-Y. Shih, A.T. Lupo Jr, R. Aslanian, S. Orlando, J.H. Piwinskii, M.J. Green, A.K. Ganguly, M.A. Clark, S. Tozzi, W. Kreutner and J.A. Hey, J. Med. Chem., 38 (1995) 1593
55. J.H.M. Lange, H.C. Wals, A. van den Hoogenband, A. van de Kuilen and J.A.J. den Hartog, Tetrahedron, 51 (1995) 13447
56. A.S. Bell, D.A. Roberts and K.S. Ruddock, Tetrahedron Lett., 29 (1988) 5013

R. Leurs and H. Timmerman (Editors)
The Histamine H_3 Receptor

Synthesis of radioligands for the histamine H_3 receptor

Albert D. Windhorst [(a)], Rob Leurs [(b)], Wiro M.P.B. Menge [(b)], Henk Timmerman [(b)] and Jacobus D.M. Herscheid [(a)].

(a) : Radionuclide Center, Vrije Universiteit, De Boelelaan 1085c, 1081 HV Amsterdam, The Netherlands.
(b) : Leiden / Amsterdam Center for Drug Research (LACDR), Division of Medicinal Chemistry, Vrije Universiteit, De Boelelaan 1083, 1081 HV Amsterdam, The Netherlands.

1. INTRODUCTION

Radiolabelled receptor ligands are of great importance for studying receptor-ligand interactions. Pharmacologists use radiolabelled compounds, not only for in-depth pharmacological research, e.g. distribution, but also for screening compounds, e.g. radioligand displacement.

In vitro techniques for the research of receptors remain important but, in addition, *in vivo* research techniques are becoming more important. In the last two decades techniques have been developed which can be used to study receptors *in vivo* in a non-invasive manner. These techniques, Single Photon Emission Computed Tomography (SPECT) and Positron Emission Tomography (PET) monitor the distribution of a radiopharmaceutical by measuring its γ-radiation outside the body. Especially in the study of brain disorders, distributions of receptors can be of great importance for diagnosing and the development of new therapies. With SPECT the maximum reachable resolution is 7 to 8 mm which implicates that this technique is not adequate to study distributions in small laboratory animals. However the newest generation of PET camera's have a resolution of less then 3 mm, which allows the visualisation of important rat brain structures like the striatum separately. In addition, absolute quantification is possible with PET which implicates that not only the distribution but also the amount of binding sites (B_{max}) and even the dissociation constant (K_d) can be determined more accurately then with SPECT. A disadvantage of PET over SPECT is the necessity to have an in house cyclotron and fast radiochemistry.

For all the above methods and techniques one needs specially synthesised compounds which are suitably labelled : ^{3}H, ^{125}I and ^{14}C for *in vitro* pharmacology, ^{123}I for SPECT and ^{18}F or ^{11}C for PET (table 1).

Table 1.
Physico-chemical properties of the isotopes described in this chapter.

Isotope	Decay	Energy (keV)	Half-life (#)	Source	Main use
^{3}H	β^-	18	12.2 y	Commercially available	Bindingassays, autoradiography
^{11}C	β^+ γ	960 511	20.4 m	On-site production	PET
^{14}C	β^-	152	5380 y	Commercially available	Bindingassays
^{18}F	β^+ γ	635 511	109.6 m	On-site production	PET
^{123}I	γ	159	13.2 h	on-site production, limited commercial availability	SPECT
^{125}I	γ	35	60 d	Commercially available	Bindingassays, autoradiography

(#) : y = year, d = days, h = hours, m = minutes.

The unique properties of each isotope make it more or less suited for different techniques. Important to consider are the type of radiation (β particles or γ rays), the energy of the radiation and the half-life of the isotope.

In this chapter an overview will be given on the synthesis of radiolabelled ligands for the histamine H_3 receptor. The physico-chemical properties and characteristics of radionuclides used for the labelling of histamine H_3 receptor radioligands will be described briefly. First some general labelling methods will be discussed: Tritiation, electrophilic and nucleophilic iodination with ^{123}I and ^{125}I, nucleophilic fluorination with ^{18}F and methylation with ^{11}C, ^{14}C or ^{3}H labelled methyliodide. Finally the application of these methods for the synthesis of the known histamine H_3 receptor radioligands will be described.

2. LABELLING METHODS USED IN THE SYNTHESIS OF RADIOLIGANDS FOR THE HISTAMINE H_3 RECEPTOR

2.1 Tritiation

Many tritium labelled receptor ligands are commercially available. The advantages of the use of tritium as an isotope for the radiolabelling of receptor ligands are its long half-life and the relatively safety of use. The low energy of the β^- particle is beneficial for autoradiography, especially for the resolution. A disadvantage of the low radiation energy is that special detection facilities are required and long exposure times are needed.

Isotopic exchange in most cases does not greatly influence the physico-chemical properties of the total compound. In the case of tritium for hydrogen exchange however, the isotope effect of tritium can influence the pK_A of a basic moiety in the vicinity of the tritium atom [1,2]. Although only a small shift in the pK_A is observed (0.05-0.1), this can be of great importance for the biological activity of a radiopharmaceutical.

Incorporation of tritium is mainly achieved by (catalytic) reductions, (catalytic) exchange reactions or methylation with [^{3}H]methyliodide. All potent H_3 receptor ligands contain a imidazole moiety and since the imidazole can be labelled in the 2 or 5 position via a Pd catalysed tritio-deiodination [3], some H_3 ligands have been labelled with tritium in these positions. A disadvantage of the use of [2,5-^{3}H]-imidazole moiety however, is that the tritium atoms in these two positions have a weakly acidic character and can be exchanged back for hydrogen during storage.

Although [^{3}H]CH_3I with a high specific activity is commercially available it can also be produced on site from tritium gas. Lee and co-workers recently published two methods for the synthesis of [^{3}H]CH_3I (scheme 1) [4].

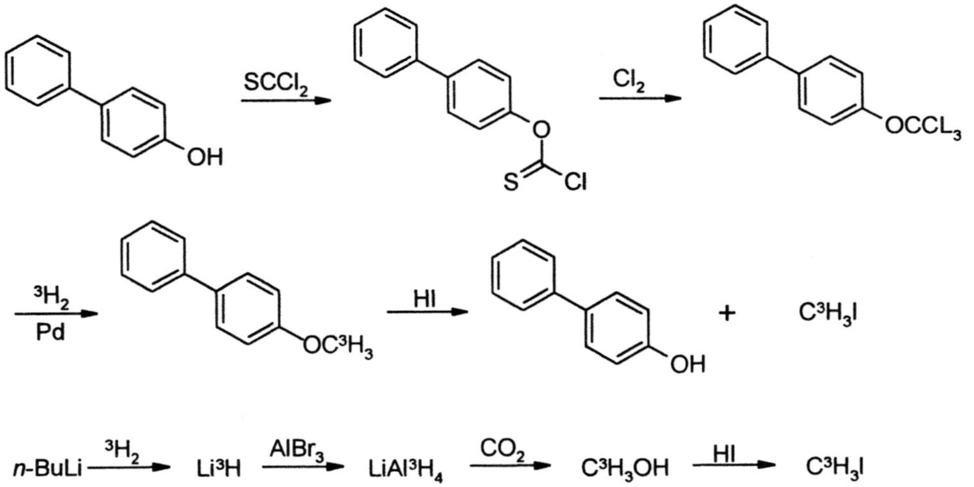

scheme 1. Synthesis of [^{3}H]methyliodide.

2.2 Iodination

Of all the possible radioisotopes of iodine ^{125}I and ^{123}I are most widely used in receptor-based research.

Binding assays and autoradiography can be performed with ^{125}I and detection is faster and easier then with tritium. For autoradiography however, tritium is still preferred, since the low energy of the β particle has a very short range which results in a higher resolution. For many receptors ^{125}I labelled ligands are commercially available. If a custom synthesis is needed, $Na^{125}I$, with a specific activity of 2 Ci/μmol in 0.01 M NaOH, is also commercially available.

For SPECT, ^{123}I is used. It has a half-life of 13.2 hours and therefore has only limited commercial applicability. It can be produced on site with a cyclotron by bombarding ^{124}Xe with protons of 24 MeV according to the ^{124}Xe (p,2n)^{123}Cs reaction. ^{123}Cs has a half-life of 5 minutes and decays to ^{123}Xe which, with a half-life of 2 hours, decays to ^{123}I. After 6 hours, ^{123}I can be isolated as $Na^{123}I$ in 0.01M NaOH with a specific activity of 200 Ci/μmol.

Basically two types of reactions can be used for iodinations : electrophilic- or nucleophilic iodination at a unsaturated carbon. Iodine attached to a saturated carbon atom is labile and is readily substituted by nucleophiles. For the same reason, radiopharmaceuticals which are labelled at a saturated carbon atom, are quickly metabolised *in vivo*.

2.2.1 Electrophilic Iodination

Labelling of receptor ligands with iodine is often performed by electrophilic aromatic iodination, because the labelling procedure is relatively easy and many compounds contain a phenyl moiety. This will always lead to several isomers since this iodination is not regioselective. If however a strong activating substituent is present an electrophilic iodination will predominantly give para-substitution or, if the para-position is already substituted, the ortho isomer. A mixture of several regio isomers should be expected and if a mixture is obtained, a tedious semipreparative HPLC purification will be necessary. Regiospecific iodination can be achieved by using a demetallation reaction of a tri(alkyl)stannane precursor. In hat case, substitution will occur at the position of the stannane group (scheme 2). A second advantage of a demetallation reaction is that in this way a meta-substituted iodine labelled compound can be obtained which is hardly possible with a standard electrophilic substitution.

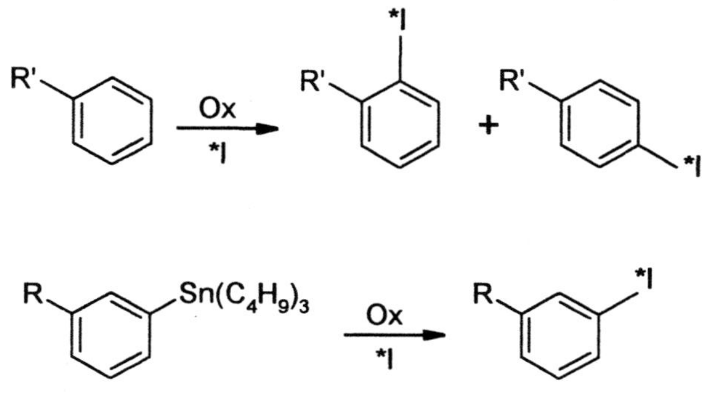

Scheme 2. Possible substitution sites for electrophilic iodinations.
R = non-activating substituent, R' = activating substituent. Ox = oxidising agent.

Two oxidising agents are commonly used in electrophilic radioiodinations of receptor ligands : Chloramine-T and peracetic acid. In both cases I^- is converted to an unknown iodine-compound, which is the reactive iodination species. Chloramine-T is a very strong oxidising agent, but has the disadvantage that it can also chlorinate the substrate in the case of radioiodination. Although the chlorination reaction is much slower than iodination, it still occurs due to kinetic effects because there is such a large excess of chloramine-T compared to the amount of radioactive I^-. For this reason reaction times with electrophilic iodinations with chloramine-T as oxidising agent should be as short as possible since the chloro-compound will make the purification more difficult. Also over-oxidation and disruption of reaction product should be avoided. Therefore reaction times hardly ever exceed a period of five minutes, whereafter the reaction is stopped by the addition of a strong reducing agent like sodium metabisulfite.

Peracetic acid is a milder oxidising agent and is prepared *in situ.* The iodination reaction is performed in 98% acetic acid and upon the addition of 30% hydrogenperoxide, peracetic acid is generated. No general rules for the selection of the oxidising agent are available, it is often a matter of trial and error.

Advantages of electrophilic iodination are speed of the reactions, which moreover, can generally be performed at ambient temperatures. Furthermore, the method is applicable to a wide variety of receptor ligands, as long as these ligands are stable under the strong oxidising conditions of the reaction and have a site where the addition or substitution can take place. Electrophilic iodination is however not applicable for an isotopic exchange reaction.

2.2.2 Nucleophilic Iodination

The most widely used method for nucleophilic iodination is the Cu^+-assisted method, which was first described by Mertens et al [5]. This method can be used in a bromine for radioiodine or an iodine for radioiodine exchange reaction, and results in a regiospecific substitution (scheme 3). A limitation of this labelling method is that it has to be performed in water. If the substrate however is insoluble in water, a mixture of 10% ethanol in water is also possible, higher concentrations of ethanol lead to the precipitation of reagents. It is therefore difficult to iodinate lipophilic compounds with this method.

Scheme 3. Nucleophilic iodination according to Mertens [5]. R' = Br or I.
i : $CuSO_4$, $SnSO_4$, citric acid, 2,5-dihydroxybenzoic acid, *I, 100-160 °C.

The mechanism of the Cu^+-assisted nucleophilic iodination was postulated by Gysemans et al in 1992 [6]. They postulated a mechanism in which the Cu^+ coördinates to the phenyl-halogen bond (scheme 4).

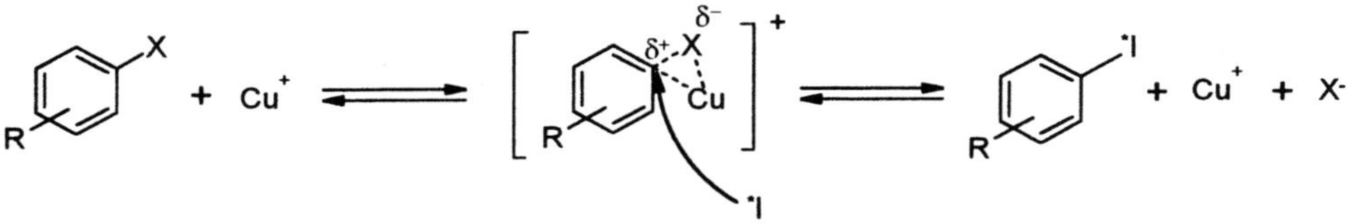

Scheme 4. Proposed mechanism for the nucleophilic iodination with the Cu^+ method according to Gysemans et al [6]. X = Br or I.

In this intermediair the phenyl-halogen is destabilised and the carbon atom of the phenyl-halogen bond is activated which is therefore more susceptible for a nucleophilic attack of $^{*}I^{-}$.

With this method one can perform a iodine for radioiodine exchange reaction resulting in a easy work-up procedure, since no semi-preparative HPLC purification is necessary : Selective filtration over a Seppak-column is usually sufficient to purify the radiolabelled compound from unreacted radioiodine and (in)organic reagents. Disadvantage of the iodine for radioiodine exchange reaction is that a high specific activity cannot be achieved, since the starting material, the non-radioactive iodinated compound, can not be separated from the radiolabelled compound. If there is a need for a radiolabelled compound with high specific activity, a bromine for radioiodine exchange reaction is the reaction of choice. In this case however a semi-preparative HPLC purification is required.

2.3. Fluorination

Many radiopharmaceuticals for PET are labelled with ^{18}F. ^{18}F decays with emission of a positron with a relative low energy which limits its range in the body and thus enhances the resolution of a PET study. A second advantage of ^{18}F is the half-life of 110 minutes which allows synthesis and PET studies, and allows use in humans, not causing radiation damage.

Many receptorligands do not contain a fluorine atom, so the introduction of ^{18}F into a receptor ligand leads to a new compound with a possible new pharmacological profile. In the past a replacement of H or OH for F has proven to be successful, but also certain problems are known, since the introduction of a fluorine atom can influence the biological activity of a compound. So the ^{19}F analogue always has to be synthesised first for a thorough pharmacological characterisation. The same is true for every radioligand which differs from the lead-compound.

Mainly two types of reactions are possible for fluorinations with ^{18}F : Electrophilic and nucleophilic fluorinations. However, electrophilic fluorinations are not suitable for PET radioligands, since in that case the ^{18}F is isolated as $[^{18}F]F_2$ with a specific activity which is too low for receptor research.

2.3.1 Nucleophilic fluorination

$^{18}F^-$ is produced with a specific activity of up to 200 Ci/µmol by bombarding ^{18}O enriched water with protons of 8-14 MeV according to the $^{18}O(p,n)^{18}F$ reaction, and isolated as $^{18}F^-$ in water [7].

Nucleophilic fluorinations can be performed at a primary, sometimes also at a secondary, aliphatic carbon atom [8]. Good leaving-groups for the aliphatic nucleophilic fluorination are trifluoromethylsuphonate (triflate)- , p-toluenesulphonate (tosylate)- and the methylsulphonate (mesylate)- groups. In this series the triflate is a very good leaving-group, but a disadvantage is its chemical instability. The tosylate is much more stable, the mesylate is relatively easy to prepare but is less reactive then the triflate or the tosylate.

Also nucleophilic aromatic substitutions can be performed with ^{18}F, however they only take place at electron deficient phenyl-groups. A quaternary ammonium or a nitro group can be substituted by ^{18}F with a nucleophilic aromatic substitution reaction [9,10].

The nucleophilic fluorinations have to be performed with the exclusion of water and under an inert atmosphere in a non-protic organic solvent. Since under those circumstances $^{18}F^-$ is insoluble, a phase transfer reagent has to be used. Often used are : Kryptofix 2.2.2 and tetra-*N*-butylammmoniumhydrogencarbonate, the first one giving the better yields [11].

A major complication in the use of ^{18}F, and other nuclides with a short half life, for the labelling of radiopharmaceuticals is that the half-life is often the limiting factor for the radiochemical yield of the synthesis. To obtain reasonable amounts of radiolabelled compound at the end of the synthesis, relatively high amounts of radioactivity (up to 6 Ci at the beginning of the synthesis) are used and therefore special safety facilities are needed. So in general, fast reactions are of great importance in the production of $^{18}F^-$ labelled compounds. For instance a chemical reaction which gives 100% chemical yield after 90 minutes can give a theoretical radiochemical yield of only 67% due to the half-life of ^{18}F. Speeding up the reaction 3 times increases the maximum theoretical radiochemical yield to 83%, see figure 1.

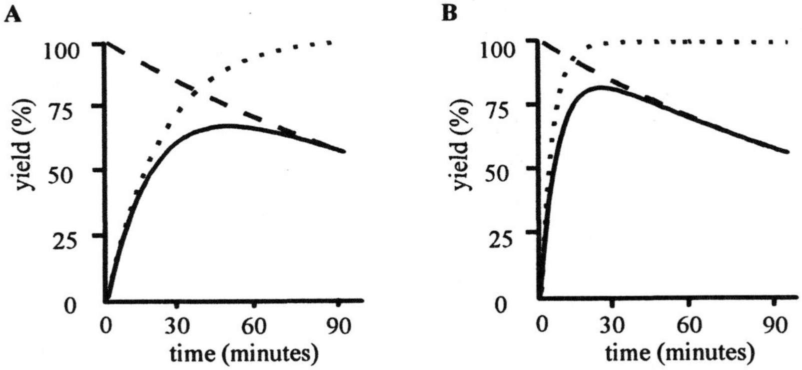

Figure 1. Influence of chemical reaction speed on the maximum achievable radiochemical yield. A: Chemical yield 100% after 90 minutes, maximum radiochemical yield 67% .
B: Chemical yield 100% after 30 minutes, maximum radiochemical yield 83%.
(— — radioactive decay, • • • • chemical yield, ——— radiochemical yield)

2.4. Methylation

Methylation reactions with [^{3}H]-, [^{14}C]- or [^{11}C]methyliodide or [^{11}C]methyltriflate are widely used for the introduction of a labelled methyl group through nucleophilic substitution reactions. [^{3}H]- and [^{14}C]methyliodide are commercially available or can be produced as has been mentioned in section 2.1 of this chapter, [^{11}C]methyliodide and [^{11}C]methyltriflate have to be prepared on site, due to the half-life of 20.4 minutes of ^{11}C.

^{11}C is produced as [^{11}C]CO_2 by bombarding nitrogen with 14 MeV protons according to the ^{14}N(p,α)^{11}C reaction. Because the nitrogen which is used as target gas contains a small amount of oxygen, the produced ^{11}C is immediately converted into [^{11}C]CO_2. After the isolation of the [^{11}C]CO_2 it is reduced with $LiALH_4$ into [^{11}C]methanol and subsequently converted into [^{11}C]methyliodide with HI [7]. [^{11}C]methyltriflate is produced by reacting [^{11}C]methyliodide with trifluoromethanesulfonic acid silver salt [12,13]. Advantages of [^{11}C]methyltriflate are that it is less volatile and far more reactive then [^{11}C]methyliodide [14].

A problem with this production of [^{11}C]methyliodide is the specific activity. Not only is there always CO_2 in the atmosphere, reducing the specific activity considerably, also the use of the reducing agent $LiAlH_4$ is a main source of CO_2 in this production. The highest specific activity which can be reached by this method is approximately 30 Ci/µmol [7], but most producers of [^{11}C]methyliodide can only reach a specific activity up to 5 Ci/µmol. A solution for this problem is to exclude $LiAlH_4$ in the production of [^{11}C]methyliodide. This can be achieved by adding 5% hydrogen to the target nitrogen gas. In that case [^{11}C]methane is produced, which can be converted to [^{11}C]methyliodide without the use of $LiALH_4$.

Most methylated radiopharmaceuticals are relatively easily labelled with radioactive methyliodide, but in some cases catalysts or special reaction conditions are required for satisfying yields. For instance in the case of JES 1798 and PF 1022A [15,16] silver oxide as a catalyst and a defined ratio between the [^{3}H]methyliodide and the substrate is required for the methylation reaction.

3. RADIOLIGANDS FOR THE HISTAMINE H_3 RECEPTOR

3.1. Radiolabelled agonists for the H_3 receptor

Two radiolabelled agonists for the H_3 receptor are known so far : [^{3}H]-[R]-α-methylhistamine and [^{3}H]- N^{α}-methylhistamine; both compounds are commercially available. [^{3}H]-[R]-α-methylhistamine is labelled at the 2 or 5 position of the imidazole moiety with a specific activity of 23 Ci/mmol and [^{3}H]- N^{α}-methylhistamine is labelled with ^{3}H in the methyl group with a specific activity of 45-90 Ci/mmol.. Though the synthesis of these two ligands has not been described in literature and the manufacturers do not give any details about the synthesis, it is very likely that [^{3}H]-[R]-α-methylhistamine has been synthesised by a Pd catalysed tritio-diiodination procedure as has been described by Bloxsidge et al [3] and [^{3}H]-N^{α}-methylhistamine by a methylation of histamine with [^{3}H]methyliodide. Both compounds have been widely used *in vitro* to investigate the H_3 receptor (see the chapter "Radioligand binding studies for the H_3 receptor" by F.P. Jansen *et al*).

3.2. Radiolabelled antagonists for the H_3 receptor

3.2.1. Synthesis of [^{125}I]iodophenpropit

Clobenpropit is a selective histamine H_3 antagonist and, with a pA_2 of 9.9, the most active H_3 antagonist known so far. Following this observation the iodo-analogue was evaluated as a H_3 antagonist and a potential radiolabelled H_3 antagonist upon labelling with ^{125}I. Iodophenpropit, the selected compound, proved also to be a very active compound with a pA_2 of 9.6 and a selective compound : It is 1000-fold more active at the H_3 receptor then at the H_1- or the H_2 receptor [17]. Subsequently it was labelled with ^{125}I and was the first radiolabelled histamine H_3 antagonist described in literature [18]. It was synthesised from the bromoprecursor according to the Cu^+ method of Mertens (scheme 5) [5].

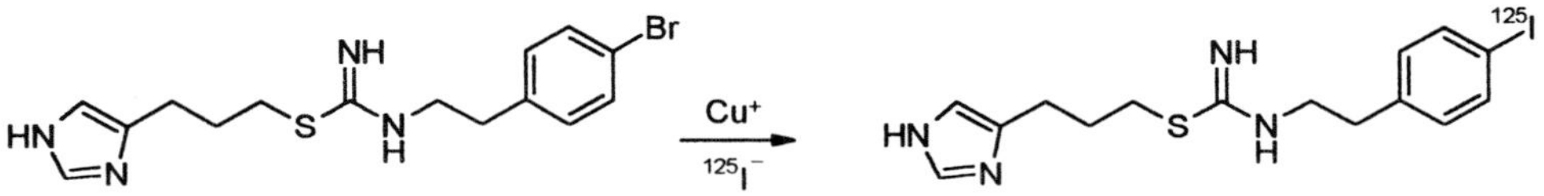

Scheme 5. Synthesis of [^{125}I]iodophenpropit.

[^{125}I]Iodophenpropit n.c.a. (no carrier added) was obtained with a specific activity of around 2000 Ci/mmol and a radiochemical purity of > 99% by a double semi-preparative HPLC. In this way the bromoprecursor could be separated completely from [^{125}I]iodophenpropit. Typically about 40 to 50 % of the radioactivity was recovered as [^{125}I]iodophenpropit. The purified product was trapped on a RP18 Seppak and reversibly eluted into 0.3 ml ethanol/H_2SO_4 (50/4, v/v). In this solution the compound is stable for at least 2 months at -20 °C.

Iodophenpropit was also labelled with ^{131}I, according to the same method. This radioligand was used to study the biodistribution of [^{131}I]iodophenpropit in rats. In these studies only a poor uptake of the compound in the brain was found. Also from whole body autoradiography studies with [^{125}I]iodophenpropit, it was concluded that [^{125}I]iodophenpropit was poorly taken up into the brain [19].

3.2.2. Synthesis of [^{123}I]GR 190028a

Although *in vitro* GR 175737 is less potent (pKi = 8.2) than clobenpropit (pK_i = 9.8), *in vivo* GR 175737 shows a higher activity (ED_{50} 1.4 mg/kg) in an *ex vivo* binding assay than clobenpropit (ED_{50} = 10.3 mg/kg) [20]. Apparently the bioavailability of GR 174737 is much better than clobenpropit and the authors postulated that this observed difference might be caused by the relative ease of blood-brain-barrier passage.

The iodo analogue in this series of compounds, GR 190028a (pK_i = 8.2), was shown to be equipotent *in vitro* with GR 175737 [20]. This compound was labelled by an isotopic exchange reaction (scheme 6) and evaluated for its potential use as a SPECT ligand for the H_3 receptor.

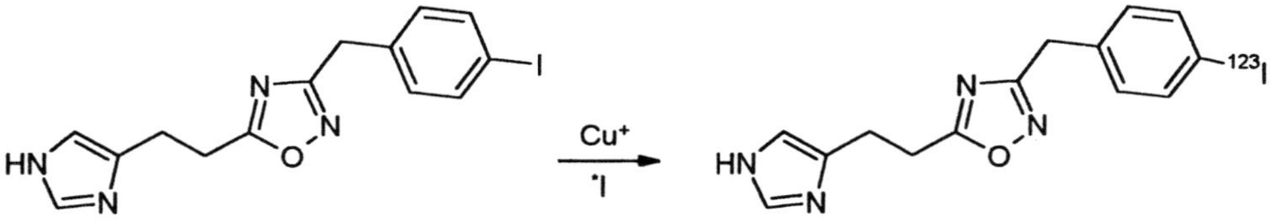

Scheme 6. Synthesis of [^{123}I]GR 1900028a.

[^{123}I]GR 190028a was purified by HPLC from a non-radioactive impurity which was formed during the reaction, and isolated in a radiochemical purity of greater than 99% and a specific activity of 18 Ci/mmol in a yield of 72 % [21].

[^{123}I]GR 190028a did show a better brain uptake than [^{131}]iodophenpropit, but the uptake still was not high enough to further study the use of [^{123}I]GR 190028a as a SPECT ligand.

3.2.3. Synthesis of [^{125}I]iodoproxyfan

Iodoproxyfan was selected from a series of iodinated histamine H_3 ligands in the search for potential iodinated radioligands for the histamine H_3 receptor. *In vitro* pharmacology showed that iodoproxyfan had the highest affinity in this series for the H_3 receptor with a K_i of 5 nM. Iodoproxyfan was then labelled with ^{125}I according to the same procedure as [^{125}I]iodophenpropit from the bromo-precursor (scheme 7) [22].

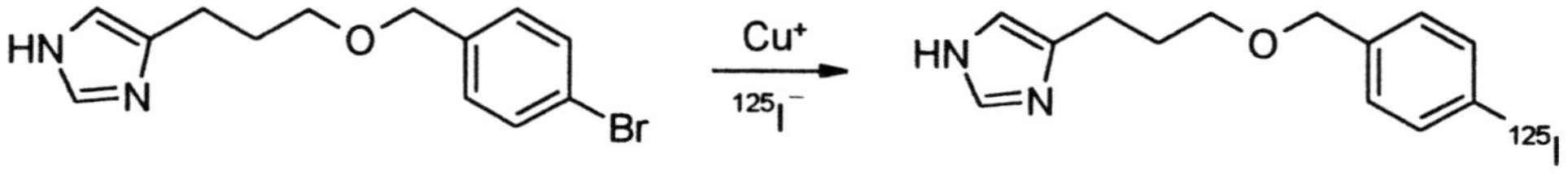

Scheme 7. Synthesis of [^{125}I]iodoproxyfan.

[^{123}I]iodoproxyfan has also been synthesised by an electrophilic iodination with Na^{125}I of the tributylstannyl precursor by a demetallation reaction with chloramine-T as the oxidising agent in a moderate yield [23].

The ^{123}I analogue of iodoproxyfan has been synthesised in our laboratory, starting from iodoproxyfan with an isotopic exchange reaction according to the same reaction conditions as used for the synthesis of [^{125}I]iodophenpropit. [^{123}I]iodoproxyfan was isolated by a simple Seppak C18 cartridge filtration procedure. The reaction had a yield of 94%, [^{123}I]iodoproxyfan had a specific activity of 120 Ci/mmol an a radiochemical purity of greater than 99% [21]. This radiopharmaceutical has been evaluated as a potential SPECT ligand for the histamine H_3 receptor; it showed a relatively good brain uptake in the rat after iv injection. No details with respect to *in vivo* H_3 receptor binding are known at this moment.

3.2.4. Synthesis of [^{18}F]VUF 5000

An analogue of the classical histamine H_3 antagonist thioperamide has been evaluated for its potential use as a PET ligand for the histamine H_3 receptor. After attempts to fluorinate thioperamide in the cyclohexane ring failed [24], a fluoromethylated analogue of thioperamide,

VUF 5000, was prepared. This compound proved to be an active ($pA_2 = 9.0$, $K_i = 2.3$ nM) and selective H_3 receptor antagonist. It was subsequently radiolabelled according to scheme 8.

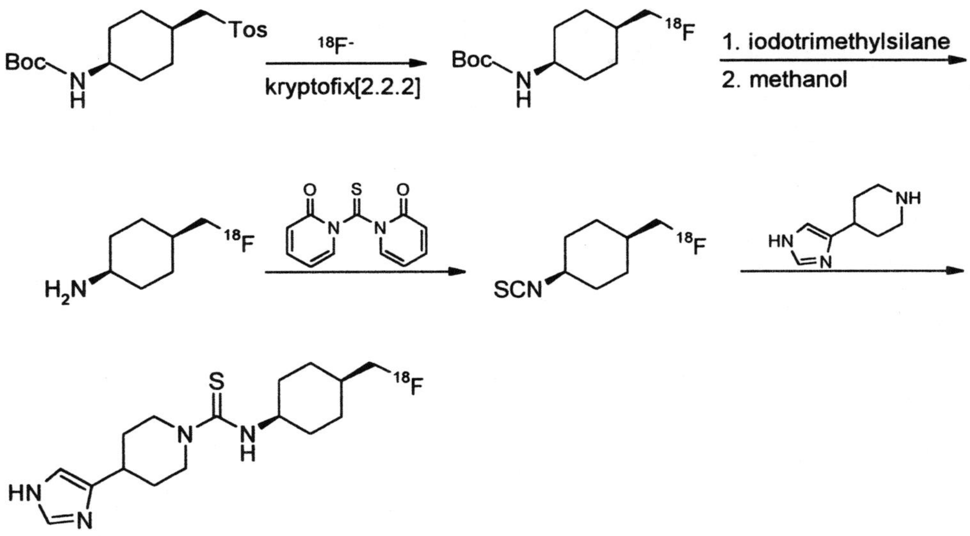

Scheme 8. Synthesis of [^{18}F]VUF 5000.

[^{18}F]VUF 5000 was obtained in 23-34% radiochemical yield (decay corrected) within 3.5 hours and purified with HPLC. It was isolated with a radiochemical purity of > 99% and a specific activity of more then 2500 Ci/μmol [25]. Biodistribution however showed a very low brain uptake after iv injection in the rat and so [^{18}F]VUF 5000 proved to be inappropriate for brain imaging with PET.

3.2.5. Synthesis of [^{18}F]FUB 272

In the search for a PET ligand for the H_3 receptor [^{18}F]FUB 272 was investigated by Ponchant et al. [26]. It was synthesised by a nucleophilic aromatic substitution of its trityl protected nitro precursor (scheme 9).

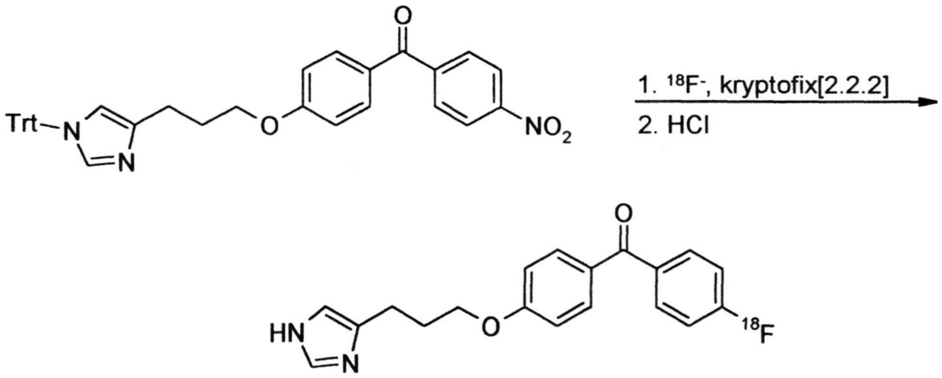

Scheme 9. Synthesis of [^{18}F]FUB 272.

In this way n.c.a. [^{18}F]FUB 272 was obtained (no specific activity mentioned) after 146 minutes in a 65% yield (corrected for decay) after a triple HPLC purification, due to impurities with a retention time very close to the desired product. Biodistribution studies showed that brain uptake was again too low for use in PET. The study also showed an uptake into the cerebellum, where no H_3 receptors are present, so [^{18}F]FUB 272 also binds to a non H_3 receptor binding site in the brain.

3.2.6. Synthesis of [^{3}H]-thioperamide

The classical histamine H_3 antagonist thioperamide has been labelled and evaluated as radioligand for the H_3 receptor by Alves-Rodrigues et al [27]. [^{3}H]thioperamide was synthesised from [^{3}H]4-(1H-imidazol-4-yl)piperidine, labelled with tritium in the 2 or 5 position of the imidazole and at the 4 position of the piperidine, and cyclohexylisothiocyanate (scheme 10) [28]. It was purified with a semipreparative HPLC method and isolated with a specific activity of 6 Ci/mmol.

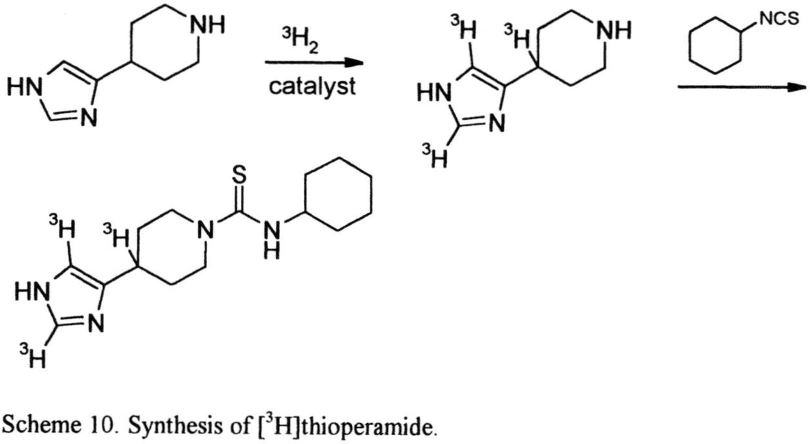

Scheme 10. Synthesis of [^{3}H]thioperamide.

As has been mentioned, a problem with tritiation at the 2 and 5 positions of the imidazole, is that that the tritium atoms are relatively acidic and can exchange back for hydrogen during storage. Indeed we have found in our laboratories that after storage of [^{3}H]thioperamide for a year, all tritium from [^{3}H]thioperamide was exchanged back with hydrogen from the stock solution. Solution for this problem is tritiation of the cyclohexane ring, but this has not been done for thioperamide.

3.2.7. Synthesis of [^{3}H]-GR 168320

[^{3}H]GR 168320 was developed as a (stably labelled) tritiated H_3 antagonist and evaluated for its *in vitro* use as a radioligand for the H_3 receptor [29]. As has been shown for thioperamide, the tritium labels at the 2 and 5 position of the imidazole moiety and at the 4 position of the piperidine moiety are sensitive to exchange with hydrogen from the storage solution. [^{3}H]GR 168320 is labelled at the cyclohexane, by a catalytic reduction with $^3H_2$, no details about the synthesis have been published (scheme 11). [^{3}H]GR 168320 was isolated with a specific activity of 4.8 Ci/mmol.

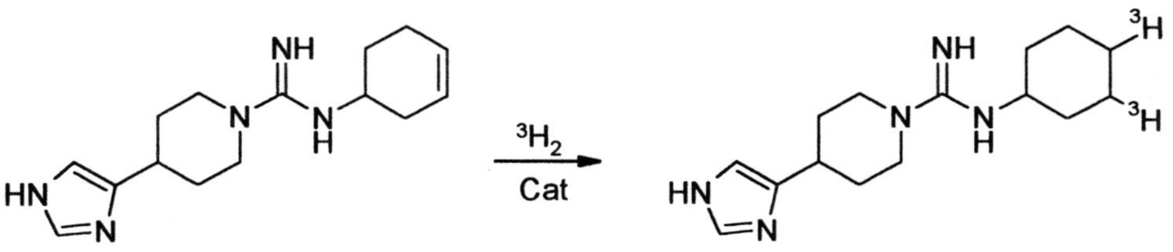

Scheme 11. Synthesis of [^{3}H]GR 168320.

The tritium at this position is much more stable than it would be if the labels were at the imidazole moiety. The compound proved to be a good radioligand for *in vitro* studies of the H_3 receptor. (see the chapter "Radioligand binding studies for the H_3 receptor" by F.P. Jansen *et al*).

3.2.8. Synthesis of [^{3}H]*S*-methylthioperamide

[^{3}H]*S*-Methylthioperamide was obtained by methylation of thioperamide with [^{3}H]CH_3I in the presence of base (scheme 12).

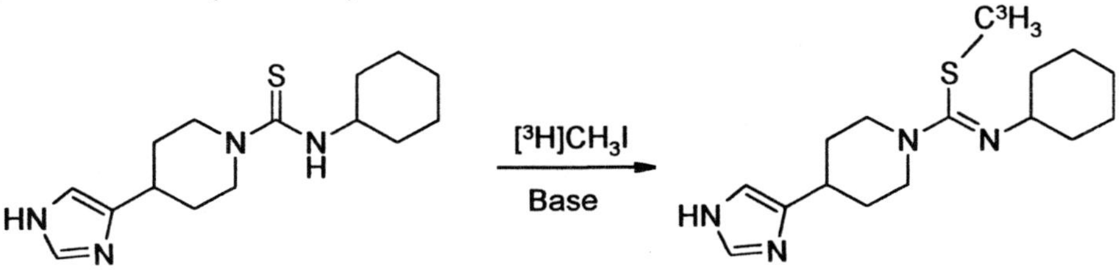

Scheme 12. Synthesis of [^{3}H]*S*-methylthioperamide.

The product was isolated with a semi-preparative HPLC method using a TSK-Gel ODS-80 column and obtained with a specific activity of 50.3 Ci/mmol [30]. It was evaluated as radioligand for the H_3 receptor. *In vitro* binding studies in the rat brain and autoradiography have been performed with this radioligand. Biodistribution experiments showed a high peripheral uptake by the liver, kidney, lung and bladder and it was suggested that [^{3}H]*S*-methylthioperamide also binds to a non-H_3 receptor site, probably hemeproteins. The ^{11}C analogue was also reported by the same authors. This compound was evaluated for use as a PET ligand for imaging the H3 receptor in the brain, [^{11}C]*S*-methylthioperamide showed however a poor uptake in the brain [31].

3.2.9. Synthesis of [^{11}C]UCL 1829.

[^{11}C]UCL 1829 was investigated for use as a PET ligand for the H_3 receptor. It has been be labelled with [^{11}C]methyliodide at the thiomethylmoiety. The labelling of the des-methyl precursor proved to be very tedious, since the nucleophilicity of the imidazole NH was close to that of the SH. After several attempts it was proven that only with a concentration of 1M of base (tributylbenzylamine), in a one to one mixture of dimethylformamide and dimethylsulfoxide at 130 °C after 7 minutes the *S*-[^{11}C]methylated was abundant in 95% against the *N*-[^{11}C]methylated in 5% (scheme 13) [26].

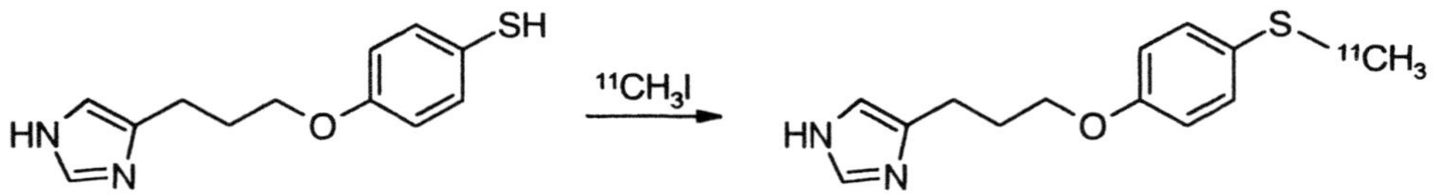

Scheme 12. Synthesis of [^{11}C]UCL 1829. i : [^{11}C]CH_3I, tributylbenzylamine, dimethylsulfoxide/dimethylformamide, 130 °C, 7 min.

[^{11}C]UCL 1829 was purified with HPLC on a RP18 column and isolated with a specific activity of 900 Ci/mmol. biodistribution in the rat showed an extremely high uptake in the lungs (25-30 % dose/gram) which was saturable with FUB 359, a H_3 antagonist. Brain penetration was however too low for brain imaging with PET.

4. CONCLUDING REMARKS

Radiolabelled compounds for the histamine H_3 receptor have been proven to be of great value in pharmacological and biochemical research. Some iodinated and tritiated compounds haven become standard research tools. In contrast, the development of PET or SPECT ligands has been proven to be troublesome. So far there has been no radiolabelled compound which is taken up into the brain in reasonable amounts to allow for PET or SPECT studies of the H_3 receptor in the brain. The search for new radiolabelled compounds is still going on however, as is research to find out why these radiolabelled compounds do not enter the brain as opposed to their non-radiolabelled equivalents. One can detect pharmacological effects which can only be caused by central nervous system activity of these non-radiolabelled compounds. In view of the potential applications of histamine H_3 receptor drugs in CNS disorders [31], it is of great importance to continue the research for a radiolabelled H_3 receptor ligand suitable for PET or SPECT in order to monitor the H_3 receptor *in vivo* in a non-invasive manner.

REFERENCES

1. P. Caldirola, H. Timmerman, J. Labelled. Compd. Radiopharm., 31 (1992) 987.
2. F.M. Kaspersen, J. van Acquoy, G.L.M. van de Laar, G.N. Wagenaars, C.W. Funke, Recl. Trav. Chim. Pays-Bas, 103 (1984) 32.
3. J.P. Bloxsidge, J.A. Elvidge. M. Gower, J.R. Jones, E.A. Evans, J.P. Kitcher, D.C. Warrel J. Labelled Compd. Radiopharm., 18 (1981) 1441.
4. S. Lee, H. Morimoto, P.G. Williams, J. Labelled. Compd. Radiopharm., 39 (1997) 461
5. J. Mertens, Eur Patent 852000798.8 (1985).
6. M. Gysemans, PhD Thesis, Vrije Universiteit Brussel (1992).
7. S.M. Qaim, J.C. Clark, C. Crouzel, M. Guilllaume, H.J. Helmeke, B. Nebeling, V.W. Pike, G. Stöcklin, in Radiopharmaceuticals for Positron Emission Tomography, G. Stöcklin, and V.W. Pike (eds), 1-43 , Kluwer Acadamic Publishers, Dordrecht, The Netherlands (1993).
8. M.R. Kilbourn, Fluorine-18 Labeling of Radiopharmaceuticals (NAS-NS-3203), National Acadamy Press (1990).
9. Y-S. Ding, J.S. Fowler, J. Gatley, S.L. Dewey, A.P. Wolf, J. Med. Chem., 34 (1991) 767.
10. G. Vaidyanathan, D.J. Affleck, M.R. Zalutsky, J. Med. Chem., 37 (1994) 3655.
11. K. Hamacher, H.H. Coenen, G. Stöcklin, J. Nucl. Med., 27 (1986) 235.
12. D.M. Jewett, Appl. Radiat. Isot., 43 (1992) 1383.
13. M. Holsbach, M. Schüller, Appl. Radiat. Isot., 44 (1993) 897.
14. K. Någren, L. Muller, C. Halldin, C.-G. Swahn, P. Lehikoinen, Nucl. Med. Biol, 22 (1995) 235.
15. U. Pleiss, A. Harder, A. Turberg, M. Londershausen, K. Iinuma, N. Mencke, P. Jeschke, G. Bonse, J. Labelled. Compd. Radiopharm., 38 (1996) 61.
16. U. Pleiss, A.Turberg, A. Harder, M. Londershausen, P. Jeschke, G. Boheim, J. Labelled. Compd. Radiopharm., 38 (1996) 651.
17. F.P. Jansen, B. Rademaker, A. Bast, H. Timmerman, Eur. J. Pharmacol., 217 (1992) 203.
18. W.M.P.B. Menge, H. van der Goot, H. Timmerman, J.L.H. Eersels, J.D.M. Herscheid, J. Labelled. Compd. Radiopharm., 31 (1992) 781.
19. F.P. Jansen, H.P. Voss, J.L.H. Eersels, J.D.M. Herscheid, B. Rademaker, A.Bast, B.S. Larsson, H. Timmerman, submitted
20. J.W. Clitherow, P. Beswick, W.J. Irving, D.I.C. Scopes, J.C. Barnes, J. Clapham, J.D. Brown, D.J. Evans, A.G. Hayes, Bioorg. & Med. Chem. Lett., 6 (1996) 833.
21. A.D. Windhorst, M.J.P.G. Kroonenbrugh, R. Leurs, W.M.P.B. Menge, H. Timmerman, J.D.M. Herscheid, submitted.(1998)
22. H. Stark, K. Purand, A. Hüls, X. Ligneau, M. Garbarg, J.-C. Schwartz, W. Schunack, J. Med. Chem, 39 (1996) 1220.
23. M. Krause, H. Stark, W. Schunack, J. Labelled. Compd. Radiopharm, 39 (1997) 601.

24. A.D. Windhorst, L. Bechger, G.W.M. Visser, W.M.P.B. Menge, R. Leurs, H. Timmerman, J.D.M. Herscheid, J. Fluorine Chem., 80 (1996) 35.
25. A.D. Windhorst, R. Leurs, W.M.P.B. Menge, H. Timmerman, J.D.M. Herscheid, 25th European Histamine Research Society Meeting, Antwerp, Belgium 1995.
26. M. Ponchant, S. Demphel, C. Fuseau, C. Coulomb, M. Bottländer, J.-C. Schwartz, H. Stark, W. Schunack, S. Athmani, R, Ganellin, C. Crouzel, XIIth International Symposium on Radiopharmaceutical Chemistry, Uppsala, Sweden (1997).
27. A. Alves-Rodrigues, R. Leurs, T.S. Wu, G.D. Prell, C. Foged, H. Timmerman, Br. J. Pharmacol., 118 (1996)2045.
28. C. Foged, Personal communications.
29. J.D. Brown, C.T. O'Shaughnessy, G.J. Kilpatrick, D.I.C. Scopes, P. Beswick, J.W. Clitherow, J.C. Barnes, Eur. J. Pharmacol., 311 (1996) 305.
30. K. Yanai, J.H. Ryu, N. Sakai, T. Takahashi, R. Iwata, T. Ido, K. Marukami, T. Watanabe, Jpn. J. Pharmacol., 65 (1994) 107.
31. K. Yanai, J.H. Ryu, T. Watanabe, M. Higuchi, T. Fujiwara, M. Itoh, R. Iwata, T. Takahasi, T. Ido, Labelling of Histamine H_1, H_2 and H_3 Antagonists with Carbon-11 using On-Line Methylation System : Potential Radiopharmaceuticals for PET Studies, *J. Labell. Compds, Radiopharm.* **1993**, *35*, 520.
32. R. Leurs, R.C. Vollinga, H. Timmerman, in : Progrees in Drug research, 45, (ed : E. Jucker), 107 - 165, Birkhauser Verlag, Basel, Switserland (1995).

R. Leurs and H. Timmerman (Editors)
The Histamine H_3 Receptor

Medicinal Chemistry of Histamine H_3 Receptor Agonists

M. Krause, H. Stark and W. Schunack

Institut für Pharmazie I, Freie Universität Berlin, Königin-Luise-Strasse 2+4, 14195 Berlin, Germany

1. INTRODUCTION

Initially, the histamine H_3 receptor was characterized as an inhibitory presynaptic autoreceptor in the central nervous system (CNS) [1,2], but more recently it has been found to occur also on non-histaminergic axon terminals modulating the release of several other important neurotransmitters [3,4].

Since its potential therapeutic value has become evident throughout the recent years, the H_3 receptor received rapidly growing interest from pharmacologists and medicinal chemists. This is mainly attributed to the high pharmacological specificity of H_3 receptors which clearly exceeds that of H_1 and H_2 receptors. This particular specificity enabled the development of highly selective H_3 receptor agonists and antagonists, and accordingly several possible indications were proposed for H_3 receptor agonists. They have been suggested to become valuable drugs for the treatment of sleep disorders, children hyperactivity, migraine, asthma, inflammation, ulcer, and depression [5–7]. This section, however, deals with the medicinal chemistry and development of the H_3 receptor agonists known at present.

2. HISTAMINE H_3 RECEPTOR AGONISTS

2.1. Histamine Derivatives

Although histamine (**1**) is the endogenous ligand of H_1, H_2, and H_3 receptors, respectively, it displays higher affinity for the latter and is thus slightly more selective for its presynaptic than for its postsynaptic sites [1]. The remarkable agonist activity of histamine at H_3 receptors was the basis for systematic structural variations, most of which, however, resulted in a dramatic decrease in potency. Particularly the 4-alkylimidazole moiety of the histamine molecule proved to be an essential partial structure for H_3 receptor agonism, and hence it is common to almost all H_3 receptor agonists of high potency known at present.

Table 1
H_3 Receptor Agonist Activity of Imidazole-methylated Histamine Derivatives

No.	Structure	Compound	pD_2 [a]	Ref.
1	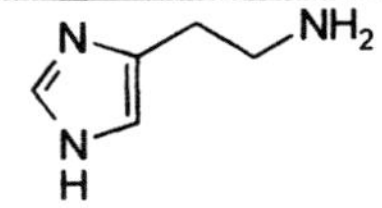	histamine	7.4	[1]
2	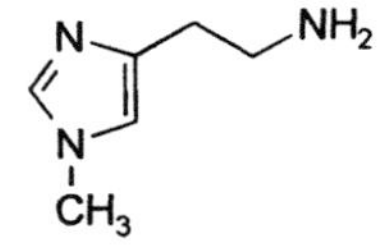	N^{τ}-methylhistamine	< 6.0	[1]
3	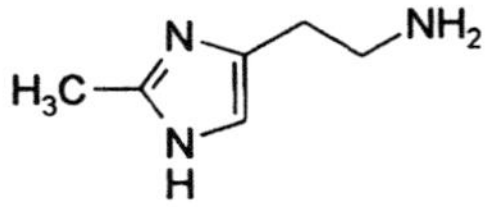	2-methylhistamine	< 4.3	[1]
4	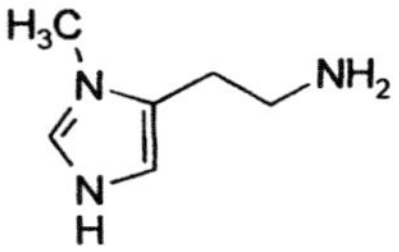	N^{π}-methylhistamine	< 6.0	[1]
5	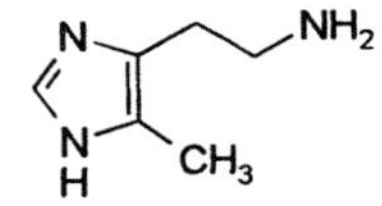	4(5)-methylhistamine	< 3.0	[1]

[a] K^+-Evoked [^{3}H]histamine release on rat brain cortex.

The replacement of the basic imidazole ring of histamine by several five- or six-membered heterocycles was one major goal in the course of H_2 receptor antagonist drug development, because imidazole is particularly critical with regard to biochemical aspects like metabolism and cytochrome P_{450} interaction. Nevertheless, during the development of H_3 receptor ligands any attempt to replace the 4-imidazolyl residue of histamine failed so far. Neither any other five- or six-membered heterocycle nor any other potential bioisosteric replacement that contains at least one nitrogen was allowed for H_3 receptor agonism [8]. At present compounds that possess H_3 receptor activity and that are devoid of an imidazole ring are only found in the field of H_3 antagonists, e. g., the H_1 receptor agonist betahistine (*N*-methyl-2-(2-pyridyl)ethanamine) and the atypical antidepressant clozapine both of which act as moderate H_3 receptor antagonists [9,10]. However, some hints for further non-imidazole H_3 receptor antagonists have been reported more recently [11,12].

Additional substitution of the imidazole of histamine is not tolerated by the H_3 receptor, too. Thus, methylation of histamine in either position of the imidazole ring

Table 2
H_3 Receptor Agonist Activity of N^{α}-Substituted Histamine Derivatives

No.	Structure	Compound	pD_2 [a]	Ref.
6	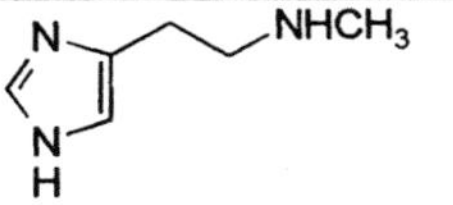	N^{α}-methylhistamine	7.8	[1]
7	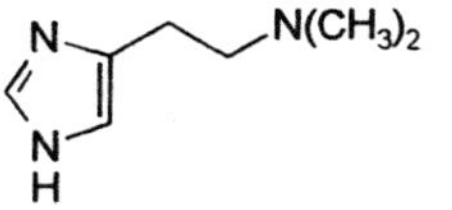	N^{α},N^{α}-dimethylhistamine	7.6	[1]
8	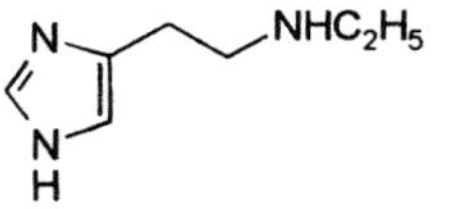	N^{α}-ethylhistamine	7.1	[14]
9	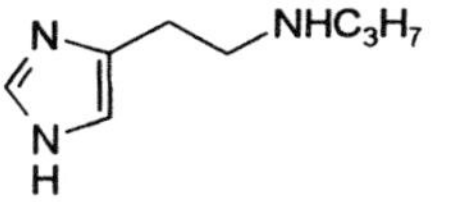	N^{α}-propylhistamine	< 5.2	[14]
10	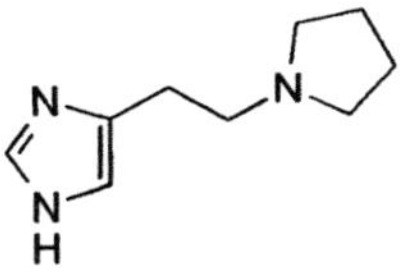	*N*-[2-(1*H*-imidazol-4-yl)ethyl]-pyrrolidine	6.2 (α = 0.56)[b]	[15]

[a] K^+-Evoked [^{3}H]histamine release on rat brain cortex. [b] Intrinsic activity (α (histamine) = 1.0).

strongly decreases H_3 receptor agonist activity (Table 1). It has to be emphasized that this behaviour is not shared by the respective postsynaptic histamine receptors, because 2-methylhistamine (**3**) is a rather selective H_1 receptor agonist, whereas 5(4)-methylhistamine (**5**) behaves as a moderate H_2 receptor agonist. Furthermore, it should be noted that N^{τ}-methylation at the imidazole nucleus is a main pathway of metabolic histamine inactivation in man, and thus formation of N^{τ}-methylhistamine (**2**) primarily accounts for the short half-life of systemically released histamine [13].

Although the 4-imidazolyl ring of histamine may not be modified in order to increase or even retain H_3 receptor agonist activity, side chain methylation emerged to be very fruitful for the development of H_3 receptor agonists. The first compound that was found to display considerable activity and moderate selectivity at H_3 receptors was the side chain methylated histamine derivative N^{α}-methylhistamine (**6**, Table 2), which is about 2.5 times more potent than the endogenous ligand [1]. However, an increase of the van der Waals volume of the N^{α}-substituent by introduction of an additional methyl in N^{α},N^{α}-dimethylhistamine (**7**) [1], or further extension of the side chain by

Table 3
H_3 Receptor Agonist Activity of Chiral Histamine Derivatives

No.	Structure	Compound	pD_2 [a]	Ref.
1	–	histamine	7.4	[1]
11	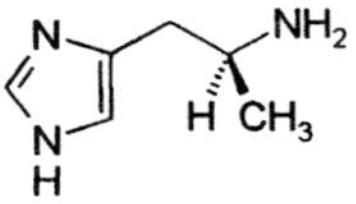	(*S*)-α-methylhistamine	6.3	[15]
12	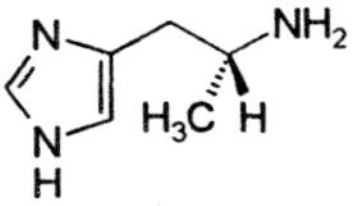	(*R*)-α-methylhistamine	8.4	[15]
13	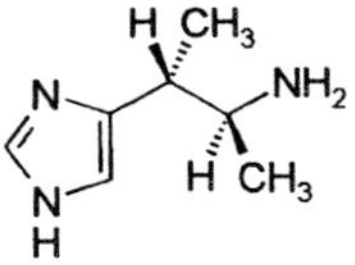	(*S*)-α,(*R*)-β-dimethylhistamine	6.5	[17]
14	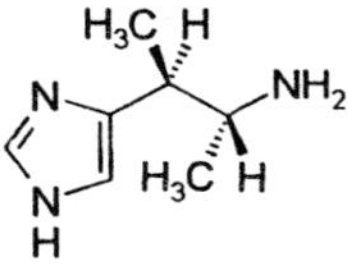	(*R*)-α,(*S*)-β-dimethylhistamine	8.5	[17]
15	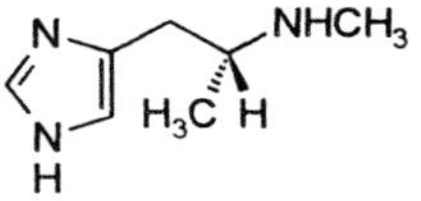	(*R*)-α,N^{α}-dimethylhistamine	5.8	[19]
16	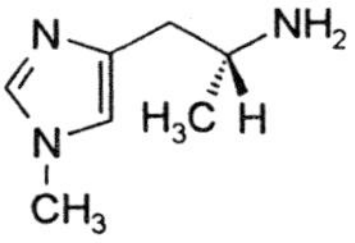	(*R*)-α,N^{τ}-dimethylhistamine	4.3	[20]

[a] K^+-Evoked [^{3}H]histamine release on rat brain cortex.

one or two methylene groups in N^{α}-ethylhistamine (**8**) and N^{α}-propylhistamine (**9**), leads to a decrease in H_3 receptor agonist potency [14]. Interestingly, inclusion of the N^{α}-nitrogen in a more rigid pyrrolidine cyclus results in a weak partial agonist, *N*-[2-(1*H*-imidazol-4-yl)ethyl]pyrrolidine (**10**), with an intrinsic activity of 0.56 versus histamine (α = 1.0) [15].

Subsequent systematic variations of the histamine side chain lead to some chiral branched derivatives which were observed to be strikingly active at and selective for the H_3 receptor. In contrast to H_1 and H_2 receptors, respectively, the H_3 receptor shows a considerably more pronounced stereoselectivity for agonistic binding, because the introduction of a single methyl group in the α-position of the histamine side chain already results in clearly differing activity of both enantiomers. While the distomer (*S*)-β-methylhistamine (**11**) displays only 13% of histamine activity, the eutomer (*R*)-α-methylhistamine (**12**, Table 3) is 15.5 times more potent than histamine at H_3 receptors [15,16]. This stereoselectivity becomes even more evident with the four possible α,β-dimethylhistamines, of which (*R*)-α,(*S*)-β-dimethylhistamine (**14**) exhibits the greatest potency by being 18 times more active at H_3 receptors than histamine, whereas its enantiomer (*S*)-α,(*R*)-β-dimethylhistamine (**13**) only shows 18% of histamine activity [17]. Accordingly, the eudismic ratio of the respective pairs of enantiomers is about 100.

In addition to their high potency, (*R*)-α-methylhistamine (**12**) and (*R*)-α,(*S*)-β-dimethylhistamine (**14**) also display a remarkable receptor selectivity with regard to H_3 vs. H_1/H_2 agonism, which is indicated by the selectivity ratios of about 20,000:1 H_3 vs. H_1/H_2 for the former and 130,000:1 for the latter [8]. However, (*R*)-α-methylhistamine (**12**) became the standard H_3 receptor agonist in pharmacological trials, because it can be prepared more easily than (*R*)-α,(*S*)-β-dimethylhistamine (**14**).

The synthesis of (*R*)-α-methylhistamine (**12**) starts with esterification of the amino acid L-histidine which is then hydrogenated and halogenated (Figure 1). Subsequent dehalogenation yielding (*R*)-α-methylhistamine (**12**) can be performed either under high pressure [16] or by transfer hydrogenolysis, which is more convenient [18].

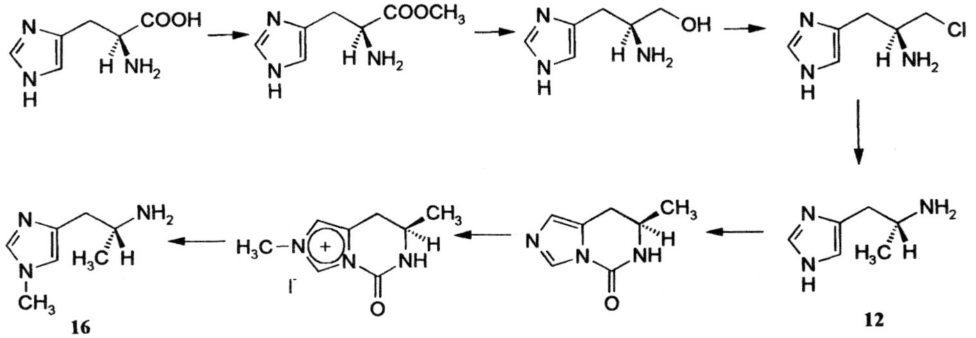

16 **12**

Figure 1: Synthesis of (*R*)-α-methylhistamine (**12**) and (*R*)-α,N^{τ}-dimethylhistamine (**16**).

In accordance with the pharmacological results above, introduction of an additional methyl in the N^{α}-position of the side chain as well as in the N^{τ}-position of the imidazole drastically decreases the H_3 receptor agonist potency of (*R*)-α-methylhistamine (**12**). In particular, (*R*)-α,N^{α}-dimethylhistamine (**15**) displays a pD_2 value of 5.8 [19] and (*R*)-α,N^{τ}-dimethylhistamine (**16**) a value of 4.3 [20].

As (*R*)-α-methylhistamine (**12**) is also inactivated by the most important histamine metabolizing enzyme in man, histamine methyltransferase (E.C.2.1.1.8), its main metabolite in man is (*R*)-α,N^{τ}-dimethylhistamine (**16**) [6], which can be prepared according to Figure 1. Simultaneous protection of the NH-functionalities of **12** by reaction with *N,N'*-carbonyldiimidazole is followed by specific methylation of one nitrogen and acidic cleavage of the protecting moiety resulting in **16**. (*R*)-α,N^{τ}-dimethylhistamine (**16**) was used for the development of a specific radioimmuno assay of (*R*)-α-methylhistamine (**12**), which allows the indirect determination of **12** in different tissues.

Interestingly, a number of histamine derivatives in which the flexible side chain is incorporated into a more rigid cyclic moiety were also reported to be potent H_3 receptor agonists. Albeit a cyclopropyl derivative of histamine, (±)-2-(1*H*-imidazol-4-yl)-cyclopropylamine (**17**), was found to be less active than (*R*)-α-methylhistamine (**12**) [21], conformationally restricted analogues of histamine which mimic three predominant conformations of **12** were recently identified as novel potent and selective H_3 receptor agonists in an approach to define the bioactive conformation of (*R*)-α-methylhistamine (**12**) and to investigate the steric requirements of H_3 receptor agonists (Table 4) [22,23]. The two conformers possessing a gauche relationship between the basic side chain nitrogen and the 4-imidazolyl moiety are (±)-*trans*-2-(1*H*-imidazol-4-yl)-1-cyclopentylamine (**18**) and (±)-*trans*-2-(1*H*-imidazol-4-yl)-1-methylcyclohexyl-amine (**19**). Their respective structures shown in Table 4 represent only one enantiomer of the diastereomers or the racemates for the sake of clarity. The anti conformer of (*R*)-α-methylhistamine (**12**), however, is imitated by the enantiomeric pyrrolidine derivatives SCH 49647 (**20**) and SCH 49648 (**21**), the latter being named immepyr. Both *trans*-3-(1*H*-imidazol-4-yl)-2-methylpyrrolidine isomers mimic the conformation to a different extend and exhibit substantial H_3 receptor activity. As a result thereof, the dextrorotatory (2*R*,3*S*)-eutomer immepyr (**21**) was found to be as active as (*R*)-α-methylhistamine (**12**) and proved to be 10-fold more active than the levorotatory (2*S*,3*R*)-distomer SCH 49647 (**20**) [22].

2.2. Imetit Derivatives

The first compound characterizing the histamine H_3 receptor was the thiourea derivative burimamide [1], but yet all of the H_3 receptor agonists with considerable activity known at present are derived from histamine, most of which contain an aliphatic or alicyclic amino group in their side chain. In the course of H_3 receptor antagonist development this amine functionality was successfully replaced by several polar moieties, and the first selective H_3 receptor antagonist was the thiourea derivative thioperamide [15]. In an effort to develop new potent agonists by replacing the N^{α}-nitrogen of histamine by several planar polar moieties, a series of amidine, guanidine

Table 4
H_3 Receptor Agonist Activity of Ring-Closed Histamine Derivatives

No.	Structure	Compound	pD_2	pK_i[a]	Ref.
17	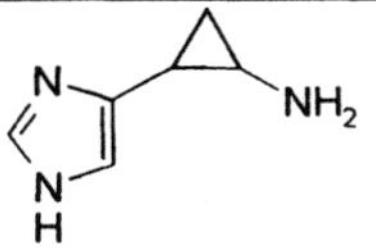	(±)-2-(1*H*-imidazol-4-yl)-cyclopropylamine	8.0[b]	7.1[c,d] 7.4[e] 8.5[f]	[21,24, 25]
18	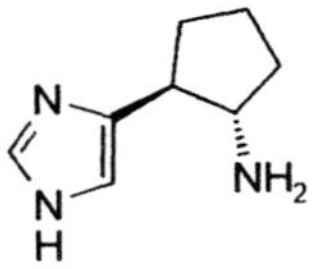	(±)-*trans*-2-(1*H*-imidazol-4-yl)-1-cyclopentylamine	–	6.1	[22,23]
19	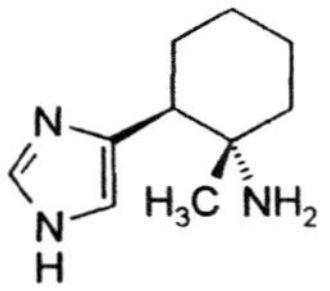	(±)-*trans*-2-(1*H*-imidazol-4-yl)-1-methylcyclohexyl-amine	–	< 5.7	[22,23]
20	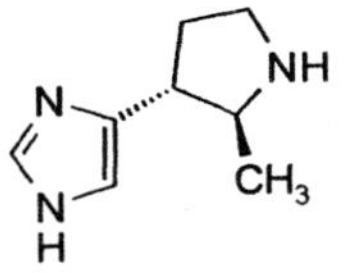	(2*S*,3*R*)-3-(1*H*-imidazol-4-yl)-2-methylpyrrolidine; SCH 49647	–	7.5	[22,23]
21	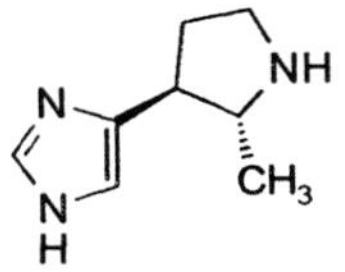	(2*R*,3*S*)-3-(1*H*-imidazol-4-yl)-2-methylpyrrolidine; SCH 49648; immepyr	7.1[c]	8.5	[22,23]
22	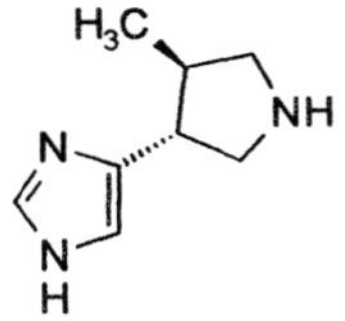	(3*R*,4*R*)-3-(1*H*-imidazol-4-yl)-4-methylpyrrolidine; SCH 50971	7.5[c]	8.6	[23]
23	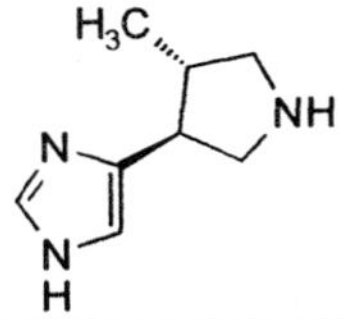	(3*S*,4*S*)-3-(1*H*-imidazol-4-yl)-4-methylpyrrolidine SCH 50972	–	7.5	[23]

[a] Displacement of [^{3}H]N^α-methylhistamine on guinea pig brain tissue. [b] K^+-Evoked [^{3}H]histamine release on rat brain cortex. [c] Electrically evoked contractions of guinea pig ileum. [d] Activity of both *trans*-configured enantiomers [24]. [e] (1*S*,2*S*) enantiomer [25]. [f] (1*R*,2*R*) enantiomer [25].

Table 5
H_3 Receptor Agonist Activity of Imetit Derivatives

No.	Structure	Compound	pD_2[a]	Ref.
1	–	histamine	7.4	[1]
24	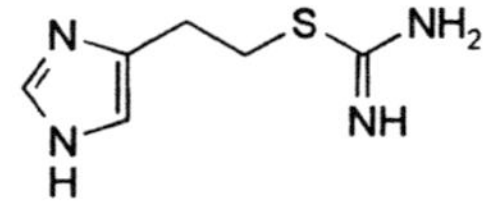	imetit	9.0, 8.1[b] 8.3[b]	[27–29]
25	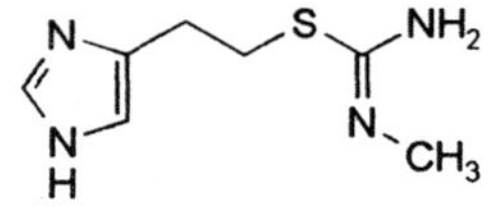	VUF 8621	7.8, 7.3[b]	[27]
26	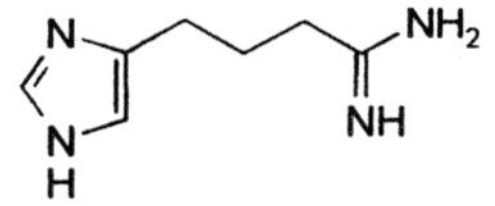	SKF 91606	9.0[b]	[28]
27	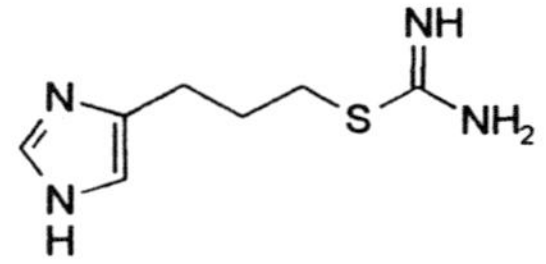	VUF 8328	(pA_2 = 8.0[b])	[27]

[a] K^+-Evoked [^{3}H]histamine release on rat brain cortex. [b] Electrically evoked contractions of the guinea pig ileum.

and isothiourea derivatives was investigated, some of which display high H_3 receptor agonist activity and selectivity [26].

The isothiourea derivative imetit (**24**, Table 5) is a highly potent compound of this series of H_3 receptor agonists being about four times as active as (*R*)-α-methylhistamine (**12**) [27–29]. However, neither methylation of the isothiourea moiety (**25**) nor introduction of an additional methylene spacer is tolerated for H_3 receptor agonism, as VUF 8328 (**27**) is even a potent H_3 antagonist. Further alkylation of the isothiourea substantiates the transition from agonist to antagonist binding, since it resulted in the highly potent H_3 antagonist clobenpropit. Replacement of the sulphur by a nitrogen leads to the corresponding guanidine, which is clearly less potent (pD_2 = 6.3) than imetit [27], whereas SKF 91606 (**26**), the corresponding amidine, displays even higher activity [28]. Showing a pD_2 value of 9.0 at the guinea pig ileum, SKF 91606 (**26**) is the most potent of the H_3 receptor agonists in this particular assay known at present.

It has to be recognized that the distance between the proton donating nitrogen and the imidazole ring in an extended conformation of imetit (**24**) is clearly different as compared to histamine, i. e., for imetit (**24**) the value is approximately 8 Å, whereas for histamine it is only 4.5 Å. According to this observation the length of the histamine side chain was lengthened by the introduction of one, two, or three methylene groups, respectively [30], but unexpectedly the agonistic activity was completely lost. However, it resulted in a series of compounds with H_3 antagonistic properties, the most potent of which was found to be impentamine which contains a pentamethylene spacer between the imidazole ring and the primary amine. Surprisingly, impentamine and related compounds behave as partial agonists in some pharmacological assays (see 2.5.).

Due to its high potency and selectivity imetit (**24**) proved to be a valuable tool in pharmacological trials related to H_3 receptors, but in spite of that the potential toxicity of its isothiourea moiety emerges to impede further pharmacological development. A similar situation arose with the H_3 antagonist thioperamide, the clinical development of which was ceased due to hepatotoxicity which is most probably related to its thiourea functionality.

2.3. Immepip Derivatives

As already discussed above, the agonist activity of histamine is abolished and turned to antagonism if the side chain is further extended [30]. However, cyclization of a lengthened side chain is allowed for H_3 receptor agonism. The racemic 3-(1*H*-imidazol-4-ylmethyl)pyrrolidine (**28**) displays only slightly less affinity for the H_3 receptor than histamine, but methylation of its secondary amine leads to an approximately 30-fold increase of affinity of the corresponding racemic *N*-methylpyrrolidine **29** [31]. Replacement of the pyrrolidine ring of **28** by a piperidine ring results in even higher affinity with a pK_i value of 9.5 for immepip (**30**), which is equipotent to (*R*)-α-methylhistamine (**12**) and imetit (**24**) in the functional test on the guinea pig ileum [32,33]. Further structural modifications, however, like methylation of the piperidine nitrogen (**31**) [32] or replacement of the piperidine by a piperazine ring (**32**) [34] decrease the affinity for the H_3 receptor 2.5-fold and 16-fold, respectively. Nevertheless, immepip (**30**) may be a useful tool for microdialyses or other experiments in which the interaction of histamine (**1**) and (*R*)-α-methylhistamine (**12**) causes problems.

2.4. Molecular Modelling

Since it is very helpful for the design of selective H_3 receptor ligands to know more about the molecular interaction of the endogenous ligand histamine and its H_3 receptor, especially in which bioactive conformation histamine interacts with the respective binding sites of the receptor, the structural and conformational requirements for receptor affinity and activity of H_3 receptor agonists were recently investigated by means of molecular modelling methods [35]. In particular, the side chain methylated

Table 6
H_3 Receptor Agonist Activity of Several Derivatives of Immepip

No.	Structure	Compound	pK_i [a]	Ref.
28	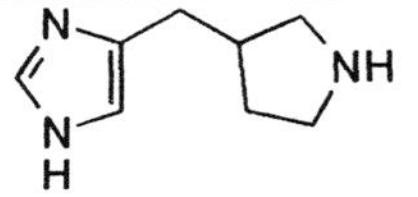	(±)-3-(1*H*-imidazol-4-yl-methyl)pyrrolidine	7.0	[31]
29	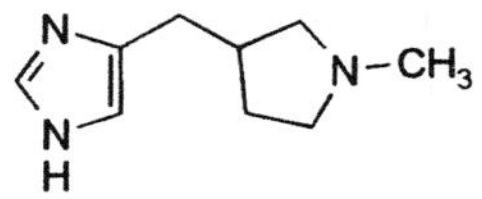	(±)-3-(1*H*-imidazol-4-yl-methyl)-1-methylpyrrolidine	8.5	[31]
30	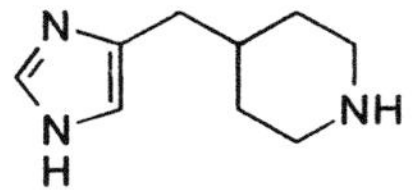	4-(1*H*-imidazol-4-ylmethyl)-piperidine; immepip	9.5, 8.0 [b]	[32,33]
31	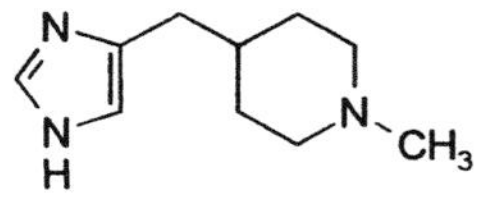	4-(1*H*-imidazol-4-ylmethyl)-1-methylpiperidine	9.1	[32]
32	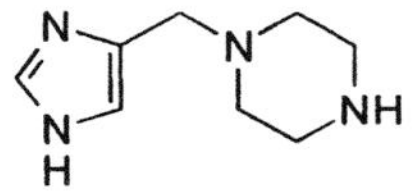	1-(1*H*-imidazol-4-ylmethyl)-piperazine	8.3	[34]

[a] Displacement of $[^3H]N^{\alpha}$-methylhistamine on guinea pig brain tissue. [b] Electrically evoked contractions of guinea pig ileum (pD_2 value).

histamine analogues N^{α}-methylhistamine (**6**), (*R*)-α-methylhistamine (**12**), and (*R*)-α,(*S*)-β-dimethylhistamine (**14**), the heterocyclic analogues (±)-2-(1*H*-imidazol-4-yl)-cyclopropylamine (**17**), immepip (**30**), immepyr (**21**), and SCH 50971 (**22**) as well as imetit (**24**) and SKF 91606 (**26**) were analyzed. Due to their rigid conformations, their similarities, and their heterogeneity these compounds served as good templates for the calculation of a possible receptor model. The calculation resulted in a single useful pharmacophore model in which both essential substructures of H_3 receptor agonists, the imidazole ring and the protonated side chain nitrogen, can be perfectly superimposed. Accordingly, the yielded pharmacophore indicated similar positions of the side chains and the imidazole rings, respectively, for all H_3 receptor agonists in this study [35].

The four possible stereoisomers of the diastereomeric mixture of (±)-2-(1*H*-imidazol-4-yl)-cyclopropylamine (**17**) have not been separated yet (except very recently both *trans* isomers (Table 4) [25]), but as a result of this molecular modelling study it was observed that a congruent pharmacophoric conformation can only be adopted by the two *trans* configured isomers [24]. The conformationally restricted pyrrolidine derivatives immepyr (**21**) and SCH 50971 (**22**) support this pharmacophore model very strikingly, because in addition to the similar positions of imidazole ring and amine functionality, the methyl group of immepyr (**21**) corresponds to the (*R*)-α-methyl of (*R*)-α,(*S*)-β-dimethylhistamine (**14**), while the pyrrolidine ring overlaps with (*S*)-β-methyl. Although the potent H_3 receptor agonists imetit (**24**) and SKF 91606 (**26**) do not contain an amine but an isothiourea and amidine functionality, respectively, they can also adopt conformations in which one nitrogen occupies the same position as the amine nitrogen of other H_3 receptor agonists [35]. The histamine side chain revealed from this study displays a gauche–trans conformation in the bioactive conformation on H_3 receptors.

Studies on the structure–activity relationship of the pyrrolidine derivatives of histamine (Table 4) established that the corresponding four-membered ring analogue of immepyr (**21**) was comparably active to (*R*)-α-methylhistamine (**12**), but less selective for the H_3 receptor, whereas the six-membered ring analogue was significantly less active [31]. The methyl substituent (**20–23**) was discovered to be essential for H_3 receptor selectivity, while increasing the size of the alkyl substituent decreases activity. Likewise, dimethylation decreases H_3 receptor agonist potency. This may give rise to the hypothesis that a lipophilic residue of a distinct volume enhances agonist affinity. Surprisingly, both *trans*-3-(1*H*-imidazol-4-yl)-4-methylpyrrolidines (SCH 50971 (**22**) and SCH 50972 (**23**)) which carry the methyl substituent in the 4-position that is *trans* orientated to the imidazole residue, exhibit a structure–activity relationship similar to that of the trans-2-isomers **20** and **21**. Again the dextrorotatory (3*R*,4*R*)-enantiomer SCH 50971 (**22**) proved to be the eutomer being more than 10-fold more potent than the corresponding (3*S*,4*S*)-antipode SCH 50972 (**23**) [23].

Thus, amino acid models of the respective H_3 receptor binding sites may be derived from this pharmacophore model combined with the molecular interaction patterns of the H_3 receptor agonists investigated. More recently, another study has suggested a conformational change of an aspartic acid residue to play an essential role in H_3 receptor stimulation [24]. However, only few studies have been performed to investigate the molecular interaction of histamine and its H_3 receptor, although the sequence of the latter is eagerly awaited.

2.5. Nonaminergic Partial Agonists

Besides the aminergic compounds *N*-[2-(1*H*-imidazol-4-yl)ethyl]pyrrolidine (**10**), impentamine, and related compounds [36] more recently novel nonaminergic compounds were found to be partial agonists at H_3 receptors. The specific radioligand iodoproxyfan as well as its deiodo analogues display partial agonism in the H_3 receptor-dependent [^{3}H]noradrenaline release assay in mouse brain slices [37], whereas in the guinea pig ileum assay only iodoproxyfan was detected to be a partial agonist [37] or full agonist [38]. This discrepancy of the two functional assays, which was also observed for impentamine and its analogues, may be caused by lower efficacy

Table 7
Nonaminergic Compounds with Partial H_3 Receptor Agonist Activity

No.	Structure	Compound	pD_2 [a]	α [b]	Ref.
1	–	histamine	7.4	1.0	[1]
33	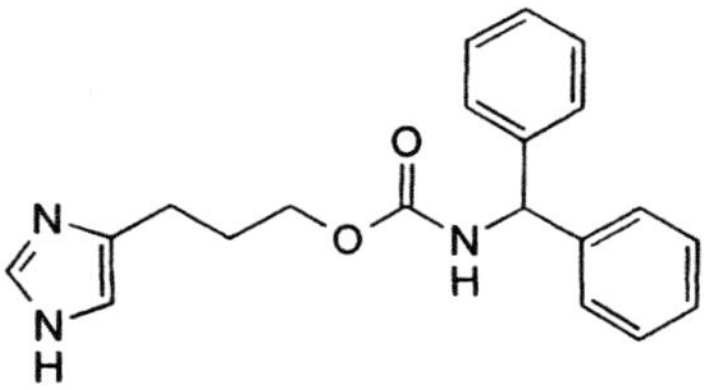	FUB 316	7.0	0.2	[39]
34	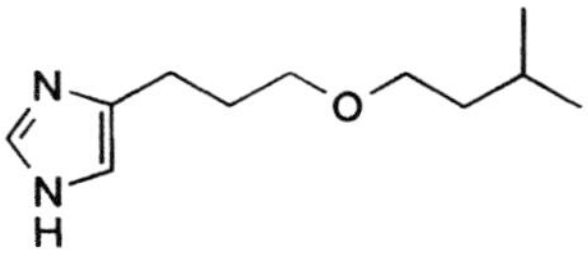	FUB 373	6.9	0.2	[39]
35	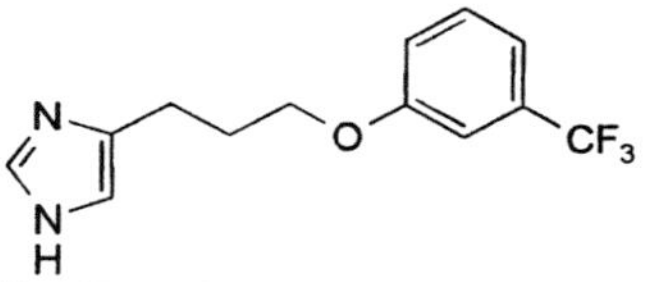	UCL 1470	7.0	0.4	[39]

[a] K^+-Evoked [^{3}H]histamine release on rat brain cortex. [b] Intrinsic activity.

of receptor coupling in the mouse brain as compared to the guinea pig ileum. It may therefore be suggested that the partial agonism of the deiodo analogue is undetectable in the guinea pig ileum assay due to its potency lower than that of iodoproxyfan and a high noise-effect ratio.

However, the observation that compounds of different structures, which were described by the groups of Schwartz and Ganellin in close co-operation with the authors, also display partial agonism is more striking (Table 7) [39]. Although compounds **44–46** belong to the general construction pattern of antagonists, they behave like partial agonists in the [^{3}H]histamine release assay on rat synaptosomes. These compounds are devoid of an amine functionality or related moieties, but though the imidazole ring is the only protonable group previous investigations suggested that it does not seem to mimic the potential aspartic receptor binding site as presumed for most of the aminergic agonists. Accordingly, other mechanisms of receptor activation than amine–carboxylic acid interaction should be taken into account for these compounds. Preliminary investigations of structure–activity relationships established that both the imidazole ring and a lipophilic residue of a defined volume have to be arranged in a distinct position to activate the H_3 receptor. This suggests that an additional lipophilic binding site might be able to activate the receptor by either changing its conformation or fixing it in an active state.

However, no H_3 receptor agonist activity of these compounds is detected in the guinea pig ileum assay. It should therefore be interesting to see which kind of effect these compounds show in other H_3 receptor assays. Although the observed difference of H_3 receptor response might be due to a heterogeneous population of H_3 receptors, this does not necessarily imply the existence of subtypes. In addition to the possible existence of subtypes, this particular discrepancy of H_3 receptor assays might either be due to differences between species, G-protein coupling, signal transduction mechanisms, or a different extent of spare receptors. A comparable discussion which came about for the H_2 receptor finally ended when cloning of the receptor undoubtedly confirmed the homogeneity of the gene and accordingly of the corresponding amino acid sequence in different tissues. Thus, future investigations will have to explain whether histamine H_3 receptor heterogeneity or other reasons account for these results.

3. AZOMETHINE PRODRUGS OF (*R*)-α-METHYLHISTAMINE

Due to its high selectivity and potency at H_3 receptors (*R*)-α-methylhistamine (**12**) is nowadays used as the standard agonist in pharmacological assays related to this receptor. With regard to its pharmacodynamic properties, which were substantiated in preclinical studies [40], (*R*)-α-methylhistamine (**12**) meets all criteria of a potential drug substance. However, in contrast to its pharmacodynamic potential (*R*)-α-methylhistamine (**12**) suffers from its pharmacokinetic drawbacks which limit its use in both pharmacological and conceivable clinical studies.

In particular, the pronounced basicity of the primary amine as well as the high polarity of the entire molecule restrict the penetration of biological membranes to a great extent regarding not only absorption from the digestive tract but also permeation of the blood-brain barrier (BBB) into the brain in which the areas of highest H_3 receptor density are found [41]. Even after *i.v.* application of high doses (*R*)-α-methylhistamine (**12**) penetrates the BBB only to a very small extent (Figure 2) [42]. The parent compound histamine itself, however, does not cross the BBB at all, but has to be synthesized, e. g., in histaminergic neurons from its amino acid precursor L-histidine [43].

Furthermore, due to its structural similarity to histamine (*R*)-α-methylhistamine (**12**) shows high affinity for histamine methyltransferase (HMT). Accordingly, methylation of the imidazole ring by HMT accounts for the short plasma half-life of (*R*)-α-methylhistamine (**12**) in man ($t_{½}$ = 3 min) [6]. It is assumed that the primary amine functionality of the side chain plays a key role in the interaction between

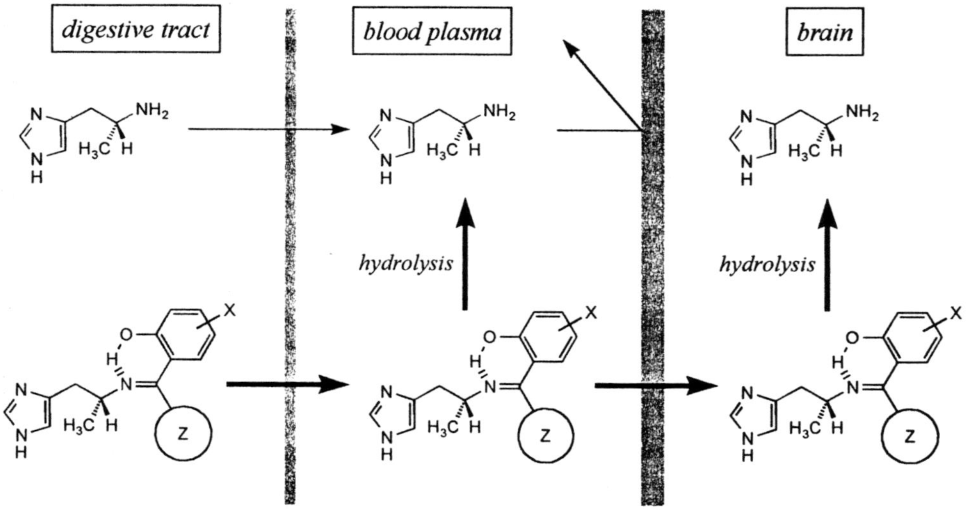

Figure 2. Prodrug conception of azomethine prodrugs of (*R*)-α-methylhistamine

enzyme and substrate, and thus its bioreversible blockade was supposed to improve metabolic stability.

To circumvent these pharmacokinetic obstacles of (*R*)-α-methylhistamine (**12**), the development of more lipophilic, less basic, and orally active prodrugs appeared to be a suitable approach. Thus, azomethine derivatives were prepared by coupling the primary amine of the side chain to several aromatic ketones in a bioreversible condensation reaction [44–46]. The azomethine double bond not only reduced the basicity of the basic amine, but also prevented the recognition of (*R*)-α-methyl-histamine (**12**) by the metabolizing enzyme HMT. Moreover, the lipophilicity of these azomethines was strikingly enhanced by the diaryl pro-moiety. With regard to the hydrolytic lability of such azomethines the phenol was found to be crucial, because it forms an intramolecular hydrogen bond that enhances the stability of the imine double bond. Therefore, it was substantiated by X-ray structure analysis. On account of the improved physico-chemical and biochemical properties of such azomethine prodrugs, the oral bioavailability and CNS penetration were strikingly increased as compared to the free base (*R*)-α-methylhistamine (**12**). Following oral administration the prodrugs are more readily absorbed from the digestive tract than the active drug. In the blood plasma azomethines partially decompose by means of non-enzymatic hydrolysis, while the intact prodrug enters the brain, the extent of which, however, strikingly depends on the substituents of pro-moiety (Figure 2) [44–46].

Several pharmacological studies were carried out to substantiate the proposed prodrug concept for (*R*)-α-methylhistamine (**12**). The parent compound of this series of azomethines is the 2-hydroxybenzophenone derivative of (*R*)-α-methylhistamine which is designated BP 2.94 (**36**). To evaluate the desired protection against HMT metabolization, BP 2.94 (**36**) was tested as a possible substrate in the presence of *S*-adenosylmethionine in comparison to histamine and (*R*)-α-methylhistamine (**12**). As a result no methylation of the prodrug **36** was observed, whereas both primary amines displayed similar K_M- and V_{max}-values [6].

Accordingly, the metabolic stability of BP 2.94 *in vivo* was strikingly enhanced as compared to the active drug. Following oral application of BP 2.94 (**36**) to healthy human volunteers, the progressive hydrolysis of the prodrug led to a parallel decrease in plasma levels of both compounds. A plasma half-life of 1 h was obtained for liberated (*R*)-α-methylhistamine (**12**) and > 24 h for BP 2.94 (**36**) as compared to 3 min for non-derivatized (*R*)-α-methylhistamine (**12**). The bioavailability in healthy human volunteers expressed as area under the curve (AUC) of (*R*)-α-methylhistamine (**12**) was about 100-fold increased after oral application of BP 2.94 (**36**) at a dose of 0.5 mg/kg (≅ 35 mg or 0.1 mmol). Even 24 h after application of this single dose the plasma level of (*R*)-α-methylhistamine (**12**) was observed to be 30 nM, which is approximately ten times higher than the EC_{50} value at H_3 receptors [6].

At present, BP 2.94 (**36**) is under clinical development in Phase II trials for the treatment of asthma, pneumoallergic diseases, and others. Preclinical studies in rodents clearly displayed anti-inflammatory as well as antinociceptive activity of BP 2.94 (**36**) given orally at low doses. These effects are mediated by inhibitory H_3 receptors located on sensory C-fibres in several different tissues. In particular, capsaicin-induced plasma protein extravasation was dose-dependently inhibited in airways, digestive tract, skin, conjunctiva, urinary bladder, nasal mucosa, and dura mater of the rat. In the *p*-phenylbenzoquinone-induced writhing test in mice, BP 2.94 (**36**) had a pronounced antinociceptive activity similar to that of acetylsalicylic acid. This effect was significantly abolished by the H_3 receptor antagonist thioperamide but not by naloxone. Furthermore, BP 2.94 (**36**) reduced zymosan-induced edema. This anti-inflammatory effect was also abolished by thioperamide [6].

The outcome of preclinical studies in rodents furthermore suggests that BP 2.94 (**36**) may also be used as an antiulcer agent, since it was found to prevent gastric mucosal lesions induced by either ethanol or non-steroidal anti-inflammatory drugs, i. e., acetylsalicylic acid and indomethacin [47-49].

Although BP 2.94 (**36**) shows pronounced peripheral activity, it enters the brain only to an insufficient extent. However, the CNS delivery of (*R*)-α-methylhistamine (**12**) at sufficient levels after oral application is a prerequisite for possible central indications like depression or sleep disorders. To optimize the pharmacokinetic profile

Table 8
Pharmacokinetic Data of Diaryl Azomethine Prodrugs of (*R*)-α-Methylhistamine (**12**)

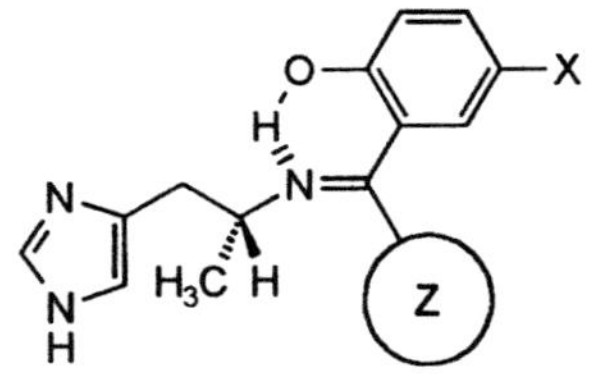

No.	X	Z	Compound	AUC$_{plasma}$ (**12**)[a] (ng × h/ml)	AUC$_{CNS}$ (**12**)[a] (ng × h/g)	Ref.
36	H	Phenyl	BP 2.94/FUB 94	186	–	[44]
37	F	Phenyl	FUB 275	1182	9	[44]
38	F	4-F-Phenyl	FUB 303	296	32	[44]
39	F	4-Cl-Phenyl	FUB 246	387	55	[44]
40	Cl	4-F-Phenyl	FUB 337	230	33	[44]
41	Cl	4-Cl-Phenyl	FUB 338	361	80	[44]
42	F	2-Pyrrolyl	FUB 307	247	97	[45]
43	F	2-Furanyl	FUB 274	481	83	[45]
44	F	3-Furanyl	FUB 353	569	94	[45]
45	F	2-Thienyl	FUB 258	726	56	[45]
46	F	3-Thienyl	FUB 306	649	24	[45]

[a] Free (*R*)-α-methylhistamine (**12**) following *p.o.* application of a defined dose of the respective prodrug equivalent to 24 μmol (**12**)/kg or 3 mg (**12**)/kg.

substitution pattern of the pro-moiety was systematically altered. The obtained novel azomethine prodrugs were orally administered to mice to determine their *in vivo* pharmacokinetic parameters by means of the above-mentioned RIA, which allowed the quantitative determination of (*R*)-α-methylhistamine (**12**) in both plasma and CNS.

In the course of the prodrug development lipophilicity-enhancing and electron-withdrawing halogen substituents were introduced into the benzophenone pro-moiety of BP 2.94 (**37–41**). This derivatization resulted in strikingly increased absorption and CNS delivery of (*R*)-α-methylhistamine (**12**) (Table 8). As expected from the estimated electronic parameters the halogenated prodrugs **34–38** displayed higher levels of liberated (*R*)-α-methylhistamine (**12**) in plasma and CNS compared to the parent prodrug BP 2.94 (**36**). Although BP 2.94 (**36**) reached high plasma levels of (*R*)-α-methylhistamine (**12**), no CNS penetration was detectable by the specific RIA. However, halogenation led to a significantly increased CNS potency, and the data presented in Table 8 illustrate the progressive increase in (*R*)-α-methylhistamine (**12**) levels in the CNS achieved by systematic fluorination and chlorination of the pro-moiety. Finally, the dichloro derivative FUB 338 (**41**) proved to be most effective for central delivery, which is mainly attributed to well balanced stability and lipophilicity. Surprisingly, the monofluorinated prodrug FUB 275 (**37**) reached by far the highest plasma levels of (*R*)-α-methylhistamine (**12**), although its brain penetration was observed to be relatively low. Accordingly, FUB 275 (**37**) may be suitable for peripheral indications like BP 2.94 (**36**), because it was also found to prevent gastric mucosal lesions induced by either ethanol or non-steroidal anti-inflammatory drugs [47,48].

On the basis of these observations, a series of heteroaryl phenyl azomethine prodrugs were developed in which the fluorinated phenolic residue of FUB 275 (**37**) was retained, whereas the second phenyl ring was substituted by several hetero-aromatic cycles. Significant differences of the *in vivo* pharmacokinetics were observed between the electron-withdrawing six-membered heterocycles and the electron-donating five-membered heterocycles (**42–46**), the latter proving to be most effective [45]. In contrast to the former which were found to be far too unstable for oral administration, the five-membered heteroaryl derivatives **42–46** appeared to penetrate easily into the brain and displayed high CNS potency, comparable to that of the halogenated benzophenone imines. Nevertheless, striking differences were observed depending on the heteroaromatic residue and its position of substitution (Table 8). With regard to the CNS delivery of (*R*)-α-methylhistamine (**12**), it is clearly indicated that the pyrrolyl derivative FUB 307 (**42**) as well as the isomeric furanyl imines FUB 274 (**43**) and FUB 353 (**44**) showed the highest CNS levels. Although the 2-thienyl prodrug FUB 258 (**45**) delivered slightly lower CNS levels of (*R*)-α-methylhistamine (**12**) than the furanyl analogues **43** and **44**, it was still more than twice as effective as compared to its 3-thienyl isomer FUB 306 (**46**) [45].

All azomethines showed the same pharmacokinetic profile with a t_{max} value of 0.5 h which drops off to zero within 3 h. However, in contrast to any other azomethine the oral application of the pyrrolyl imine FUB 307 (**42**) led to an incomparably prolonged CNS delivery of (*R*)-α-methylhistamine (**12**) with a moderate C_{max} value.

Thus, FUB 307 (**42**) may be regarded as a kind of 'retard prodrug' and represents the first reported compound of the azomethine prodrug type and particularly the first histamine H_3 receptor agonist to possess such prolonged *in vivo* pharmacokinetics [45].

The ratio of CNS and plasma values of liberated (*R*)-α-methylhistamine (**12**) underlines the suitability of these azomethines to realize the above-mentioned prodrug concept (Figure 3). In conclusion, the pharmacokinetic properties of the prodrugs can be shifted to the desired direction by varying the substitution pattern of the pro-moiety.

As the pharmacokinetic properties of the azomethine prodrugs of (*R*)-α-methylhistamine (**12**) were found to strikingly depend on their physico-chemical properties, it was attempted to quantify the relationship between lipophilicity and CNS penetration of the benzophenone derived imines [50]. The main obstacle to the experimental measurement of the azomethines log *P* values was the competing hydrolysis of the

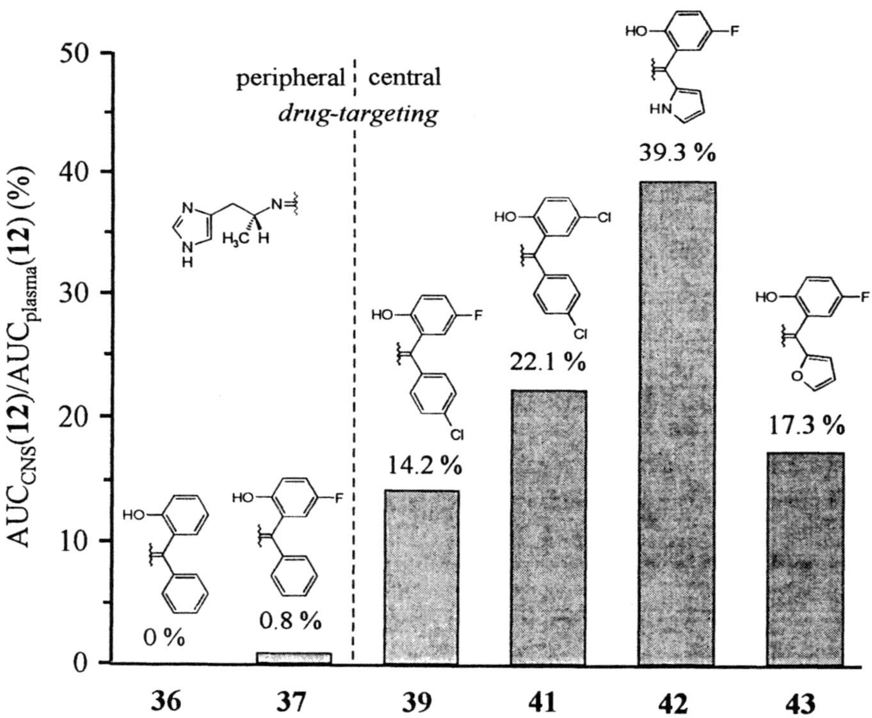

Figure 3: Ratio of CNS and plasma levels of liberated (*R*)-α-methylhistamine (**12**) defined as AUC$_{CNS}$(**12**)/AUC$_{plasma}$(**12**) after *p.o.* application of the respective azomethine prodrugs.

imine double bond. Therefore, the R_f values were determined by means of reversed phase TLC and subsequently converted to R_{m0} values which reflect the theoretical values in pure water. These data correlated well with calculated log *P* values according to Rekker's revised *f*-system. A computer assisted calculation of log *P* led to comparable results thereby proving the excellent correlation of experimentally and theoretically obtained data. It was previously suggested that the relationship between lipid solubility and brain uptake might be a sigmoidal one, and this was also observed with the more hydrophobic compounds lying on the more linear portion of the curve. The final linear regression analysis of the logarithm of the brain penetration indices, which were defined as the ratio of prodrug in the CNS to the AUC of prodrug in the plasma, versus the calculated log *P* values resulted in a significant correlation. These results clearly point out the relationship between lipophilicity and brain uptake of diphenyl azomethine prodrugs of (*R*)-α-methylhistamine (**12**), and they furthermore suggest that these prodrugs enter the brain mainly by passive diffusion [50].

4. SUMMARY

To date there is a number of structurally different histamine H_3 receptor agonists available which display high potency as well as distinct selectivity. Depending on the nature of the pharmacological experiment either (*R*)-α-methylhistamine (**12**) or imetit (**24**) are used as agonists of choice, although in some tests immepip (**30**) or other H_3 receptor agonists may have some advantages. At present we know only very little about the molecular interaction of histamine with its H_3 receptor. However, when the amino acid sequence will be known in the future, the small number of molecular modelling studies to clarify the mechanism of receptor activation will presumably increase, particularly if we take into account the results of nonaminergic partial agonists. Among the various agonists described so far, the most promising approach to achieve therapeutic application seems to be the orally active prodrugs of (*R*)-α-methylhistamine (**12**), of which one, BP 2.94 (**36**), is already under investigation in clinical phase II trials. The outcome of these studies will thus offer a clear perspective of possible therapeutic indications and the potential market of a H_3 receptor agonist development.

REFERENCES

1. J.-M. Arrang, M. Garbarg and J.-C. Schwartz, Nature (London), 102 (1983) 832.
2. J.-M. Arrang, M. Garbarg and J.-C. Schwartz, Neuroscience, 23 (1987) 149.
3. J. L. Burgaud and N. Oudart, J. Pharm. Pharmacol., 45 (1993) 955.

4. E. Schlicker, B. Malinowska, M. Kathmann and M. Göthert, Fundam. Clin. Pharmacol., 8 (1994) 128.
5. J.-C. Schwartz et al., in The Histamine H_3 Receptor as Target for New Drugs, R. Leurs and H. Timmerman (eds.), Elsevier Science Publishers B. V., Amsterdam, 1998.
6. A. Rouleau, M. Garbarg, X. Ligneau, C. Mantion, P. Lavie, C. Advenier, J.-M. Lecomte, M. Krause, H. Stark, W. Schunack and J.-C. Schwartz, J. Pharmacol. Exp. Ther., 281 (1997) 1085.
7. R. Leurs, R. C. Vollinga and H. Timmerman in Progress in Drug Research, B. Jucker (ed.), Birkhäuser Verlag, Basel, 1992, pp 127–165.
8. R. Lipp, H. Stark and W. Schunack in The Histamine Receptor, Receptor Biochemistry and Methodology, J.-C. Schwartz and H. L. Haas (eds.), Wiley Liss, Inc., New York, 1992, pp 57–72.
9. J.-M. Arrang, M. Garbarg, T. T. Quach, M. Dam Trung Tuong, E. Yeramian and J.-C. Schwartz, Eur. J. Pharmacol., 111 (1985) 73.
10. M. Kathmann, E. Schlicker and M. Göthert, Psychopharmacology, 116 (1994) 464.
11. S. Gobbi, W. Menge and H. Timmerman, Poster presented at the 11th Noordwijkerhout-Camerino Symposium 1997, Noordwijkerhout, The Netherlands.
12. C. R. Ganellin, personal communication.
13. D. G. Cooper, R. C. Young, G. J. Durant, and C. R. Ganellin in Comprehensive Medicinal Chemistry: the Rational Design, Mechanistic Study & Therapeutic Application of Chemical Compounds, C. Hansch (ed.), Pergamon Press, Oxford, U.K., 1990, pp 323–421.
14. J.-C. Schwartz, J.-M. Arrang, M. Garbarg and W. Schunack in Innovative Approaches in Drug Research, A. F. Harms (ed.), Elsevier Science Publishers B. V., Amsterdam, 1986, pp 73–89.
15. J.-M. Arrang, M. Garbarg, J.-C. Lancelot, J.-M. Lecomte, H. Pollard, M. Robba, W. Schunack and J.-C. Schwartz, Nature (London), 327 (1987) 117.
16. G. Gerhard and W. Schunack, Arch. Pharm. (Weinheim), 313 (1980) 709.
17. R. Lipp, J.-M. Arrang, M. Garbarg, P. Luger, J.-C. Schwartz and W. Schunack, J. Med. Chem., 35 (1992) 4434.
18. R. J. Friary, P. Mangiaracina, M. Nafissi, S. O. Orlando, S. Rosenhouse, V. A. Seidl and N.-Y. Shih, Tetrahedron, 49 (1993) 1993.
19. J.-M. Arrang, J.-C. Schwartz and W. Schunack, Eur. J. Med. Pharmacol., 117 (1985) 109.
20. R. Lipp, H. Stark, J.-M. Arrang, M. Garbarg, J.-C. Schwartz and W. Schunack, Eur. J. Med. Chem., 30 (1995) 219.

21. J.-M. Arrang, M. Garbarg, W. Schunack, J.-C. Schwartz and R. O. Lipp, European Patent Application 0 214 058 (1987).
22. N.-Y. Shih, A. T. Lupo, R. Aslanian, S. Orlando, J. J. Piwinski, M. J. Green, A. K. Ganguly, M. A. Clark, S. Tozzi, W. Kreutner and J. A. Hey, J. Med. Chem., 38 (1995) 1593.
23. N.-Y. Shih, R. Aslanian, A. T. Lupo, L. Duguma, S. Orlando, J. J. Piwinski, M. J. Green, A. K. Ganguly, M. Clark, S. Tozzi, W. Kreutner and J. A. Hey, Poster presented at New Perspectives in Histamine Research 1994, Winnipeg, Manitoba, Canada.
24. I. J. P. de Esch, W. M. P. B. Menge, P. H. J. Nederkorn and H. Timmerman, Poster presented at the 11th Noordwijkerhout-Camerino Symposium 1997, Noordwijkerhout, The Netherlands.
25. M. A. Khan, S. L. Yates, C. E. Tedford, K. Kirschbaum and J. G. Phillips, submitted.
26. L. B. Hough, J. K. Khandelwal and T. W. Mittag, Agents Actions, 11 (1981) 425.
27. H. van der Goot, M. J. P. Schepers, G. J. Sterk and H. Timmerman, Eur. J. Med. Chem., 27 (1992) 511.
28. W. Howson, M. E. Parsons, P. Raval and G. T. G. Swayne, Bioorg. Med. Chem. Lett., 2 (1992) 77.
29. C. R. Ganellin, B. Bang-Andersen, Y. S. Khalaf, W. Tertiuk, J.-M. Arrang, M. Garbarg, X. Ligneau, A. Rouleau and J.-C. Schwartz, Bioorg. Med. Chem. Lett., 2 (1992) 1231.
30. R. C. Vollinga, W. M. P. B. Menge, R. Leurs and H. Timmerman, J. Med. Chem., 38 (1995) 266.
31. N.-Y. Shih, R. Aslanian, A. Lupo, J. J. Piwinski, M. J. Green and A. K. Ganguly, Imidazolyl or Imidazolylalkyl Substituted with a Four or Five Membered Nitrogen Containing Heterocyclic Ring, PCT Int. Patent Appl. WO 93/12108 (1993).
32. N.-Y. Shih and M. J. Green, Imidazolylalkyl Substituted with a Six Membered Nitrogen Containing Heterocyclic Ring, PCT Int. Patent Appl. WO 93/12107 (1993).
33. R. C. Vollinga, J. P. de Koning, F. P. Jansen, R. Leurs, W. M. P. B. Menge and H. Timmerman, J. Med. Chem., 37 (1994) 332.
34. N.-Y. Shih, Imidazolyl-Alkyl-Piperazine and -Diazepine Derivatives as Histamine H_3 Agonists/Antagonists, PCT Int. Patent Appl. WO 93/12093 (1993).
35. W. Sippl, H. Stark, H.-D. Höltje, Quant. Struct.-Act. Relat., 14 (1995) 121.
36. R. Leurs, M. Kathmann, R. C. Vollinga, W. M. P. B. Menge, E. Schlicker and H. Timmerman, J. Pharmacol. Exp. Ther., 276 (1996) 1009.
37. E. Schlicker, M. Kathmann, H. Bitschnau, I. Marr, S. Reidemeister, H. Stark and W. Schunack, Naunyn-Schmiedeberg's Arch. Pharmacol., 353 (1996) 482.

38. G. F. Watt, D. A. Sykes, S. P. Roberts, N. P. Shankley and J. W. Black, Poster presented at the Meeting of the British Pharmacological Society 1997, Edinburgh, UK.
39. J.-C. Schwartz, J.-M. Arrang, M. Garbarg, A. Quemener, J.-M. Lecomte, X. Ligneau, W. G. Schunack, H. Stark, K. Purand, A. Hüls, S. Reidemeister, S. Athmani, C. R. Ganellin, N. Pelloux-Leon and W. Tertiuk, FR Patent FR 2 732 017 - A1 (1995).
40. J.-M. Arrang, M. Garbarg, J.-C. Schwartz, R. Lipp, H. Stark, W. Schunack and J.-M. Lecomte, Agents Actions, Suppl., 33 (1991) 55.
41. X. Ligneau, M. Garbarg, M. L. Vizuete, J. Diaz, K. Purand, H. Stark, W. Schunack and J.-C. Schwartz, J. Pharmacol. Exp. Ther., 271 (1994) 452.
42. S. Yamazaki, E. Sakurai, N. Hikichi, N. Sakai, K. Maeyama and T. Watanabe, J. Pharm. Pharmacol., 46 (1994) 371.
43. J.-C. Schwartz, H. Pollard, S. Bischoff, M. C. Rehault and M. Verdière-Sahuque, Eur. J. Pharmacol., 16 (1971) 326.
44. M. Krause, A. Rouleau, H. Stark, P. Luger, R. Lipp, M. Garbarg, J.-C. Schwartz and W. Schunack, J. Med. Chem., 38 (1995) 4070.
45. M. Krause, A. Rouleau, H. Stark, P. Luger, M. Garbarg, J.-C. Schwartz and W. Schunack, Arch. Pharm. Pharm. Med. Chem., 329 (1996) 209.
46. M. Krause, A. Rouleau, H. Stark, M. Garbarg, J.-C. Schwartz and W. Schunack, Pharmazie, 51 (1996) 720.
47. G. Morini, D. Grandi and G. Bertaccini, Digestion, 56 (1995) 145.
48. G. Morini, D. Grandi, M. Krause and W. Schunack, Inflamm. Res., 46 Suppl. 1 (1997) S101.
49. M. Belcheva, K. Marazova, V. Lozeva and W. Schunack, Inflamm. Res., 46 Suppl. 1 (1997) S113.
50. M. Krause, A. Rouleau, H. Stark, M. Garbarg, J.-C. Schwartz and W. Schunack, Sci. Pharm., 64 (1996) 503.

R. Leurs and H. Timmerman (Editors)
The Histamine H_3 Receptor

Medicinal Chemistry of Histamine H3 Receptor Antagonists

James G. Phillips and Syed M. Ali
Gliatech Inc., 23420 Commerce Park Road, Cleveland, Ohio 44122

1. INTRODUCTION

Since the disclosure of thioperamide **1** (K_i = 4 nM) as a potent and selective histamine H_3 receptor antagonist [1], there has been a large number of 4(5)-substituted imidazole derivatives prepared and evaluated for their H_3 receptor affinity. Most of the efforts directed towards the design of new H_3 antagonists have

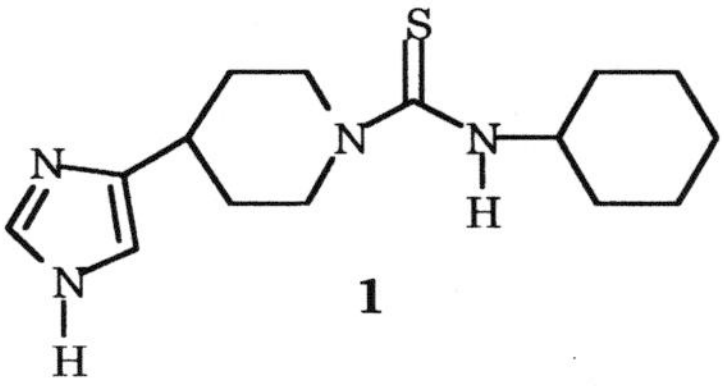

1

been summarized recently by Leurs et al. [2] and Stark et al. [3]. Prominent H_3 receptor antagonists that have emerged from these studies are clobenpropit **2** (K_i = 0.5 nM) [4], iodoproxyfan **3** (K_i = 5 nM) [5], GR 175737 **4** (K_i = 6.3 nM) [6], and GT-2016 **5** (K_i = 40 nM) [7] (Figure 1). Despite all of the efforts to date, no H_3

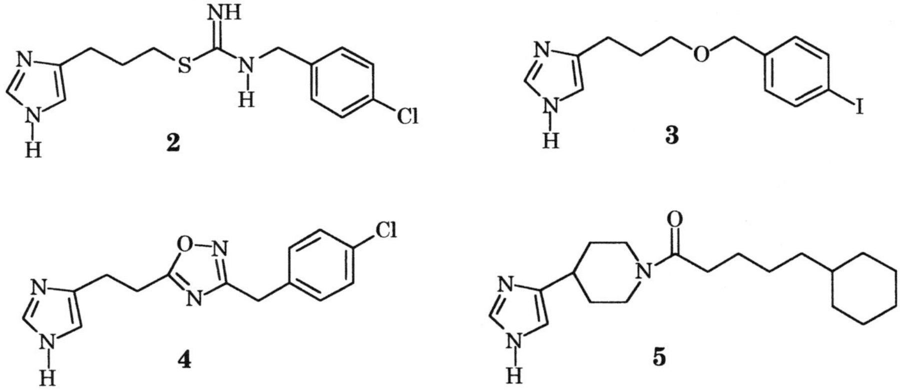

Figure 1. Prominent histamine H_3 receptor antagonists

antagonist has reached clinical trial evaluation. However, several improved clinical candidates have emerged and are progressing towards human clinical trials. The design of clinically useful H_3 antagonists must reckon with the limitations of earlier compounds. Some of the more potent H_3 receptor antagonists did not possess adequate ability to cross the blood-brain barrier [8]. Other compounds with good H_3 affinity displayed H_2 activity as well [9]. Comparison of pharmacological data of new compounds must also take into consideration that the previous classification of some known H_3 ligands as pure antagonists has come under closer scrutiny. For example, the radiolabeled ligand iodoproxyfan **3**, originally reported as a H_3 antagonist, has very recently been disclosed as behaving as a partial agonist in mouse brain cortex and guinea-pig ileum [10]. It will be the intent of this review to briefly summarize the medicinal chemistry efforts prior to 1995, and provide an update on the recent progress being made.

2. EARLY MEDICINAL CHEMISTRY OF H_3 ANTAGONISTS

Several strategies have been employed in carrying out SAR studies leading to the development of selective and potent H_3 receptor antagonists. The discovery of thioperamide **1** comes from synthetic efforts that were initiated with the knowledge that some of the known H_2 receptor ligands (impromidine **6** (K_i = 63 nM) and burimamide **7** (K_i = 63 nM)) displayed significant H_3 antagonist activity (Figure 2). In seeking a specific antagonist of histamine at cerebral H_3 receptors able to cross the blood-brain barrier, SAR studies leading to thioperamide were designed to incorporate some of the structural features of **6** and **7**, yet reduce their poor CNS penetration and eliminate H_2 activity. Thus, 4-[4(5)-imidazolyl]piperidine **8** was

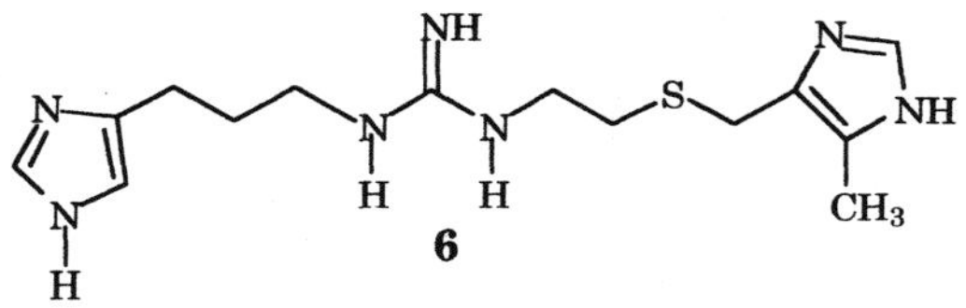
6

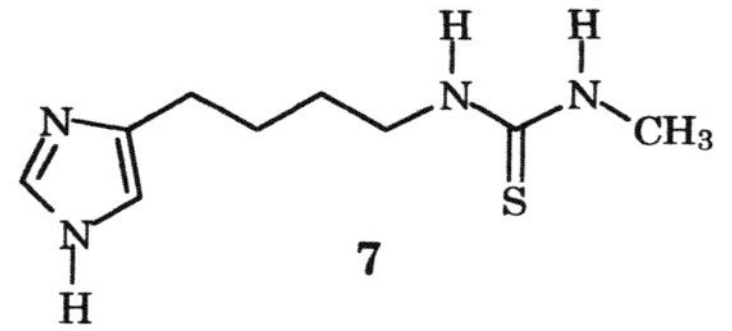
7

Figure 2. H_2 Receptor ligands with significant H_3 receptor affinity

employed as an integral building block and lead optimization studies resulted in selection of N-cyclohexylcarbothioamide substitution [11] (Figure 3). SAR efforts were subsequently directed toward the development of new H_3 receptor antagonists using **8** as a synthetic intermediate and thioperamide as a template. The emphasis of these studies has been on exploring the viability of replacing the thiourea moiety of thioperamide and obtaining good blood-brain barrier penetration. It has been reported that some thiourea containing compounds, for example, metiamide, have been associated with toxic side effects [12]. 4-[4(5)-Imidazolyl]piperidine-amide derivatives **9** (Figure 4) prepared by the Schwartz group [13] and our group [14] were the focus of early studies to circumvent the limitations of thioperamide for drug development. GT-2016 **5** is the most well characterized prototype amide

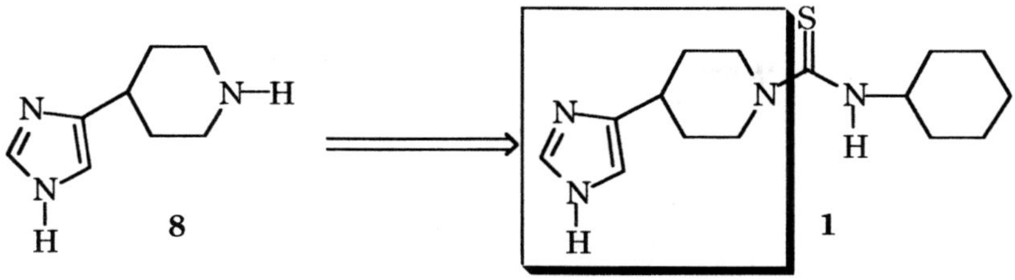

Figure 3. Early SAR studies of H_3 antagonists leading to discovery of thioperamide

analog developed, and exhibits high affinity and selectivity for the histamine H_3 receptor, as well as excellent blood-brain barrier penetration, and drug induced increases in histamine release in the cerebral cortex [7]. 4-[4(5)-Imidazolyl]piperidine containing urea, guanidine, amidine, and carbamate analogs have been mentioned in the patent literature, but very little information other than their receptor binding affinities is available [2,3,11,13]. The tritium labeled S-

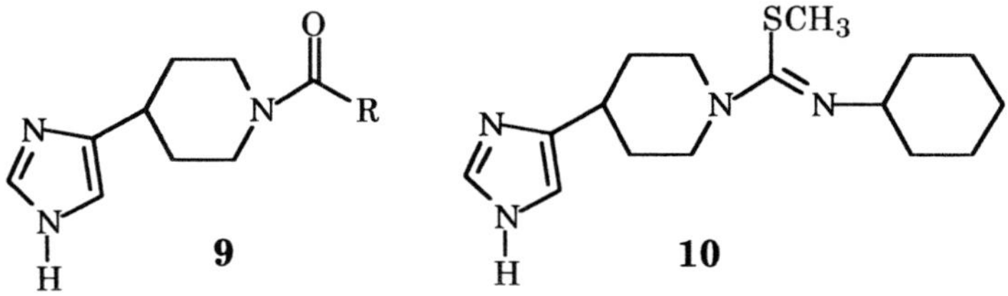

Figure 4. Early derivatives prepared to circumvent the limitations of thioperamide

methylated derivative of thioperamide **10** (K_D = 2.1 nM) [15] has been described as an H_3 antagonist, but has been shown to have high affinity for non-H_3 receptors in peripheral tissues (Figure 4).

Many of the early SAR efforts concerning the preparation of new H_3 antagonists have made use of the readily available 4(5)-substituted imidazole containing scaffolds: histamine, homohistamine, or their isothiourea or alcohol analogs. These synthetic intermediates have provided additional opportunities for the preparation of structurally diverse H_3 antagonists. SAR studies with these intermediates have focused their attention on optimizing receptor affinity by studying the influence of chain length between the imidazole ring and a polar spacer group as well as the chain length between the spacer group and a lipophilic ring. In this regard, Lipp et al. [16] prepared amides and amines of histamine. This strategy was continued using homohistamine or homologs of histamine as the scaffold, but studied the influence of amides, guanidines, thioamides, and carbamates as polar spacer groups. Some of the potent H_3 antagonists made using this approach are the disubstituted guanidine **11** (K_i = 0.74 nM) [17] and the homohistamine amide **12** (K_i = 5 nM) [18] (Figure 5). In similar fashion, 4-(2-hydroxyethyl)imidazole and 4-(3-hydroxypropyl)imidazole have been used to prepare potent carbamate containing H_3 antagonists. A representative of this series is carbamate **13** (K_i = 4 nM) which has good CNS penetration [19] (Figure 5). Van der Goot et al. [4] have made productive

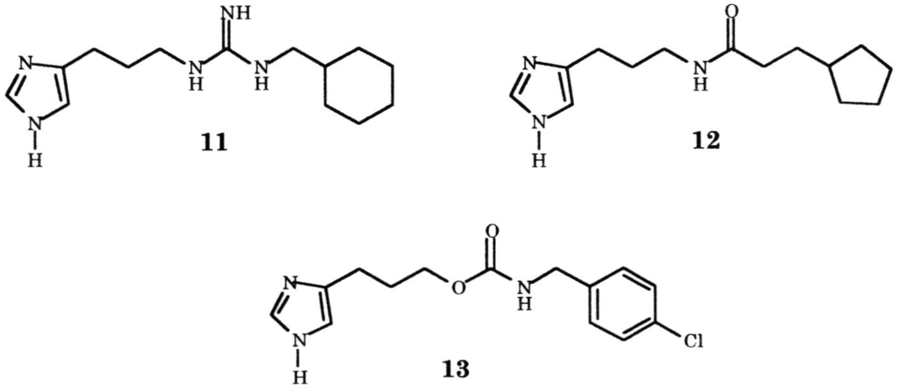

Figure 5. H_3 antagonists using histamine, homohistamine or analogous alcohol scaffolds and polar spacer groups

use of the scaffold represented by the known H_3 agonist, imetit **14**. Their studies examined the influence of chain length between the imidazole ring and an isothiourea or guanidine spacer as well as the chain length between those spacer groups and various lipophilic rings. These efforts led to the discovery of a new series of antagonists exemplified by clobenpropit **2** and the ^{125}I radiolabeled ligand, iodophenpropit **15** [20] (figure 6).

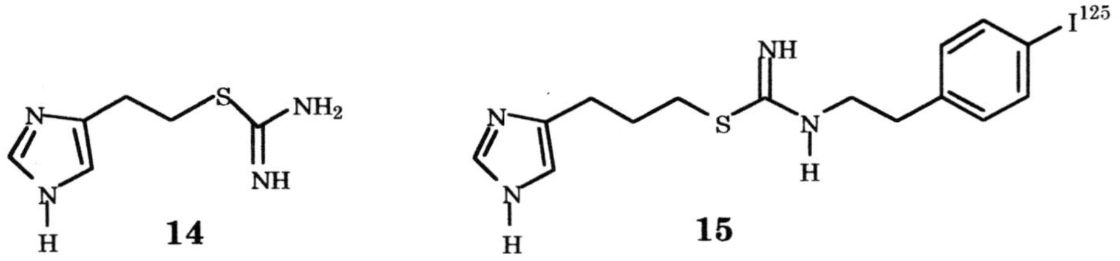

Figure 6. H_3 antagonists from derivatization of the H_3 agonist, imetit

In 1993 our group at Gliatech initiated SAR studies directed towards the development of new H_3 antagonists using 4-[4(5)-imidazolyl]piperidine as a scaffold, but used GT-2016 as a template [21]. These efforts were an attempt to further delineate some of the structural features of this antagonist that are important for potent H_3 receptor binding activity. Several prior studies had emphasized the influence of chain length between the imidazole ring and a polar spacer as well as the chain length between those spacer groups and various lipophilic rings. Our studies demonstrated that several factors besides the distances of the imidazole head group to the polar spacer and polar spacer to lipophilic tail group (cyclohexyl in these cases) are involved in determining the level of binding activity. Table 1 shows the binding data and distances for several derivatives of GT-2016.

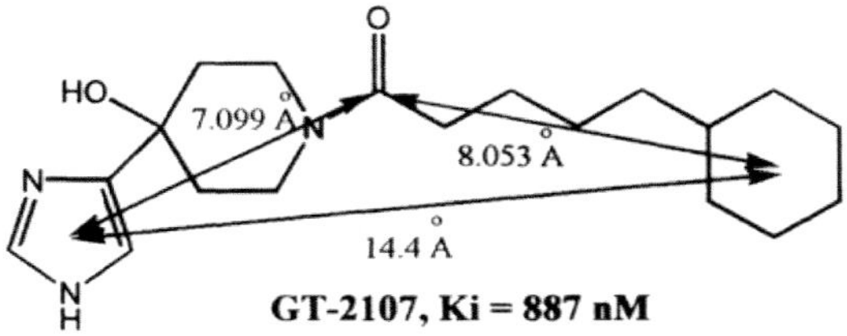

GT-2107, Ki = 887 nM

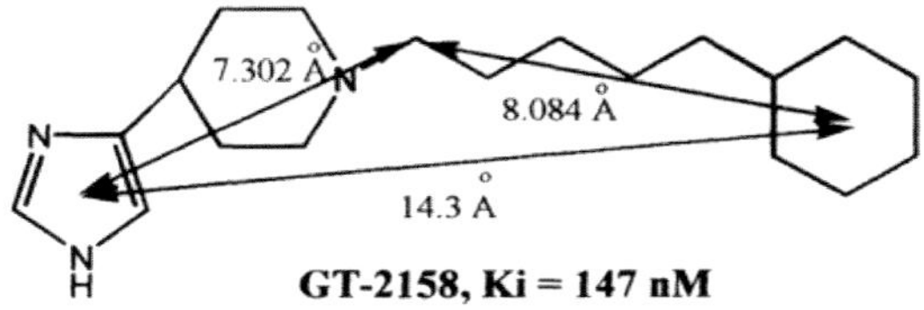

GT-2158, Ki = 147 nM

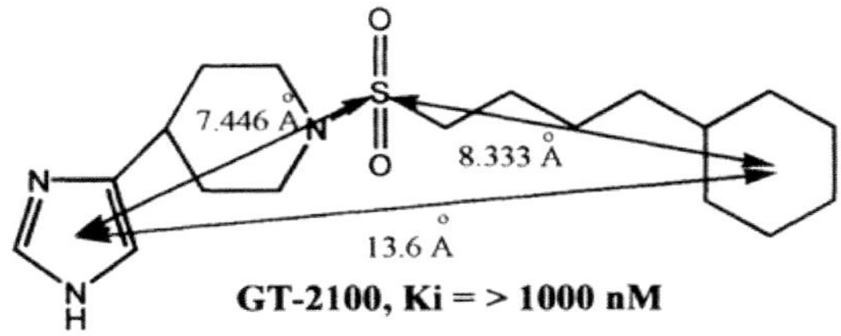

GT-2100, Ki = > 1000 nM

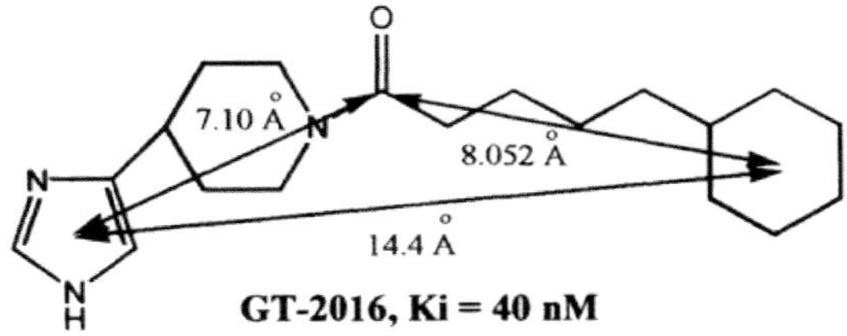

GT-2016, Ki = 40 nM

Table 1. Histamine H_3 binding data for GT-2016 derivatives

Figure 7 shows an overlay of the energy minimized structures for each of the compounds in Table 1. GT-2016, 2107, and 2158 all show good overlay of their energy minimized structures. They also exhibit comparable distances between the imidazole head group and the cyclohexyl tail, as well as comparable distances between the head group and spacer, and spacer and tail. Yet, the binding activities for GT-2016 and 2158 are distinctly better than GT-2107. Apparently, the possibility of a strong hydrogen bonding interaction between the hydroxy group on the piperidine ring of GT-2107 and the nitrogen of its imidazole head group is detrimental to ligand-receptor interaction. Moreover, the planar amide functionality of GT-2016 provides higher affinity than the more flexible tertiary

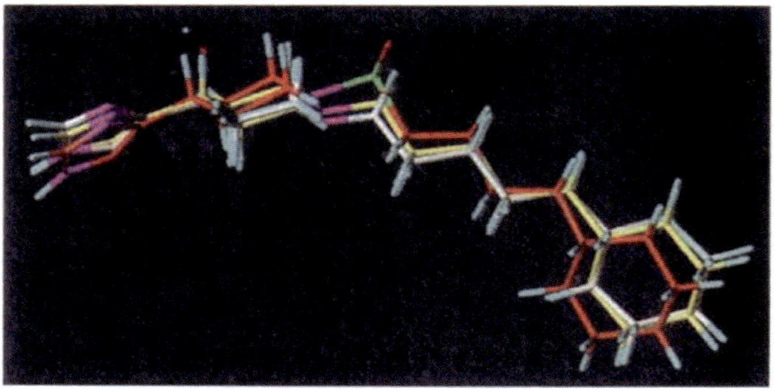

Figure 7. Overlay of energy minimized structures of GT-2016, 2100, 2107 and 2158

amine of GT-2158. The sulphonamide GT-2100 shows poorer overlap with the other 3 congeners and much weaker affinity. However, the differences of distances between head and spacer, and spacer and tail are not enough to explain differences in binding affinity. There is a bit of difference between imidazole head and lipophilic tail for GT-2100 (13.6 ang.) vs the other 3 analogs (14.3-14.4 ang.). It is important to distinguish that the sulfur atom of the polar sulphonamide functionality in GT-2100 is hybridized trigonal bipyramidal, whereas the carbonyl of the amide group of GT-2016 and 2107 is hybridized sp^2 and the carbon of the amine of GT-2158 is sp^3.

Two different research groups have investigated the replacement of the N-cyclohexylcarbothioamide portion of thioperamide with an aromatic nitrogen containing heterocycle. Successful use of this synthetic strategy in which an NH-R group (R = aromatic nitrogen containing heterocycle) served as a thiourea or urea equivalent had been employed in the design and development of brain-penetrating H_2 receptor antagonists [22]. Noteworthy contributions to the H_3 receptor field from this perspective are UCL 1283 **16** (K_i = 42 nM) [23] and the benzothiazole derivative **17** [24] (Figure 8).

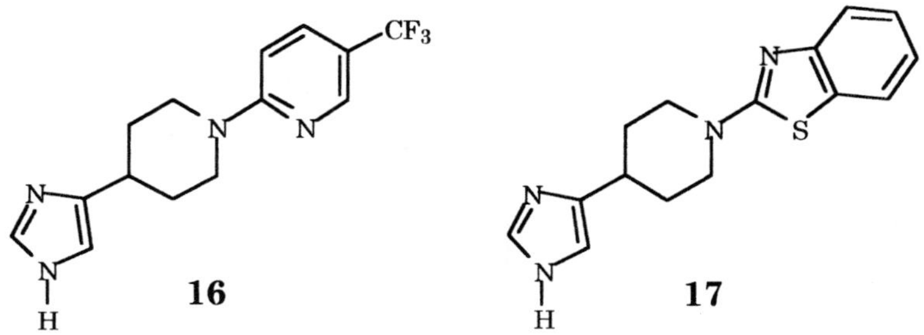

Figure 8. H_3 antagonists with the NH-R group "urea equivalent"

Ganellin et al. [23] have combined histamine or its sulfur analog with nitrogen containing heterocycles ("urea equivalent" strategy) to prepare a unique series of potent H_3 antagonists related to UCL 1283 **16**. Examples of compounds from these efforts are the pyridine **18** (K_i = 17 nM) and UCL 1199 **19** (K_i = 4.8 nM) (Figure 9).

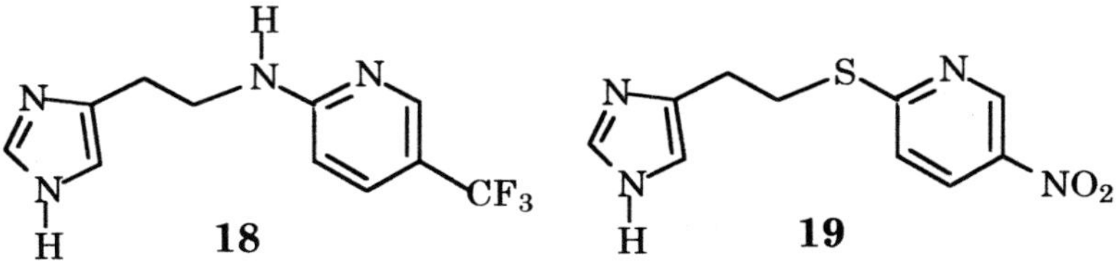

Figure 9. H_3 antagonists derived from histamine-like scaffolds and substituted nitrogen containing aromatic heterocycles.

Shih and coworkers have disclosed a new series of H_3 antagonists in which they have replaced the $-CH_2CH_2S-$ linker of clobenpropit **2** with a *para*-substituted phenyl ring. Only patent information has appeared regarding this series [25]. An example of this series is methanimidamide **20** (K_i = 7.2 nM) (Figure 10).

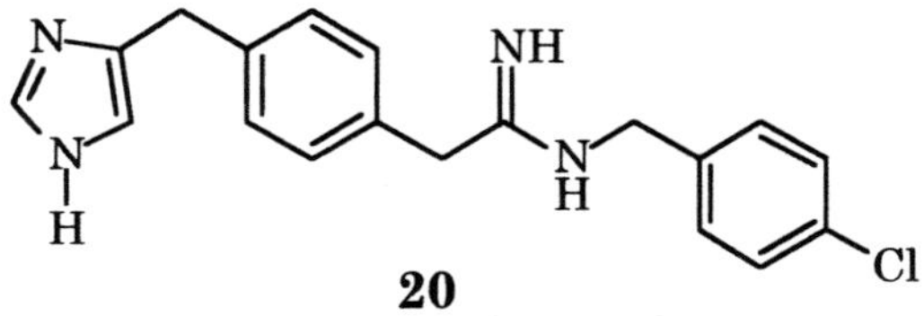

20

Figure 10. H_3 antagonists with a 4-(methylphenyl)imidazole scaffold

Glaxo scientists have reported on a new series of H_3 antagonists that contain a 1,2,4-oxadiazole ring as a bioisostere equivalent of the isothiourea functionality of clobenpropit. The most publicized representative of this series, GR 17537 **4** (K_i = 7.9 nM) [6] (figure 2), shows significantly better CNS penetrability than clobenpropit.

Timmerman's group has reported that certain homologs of histamine behave as potent and selective H_3 antagonists [26]. The 4(5)-(ω-aminoalkyl)-1*H*-imidazole derivative, impentamine **21** (pA_2 = 8.4) with a 5-carbon chain between the imidazole head group and the terminal primary amino group was the most potent compound of the series (Figure 11). However, it has been subsequently shown that impentamine behaves as a pure antagonist in the guinea-pig jejunum, but as a partial agonist in mouse brain cortex [27].

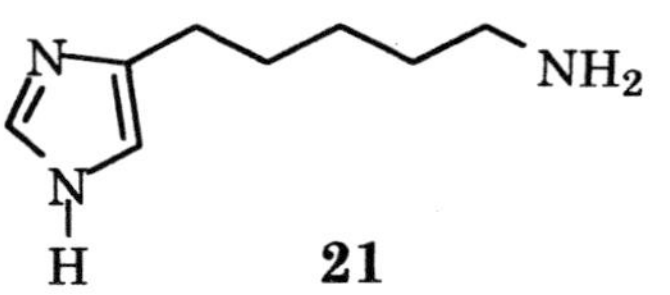

21

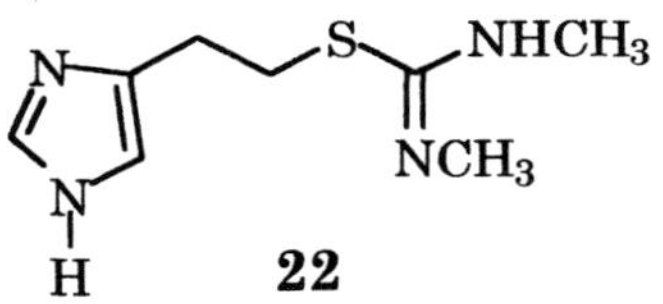

22

Figure 11. Amines and isothioureas that are H_3 antagonists

Ganellin and co-workers have also used the imetit template to prepare H_3 antagonists. The di-N-methyl derivative **22** exhibits a K_i = 51 nM (Figure 11) [28].

3. RECENT MEDICINAL CHEMISTRY OF H_3 ANTAGONISTS

Much of the more recent medicinal chemistry efforts devoted towards the development of new H_3 antagonists provide improvements over earlier studies. The availability of ^{125}I radiolabeled ligand analogs of iodoproxyfan and iodophenpropit, and advances in H_3 pharmacology, have served to intensify efforts directed towards selection of an H_3 antagonist clinical candidate. For example, Ganellin and co-workers have recently disclosed a new series of phenoxyalkyl imidazoles that are ether analogs of UCL 1199 **19** and UCL 1283 **16**. These derivatives are prepared from the readily synthesized imidazole scaffolds, 4-(2-hydroxyethyl)imidazole or 4-(3-hydroxypropyl)imidazole. Potent examples of this series [29] are UCL 1390 **23** (K_i = 12 nM) and UCL 1409 **24** (K_i = 14 nM) (Figure 12). These derivatives are reported to possess ED_{50} values that are sub 1 mg/kg. At the same time, a series of (phenylalkyloxy)propyl imidazoles were disclosed as potent H_3 antagonists [5]. The most important compound of this class is iodoproxyfan **3**. These ether derivatives are reported to possess excellent oral bioavailability, as well as preferred pharmacodynamic and pharmacokinetic profiles. However, recently Schlicker et al. [10], as well as Black's group [30], have suggested that iodoproxyfan elicits an

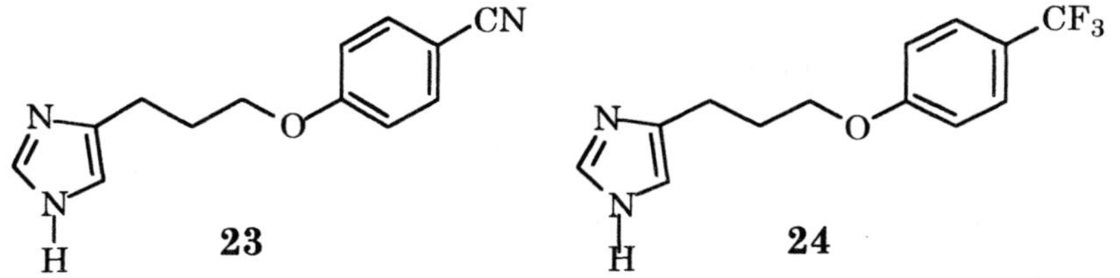
23
24

Figure 12. Phenoxyalkyl imidazole H_3 antagonists

agonist response in a selective H_3 receptor bioassay of guinea pig ileum. Other substituted aromatic derivatives of iodoproxyfan elicited agonist responses of varying magnitude that appeared to be related to the nature of the substituent in the 4-position of the aromatic ring. Conformational analysis of these compounds led to the proposal that the gradual loss of agonism through the series was associated with an increased preference for adopting folded conformations because of a possible π-stacking interaction between the imidazole ring and the remote aromatic ring. Other non aromatic ether derivatives of this series behave as full antagonists.

Mor et al. [31] have reported a QSAR study on a series of *para*- and *meta*-substituted 4(5)-phenyl-2-[[2-[4(5)-imidazolyl]ethyl]thio]imidazoles in which the carbothioamide fragment of thioperamide **1** has been replaced by a substituted imidazole ring, and the piperidine ring has been replaced by a $-CH_2CH_2S-$ linker. Potent representatives of these studies are the thiolimidazoles **25** (K_i = 10 nM) and **26** (K_i = 3.1 nM) shown in figure 13. The objective of these studies was to obtain information for optimizing the pharmocokinetic properties such as protein binding and CNS penetration of polar group containing H_3 antagonists.

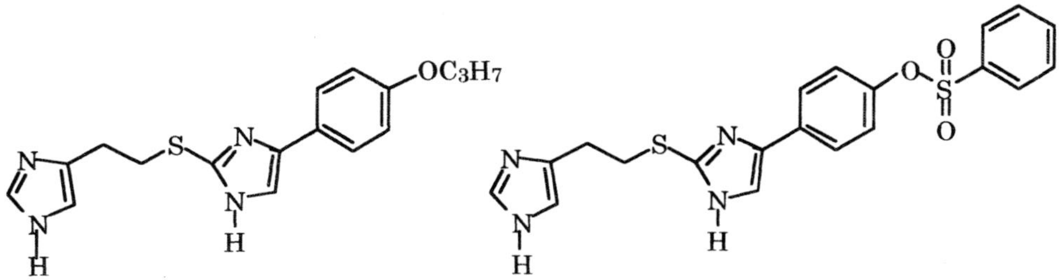
25
26

Figure 13. New thiolimidazole H_3 receptor antagonists

Other scaffolds and spacer groups continue to be used effectively for the synthesis of new H_3 antagonists. *In vitro* and *in vivo* data on a series of oxygen-containing H_3 antagonists that includes straight chain esters, ketones, and alcohols, as well as a series of straight chain amine containing compounds that includes amides, thioamides, ureas, and thioureas has been described [32-33]. The ketone **27** (K_i = 23 nM) which exhibits an oral ED_{50} = 3.5 mg/kg and the urea **28** (K_i = 8 nM) (Figure 14) are of particular interest (Figure 14).

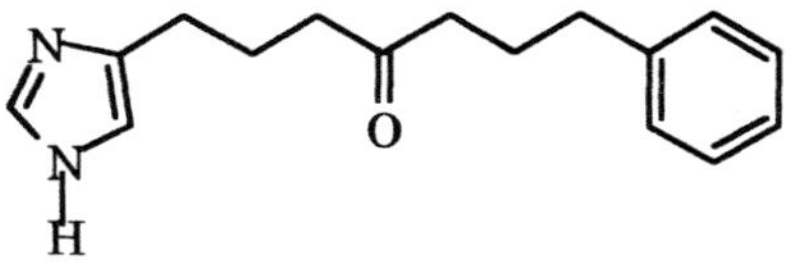

27

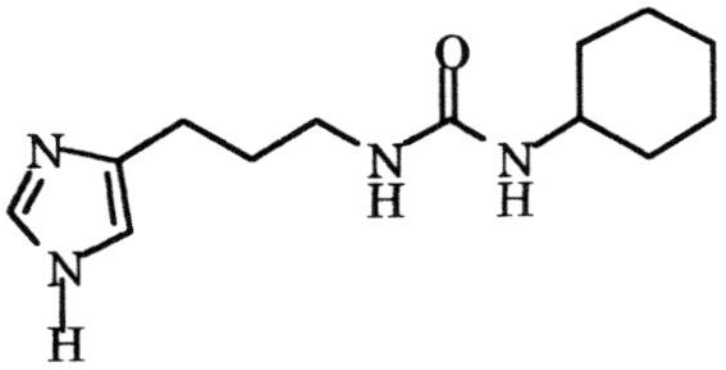

28

Figure 14. Representative acyclic oxygen and amine containing H_3 receptor antagonists

Studies directed towards the design and development of new and potent H_3 receptor antagonists have revealed three essential features required for good binding affinity: imidazole head group, spacer, and hydrophobic tail group. A variety of polar spacer groups such as amide, thioamide, guanidine, urea, thiourea, ester, and carbamate had been investigated in synthesizing potent H_3 antagonists. These studies have in large part focused their efforts by making structural modifications of the reference antagonist **1**.

Recent studies at Gliatech directed towards the development of new H_3 antagonist ligands began with the recognition that verongamine **29** (figure 15) possesses a rather unique template. Verongamine isolated from the marine sponge, *Verongula gigantea*, has been reported as an histamine H_3 receptor antagonist with an IC_{50} of 0.5 μM [34]. Viewing the structural features of **29** from the perspective of previous studies, it was easy to recognize that it contains an imidazole head group, a polar and planar amide-oxime spacer group, and an aromatic hydrophobic tail. However, consideration of these features from a conformational and perhaps a stereochemical viewpoint provided us with the impetus to embark on medicinal chemistry SAR studies that we felt would add some new insights to the H_3 receptor field [35]. It had already been demonstrated that potent and selective histamine H_3 receptor agonists possess distinct stereochemical features [36-37]. None of the reported H_3 receptor antagonists exhibit any stereochemical presentations. Furthermore, no studies concerning the effect of a stereochemical feature on antagonist binding had been reported. First, we considered that **29** contains a 2-carbon straight chain between a 4(5)-substituted imidazole head group and the amide-oxime spacer (figure 15). We envisioned that replacement of these first two carbons of the 2-carbon straight chain of **29** with a *trans* cyclopropane ring directly attached to the imidazole ring at the 4(5)-position would produce conformationally restricted analogs. At the same time, this cyclopropane incorporation would introduce the chiral element of cyclopropyl ring orientation. Next, we contemplated that verongamine is most likely derived biosynthetically from the coupling of histamine with tyrosine. The histamine-tyrosine amide **30** (figure 16) derived

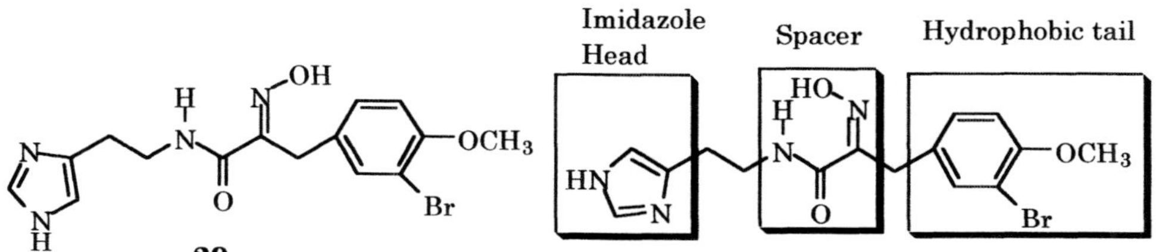

29

Figure 15. Verongamine, the only natural product disclosed as an H_3 antagonist

from this premise has a chiral amino substituent. We anticipated that the amide-amine moiety of **30** or an olefin-amine isostere as shown in **31** (figure 16) could serve as functional equivalents of the amide-oxime array of **29**. It is well established from efforts in the preparation of peptide mimetics that a *trans* olefin functional group can serve as an isostere of an amide bond. We felt that the chiral

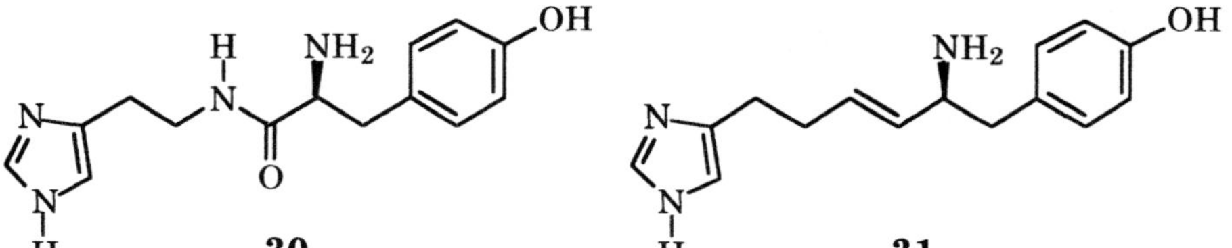

30 **31**

Figure 16. Hypothetical chiral analogs of Verongamine

amino substituent might provide additional conformational alignment with the H_3 receptor through its potential ability to interact with the receptor via a hydrogen bonding or an ionic interaction. Certainly, we also considered that these analogs could be prepared with diligence from available amino acids. For completeness, we also decided to investigate the functional equivalence of the *trans* and *cis* olefin spacers, as well as the acetylene (Figure 17).

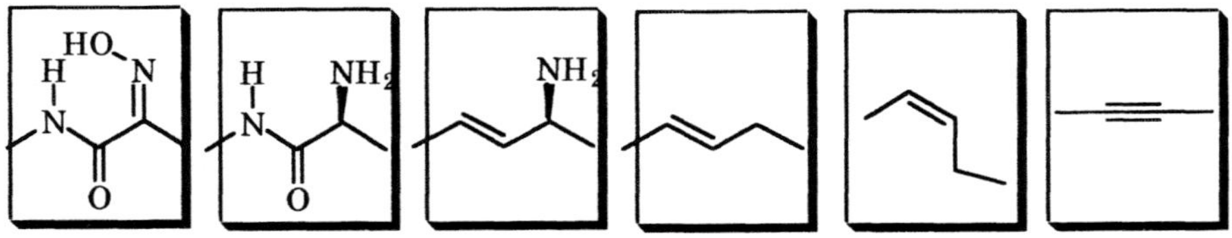

Figure 17. New spacers considered for use in synthesis of H_3 antagonists

Thus, our studies examined the effect of these incorporations on H_3 receptor affinity by making ligands that could be envisioned synthetically by modifying the two structural features A and B of **29** illustrated in figure 18: A: 2-carbon straight chain vs *trans* cyclopropyl ring and B: amide-oxime array vs the other spacers depicted in figure 17. For the purpose of these SAR studies, we used the cyclohexyl or phenyl tail which were available from commercially available (S)-N-Boc-cyclohexylalanine or (S)-N-Boc-phenylalanine.

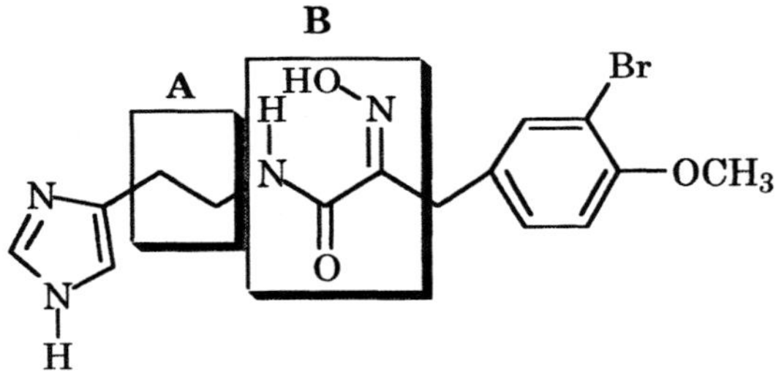

Figure 18. SAR studies of new H_3 antagonists using Verongamine template

The implementation of our strategies to prepare new and potent H_3 receptor antagonists first examined the effect of the chiral amino substituent and olefin incorporation (feature B) on binding affinity. Table 2 shows the binding data for several of these amino containing derivatives. The compounds containing the natural configuration of the chiral amino substituent were all more potent than verongamine. GT-2231 which contains both the chiral (*S*) amino substituent and the *trans* olefin isostere was approximately 500 times more potent than the natural product. The reduced affinity of GT-2136 (D-amino derivative) demonstrated the importance of the amino functionality in the ligand-receptor interaction.

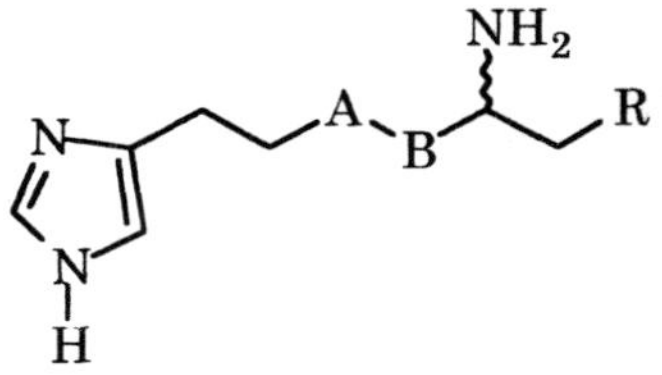

GT #	R	Chirality of Amino Group	A—B	K_i (nM)
2130	Phenyl	S	N—C(O)	104± 14.0
2136	Phenyl	R	N—C(O)	2418±112
2140	Cyclohexyl	S	N—C(O)	30.8 ± 2
2231	Cyclohexyl	S	Trans C=C	1 ± 0.1

Table 2. New H_3 antagonists containing a chiral amino substituent

The cyclopropyl analogs of interest to us were prepared from either the acids **32** and **33** or the cyclopropylamines **34** and **35** [38]. The *n*-butyl esters of **32** and **33** were separated by chiral column technology. The absolute configuration of the

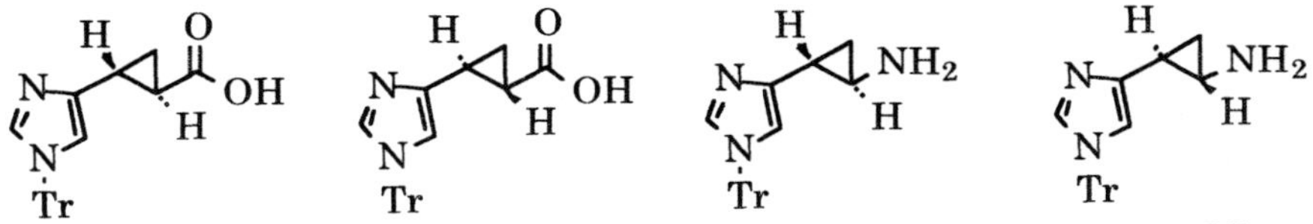

32 **33** **34** **35**

Figure 19. Key intermediates for preparation of cyclopropyl H_3 antagonists.

cyclopropyl ring of these intermediates was established from X-ray crystallographic studies of a derivative prepared by a new diastereoselective cyclopropanation method [39]. An ORTEP diagram of the derivative **36** is shown in figure 20.

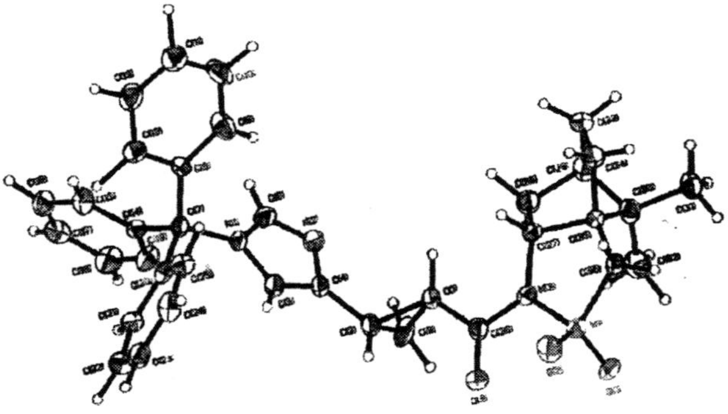

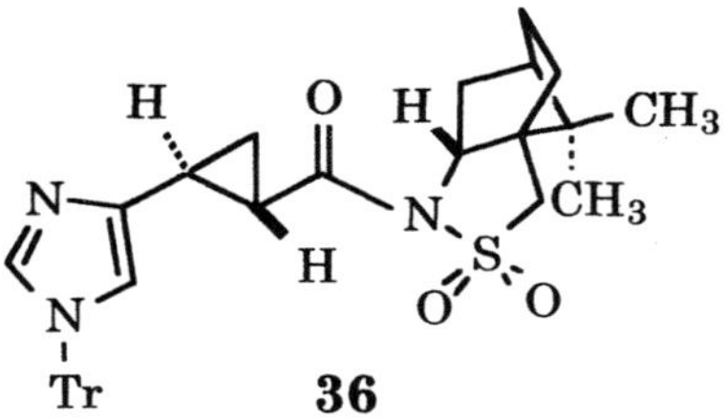

36

Figure 20. ORTEP diagram of derivative **36** used to establish cyclopropane ring stereochemistry of new H_3 antagonists

The cyclopropyl analogs of GT-2140 (Table 1), GT-2163 **37** and GT-2164 **38** (figure 21), were prepared by coupling **34** and **35** with (S)-N-BOC-cyclohexylalanine followed by deprotection. The binding affinities obtained for these derivatives

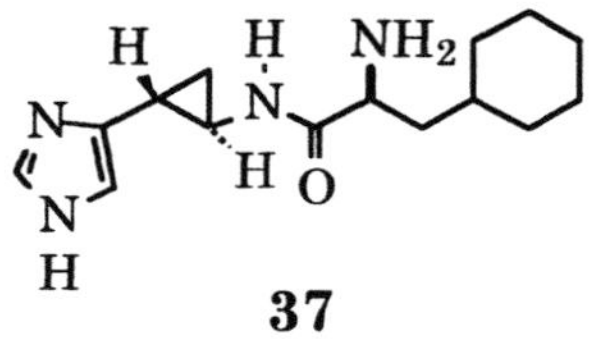

37

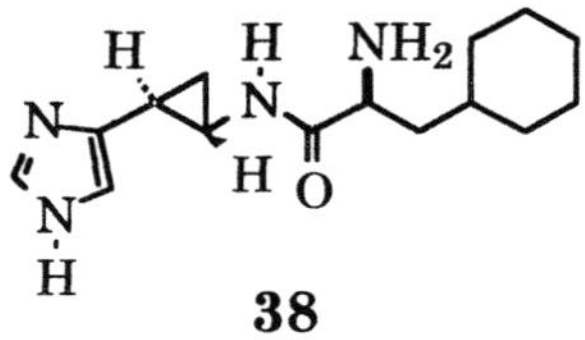

38

Figure 21. Cyclopropylamide-amino H_3 antagonists

established that there is a preference for the (1*R*,2*R*) configuration of the cyclopropane ring. Furthermore, there was an order of magnitude greater activity for the (1*R*,2*R*) cyclopropane compound in comparison to its straight chain congener (Table 3).

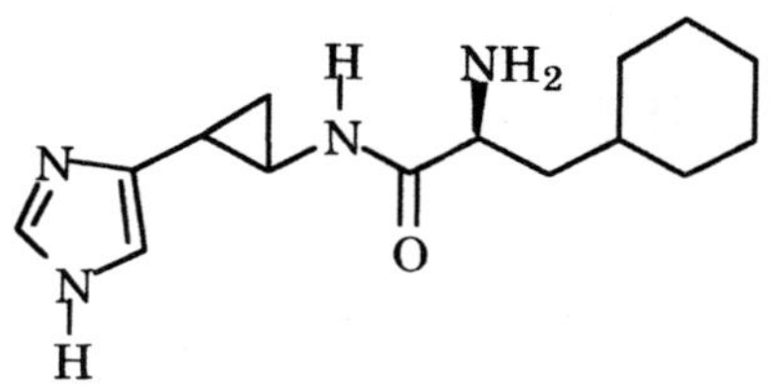

GT #	Cyclopropane configuration	K_i (nM)
2140	None	30.8 ± 2
2163	1*R*, 2*R*	1.85 ± 0.5
2164	1*S*, 2*S*	21.5 ± 1.8

Table 3. Demonstration of stereo preference of cyclopropylamide-amino H_3 antagonists

Cyclopropane compounds containing the olefin isostere replacement for the amide bond were prepared using Julia olefination chemistry. Aldehydes **39** and **40** were obtained by $LiAlH_4$ reduction of the chiral *n*-butyl esters of **32** and **33**, respectively, followed by swern oxidation of the corresponding alcohols (Figure 22). Condensation of the (S)-N-BOC-cyclohexylalanine sulfone **41** with aldehyde **39** gave after treatment with 2% Na(Hg) and deprotection, the *trans* and *cis* olefin-amines

39 **40**

Figure 22. Intermediates for synthesis of cyclopropyl-olefin-amine H_3 antagonists

42 and **43** in a ratio of 7:3, respectively (Scheme 1). GT-2232 **42** with the (1*R*, 2*R*) configuration of the cyclopropane ring, the *trans* olefin orientation, and the (*S*) configuration of the primary amino substituent is one of the most potent H_3 antagonists ever prepared. Replacement of the amide bond with the *trans* olefin isostere in this series increased H_3 binding affinity by almost an order of magnitude (Table 4).

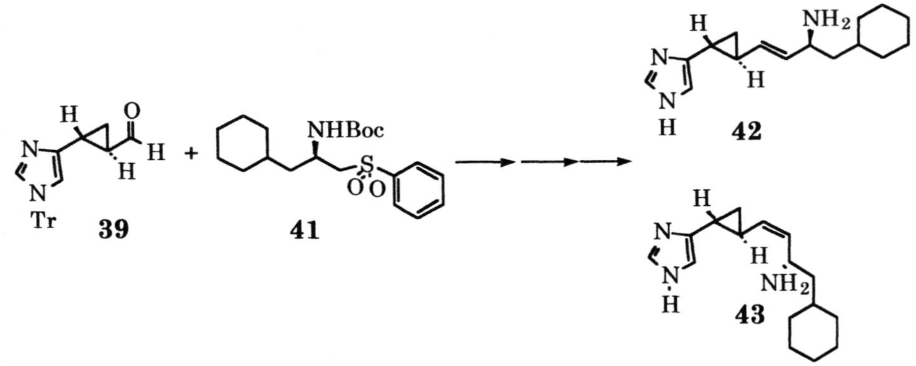

Scheme 1. Synthesis of cyclopropyl-olefin-amine H_3 antagonists

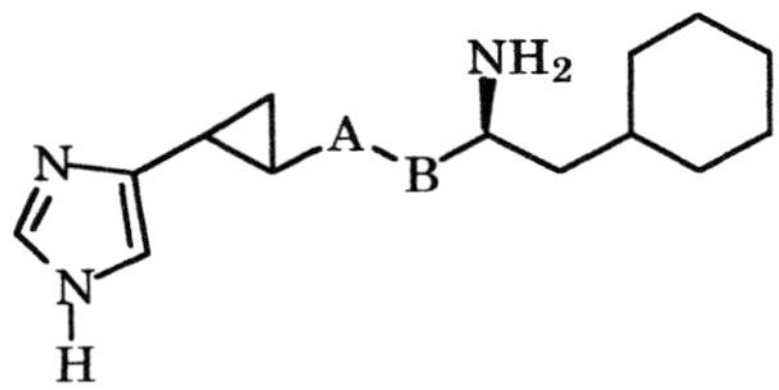

GT #	Cyclopropane configuration	A—B	K_i (nM)
2163	1*R*, 2*R*	Amide	1.85 ± 0.5
2232	1*R*, 2*R*	Trans olefin	0.37 ± 0.2
2252	1*R*, 2*R*	Cis olefin	97.7± 28

Table 4. Cyclopropane-chiral amino H_3 antagonists

The synthesis of cyclopropyl compounds containing the olefin replacement for the amide bond but without the additional primary chiral amino substituent were prepared by modified Julia olefination procedures. Addition of the benzothiazole sulfone **44** to aldehyde **39** gave trityl protected olefins in a 1:1 ratio. These derivatives were separated by HPLC and subsequently deprotected to provide cyclopropyl olefins **45** (GT-2208) and **46** (GT-2209) (scheme 2). In similar fashion,

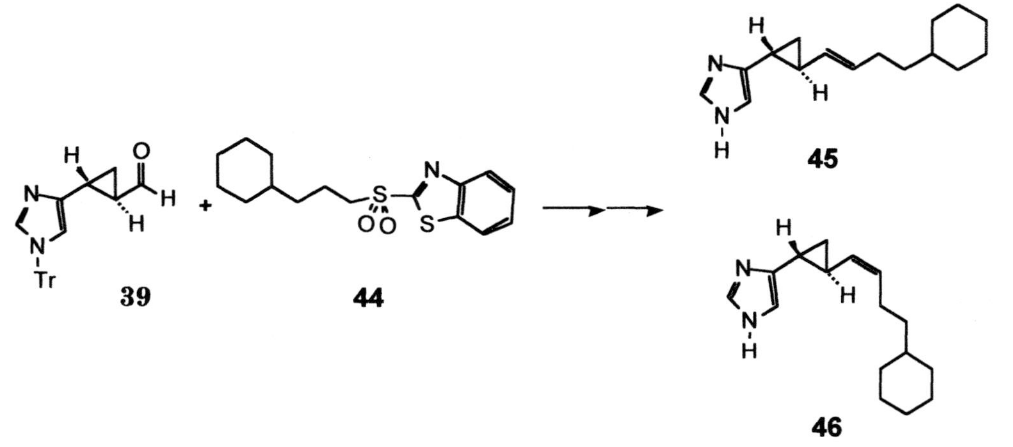

Scheme 2. Synthesis of cyclopropyl-olefin H_3 antagonists with the (1*R*,2*R*) configuration

aldehyde **40** was converted to cyclopropyl olefins **47** (GT-2207) and **48** (GT-2210) (figure 23). Table 5 shows the binding data for these four analogs. The derivatives with the (1*R*, 2*R*) cyclopropyl configuration are an order of magnitude more potent than their (1*S*, 2*S*) analogs. However, derivatives **45** and **46** are yet an order of magnitude less potent than GT-2232 which contains the additional structural feature of a chiral amino group which is five bonds away from the imidazole functionality.

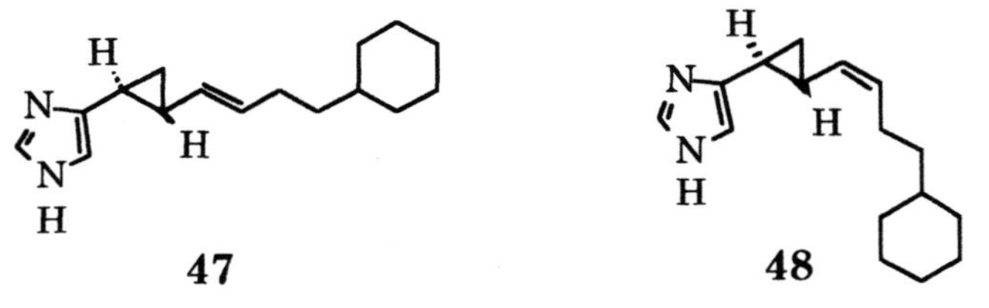

Figure 23. Cyclopropyl-olefin H_3 antagonists with the (1*S*, 2*S*) configuration

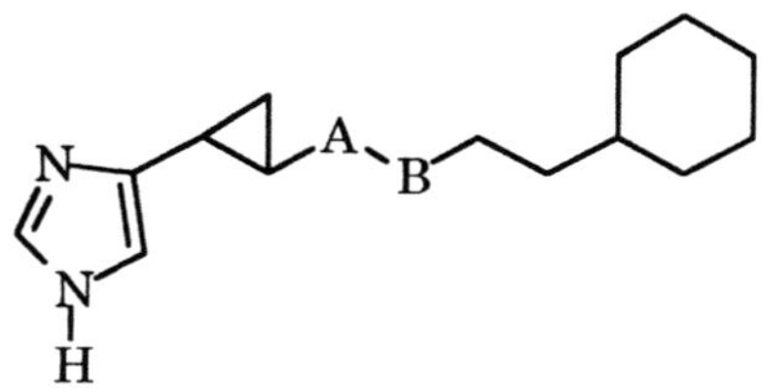

GT #	Cyclopropane configuration	A—B	K_i (nM)
2208	1*R*, 2*R*	Trans olefin	2.4 ± 0.2
2209	1*R*, 2*R*	Cis olefin	6.2 ± 0.8
2207	1*S*, 2*S*	Trans olefin	56 ± 9
2210	1*S*, 2*S*	Cis olefin	108 ± 3.9

Table 5. Binding data for cyclopropyl-olefin H_3 antagonists

The straight chain congeners of GT-2208 and GT-2209 were synthesized from aldehyde **49** which is readily available from urocanic acid. Wittig chemistry provides the *cis* olefin **50** from **49** (Figure 24). The *trans* olefin **51** is obtained from Na (lig. NH_3) reduction of acetylene **52** (Figure 24).

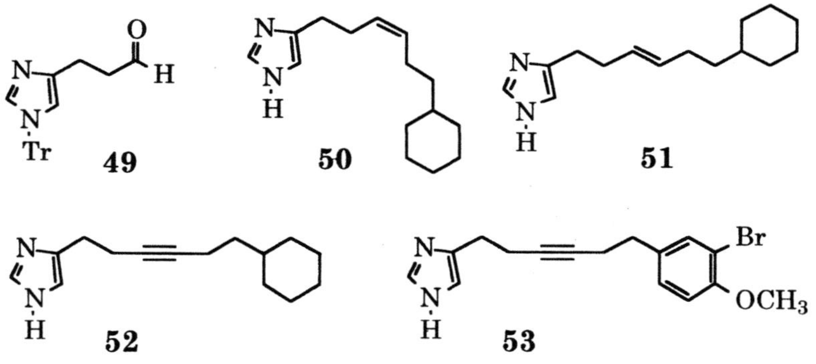

Figure 24. Straight chain scaffold that provides olefin and acetylene H_3 antagonists

The binding activities exhibited for these straight chain compounds are not substantially different than their (1*R*, 2*R*) cyclopropyl analogs (Table 6). It appears that the conformational restriction imparted by the (1*R*, 2*R*) cyclopropyl ring provides about a factor of 2 increase in binding activity when the chiral amino substituent is not included. Most importantly, it is the particular combination of three features: (1*R*, 2*R*) cyclopropyl ring configuration, trans olefin geometrical orientation, and (*S*) stereochemistry of an additional primary amino substituent

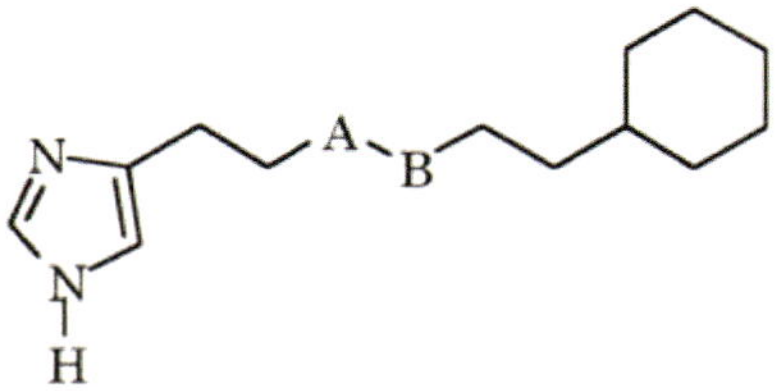

GT #	A—B	K_i (nM)
2228	Trans olefin	15.2 ± 2.4
2227	Cis olefin	4.2 ± 0.6
2260	Acetylene	2.9± 0.2

Table 6. Binding data for straight chain olefin or acetylene H_3 antagonists

that affords the best ligand-receptor interaction (GT-2232). Figure 25 shows the overlay of the energy minimized conformers of GT-2130, 2140, 2163, and 2232. The distances from the imidazole ring to the lipophilic tail of all four compounds is not significantly different (11.438 ang- 11.507 ang.) and suggests that subtle conformational effects are important in the ligand-receptor interaction.

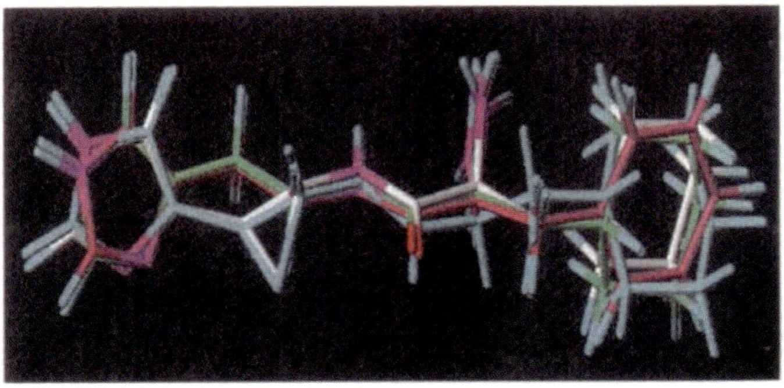

Figure 25. Overlay of the H_3 receptor antagonists: GT-2130, 2140, 2163, and 2232.

The binding data obtained for acetylene **52** encouraged us to pursue a more detailed study of acetylene-based H_3 antagonists [40]. Molecular modeling studies performed with **29** and the acetylene congener **53** (figure 24) show a good overlay of their energy minimized conformations (Figure 26). The best overlay of these two structures was obtained when a 2-carbon linker between the acetylene moiety and the aromatic hydrophobic tail was employed for **53**. Employing a Topliss operational scheme for aliphatic side chain substitution [41], we evaluated ligands having the

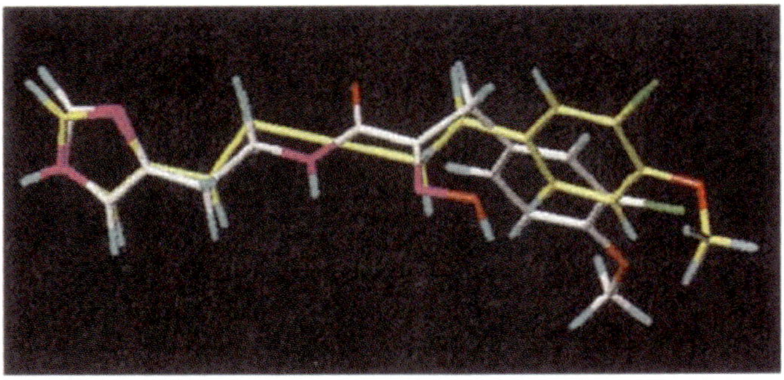

Figure 26. Overlay of energy minimized conformations of **29** and **53**.

general structure **54** which were prepared using the acetylene derivative **55** as the base compound (Figure 27).

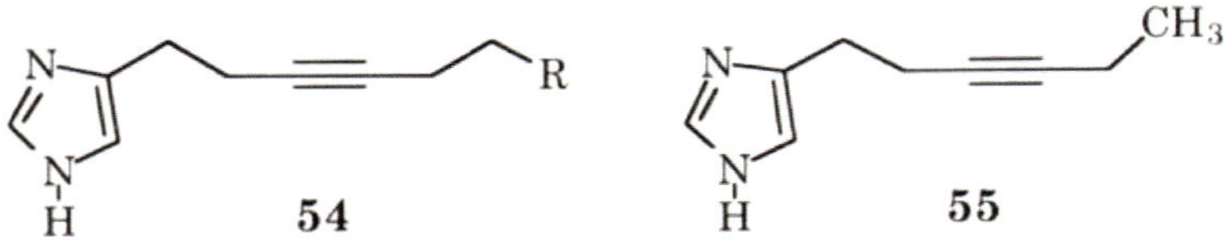

54 **55**

Figure 27. Studies of acetylenes as H_3 antagonists

56 **54**

Scheme 4. Synthesis of acetylene H_3 antagonists

Several acetylene derivatives were prepared by alkylation of the terminal acetylene **56** with the requisite alkyl iodides followed by trityl deprotection (Scheme

4). Table 7 shows the H_3 binding data of the acetylenes prepared in the study. The acetylenes GT-2293 and GT-2286 were the most potent compounds prepared. GT-2286 crosses the blood-brain barrier efficiently, exhibiting a 50% H_3 cortical receptor occupancy level at 60 minutes with an ip dose of between 0.3-1.0 mg/kg in rats.

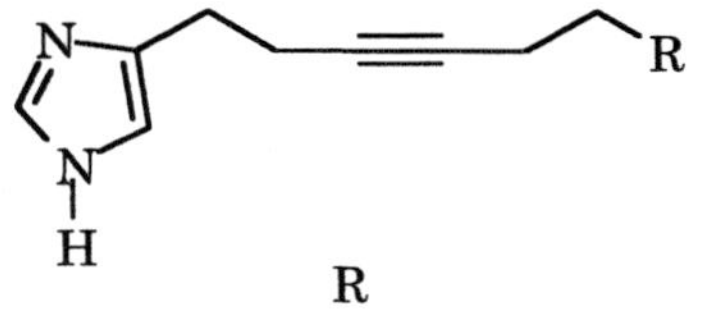

GT #	R	K_i (nM)
2228	H	79 ± 22
2283	CH_3	27 ± 7.5
2284	*i*-C_3H_7	3.7 ± 1.1
2286	cyclo-C_5H_9	0.95± 0.3
2260	cyclo-C_6H_{11}	2.9± 0.2
2330	$(CH_2)_4Ph$	3.5 ± 1.0
2293	*tert*-butyl	0.8 ± 0.04

Table 7. Binding data for acetylene based H_3 antagonists

The (1*R*, 2*R*) cyclopropyl derivatives of all of our compounds were consistently more potent than their straight chain analogs. We therefore prepared the (1*R*, 2*R*) derivative GT-2331 and the (1*S*, 2*S*) analog GT-2342 (Figure 28). GT-2331 is one of the most potent H_3 receptor antagonists ever prepared (K_i = 0.15 ± 0.04 nM), whereas GT-2342 has a K_i = 5.3 ± 0.5 nM [42].

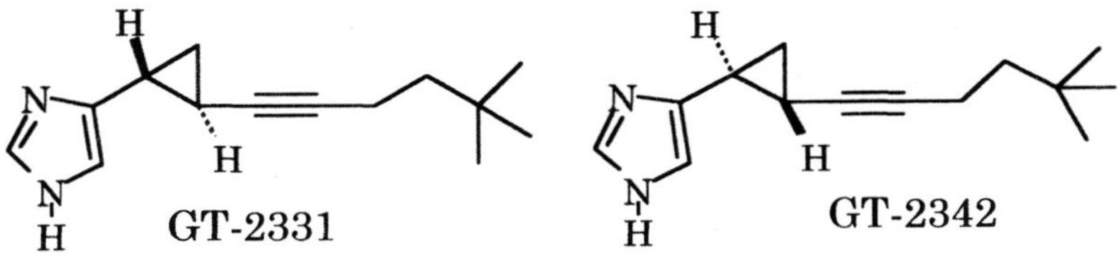

GT-2331 GT-2342

Figure 28. New cyclopropyl-acetylene H_3 antagonists

5. MEDICINAL CHEMISTRY OF H_3 RECEPTOR ANTAGONISTS-FUTURE DIRECTIONS

There are several important issues that are presently being addressed in order that an H_3 receptor antagonist be more confidently advanced for clinical evaluation. West et al. [43] have shown that two H_3 receptor antagonists, thioperamide and burimamide could identify two affinity sites termed H_{3A} and H_{3B}. Although, these data implicate possible subtypes of the H_3 receptor, interpretation of these data is confounded by the use of an agonist radioligand. It has not been determined whether these two sites represent two affinity states or two distinct receptor sites. Furthermore, most of the binding affinity data reported in the literature for H_3 antagonists has been determined using agonist radioligands, i.e. [^{3}H]N$^{\alpha}$-methylhistamine. Therefore, the development of radiolabeled H_3 receptor antagonists has become a priority. The radiolabeled ligands, [^{125}I]-iodophenpropit, [^{125}I]-iodoproxyfan, and [^{3}H]-S-methylthioperamide, all have some limitations. Recently, Glaxo has reported a new radiolabeled H_3 receptor antagonist [^{3}H] GR-168320 **57** (K_d = 0.12 nM) [44] (Figure 29). Saturation studies with this ligand suggested the involvement of a single binding site. Moreover, both agonist and antagonist affinities of several known H_3 and H_2 ligands correlated well when their affinities were compared with affinities obtained using [^{3}H]N$^{\alpha}$- methylhistamine. Several of the new potent and selective H_3 ligands described would be suitable for radiolabelling.

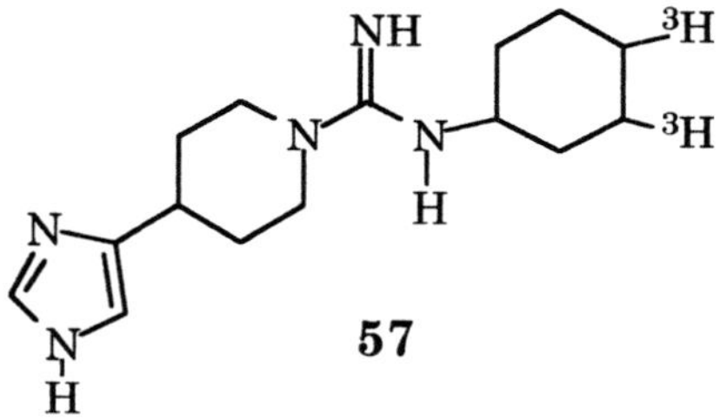

57

Figure 29. New radiolabeled H_3 receptor antagonist ligands

Attempts are also being made to prepare or discover through screening, non-imidazole containing H_3 receptor antagonists. It has been disclosed that the anti-psychotic clozapine **58** (Figure 30) exhibited high affinity for the H_3 receptor (K_i = 236 nM) in competitive binding experiments using the H_3 antagonist [^{125} I]-iodophenpropit [45]. Previously, it was thought that the 4(5)-substituted-1*H*-imidazole moiety was required for high affinity [46].

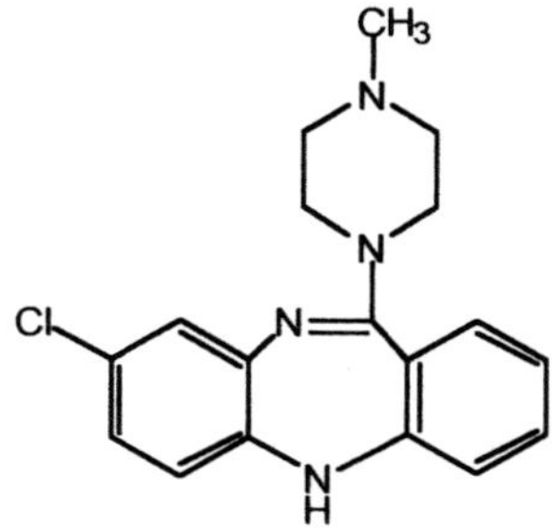

58

Figure 30. Non-imidazole containing H_3 receptor antagonist (Clozapine)

In conclusion, dramatic progress has been made in the development of new and improved H_3 receptor antagonists. These advances provide better tools to establish the pharmacology of the H_3 receptor. Moreover, several of these new agents are under clinical development and should provide for a new class of therapeutics.

REFERENCES

1. Arrang, J. M.; Garbarg, M.; Lancelot, J. C.; Lecomte, J. M.; Pollard, H.; Robba, M. F.; Schunack, W.; Schwartz, J. C. *Nature* **1987**, *327*, 117.
2. Leurs, R.; Vollinga, R. C.; Timmerman, H. *Progress in Drug Research*, Jucker, E., Ed.; Birkhauser Verlag: 1995; pp. 107-165.
3. Stark, H.; Schlicker, E.; Schunack, W. *Drugs of the Future* **1996**, 21(5), 507-520.
4. (a) Van der Goot, H.; Schepers, M. J. P.; Sterk, G. J.; Timmerman, H. *Eur. J. Med.Chem.* **1992**, 27, 511. (b) Barnes, J. C.; Brown, J. D.; Clarke, N. P.; Clapham, J.; Evans, D. J.; O'Shaughnessy, C. T. *Eur. J. Pharmacol.* **1993**, 250, 147-152. (c) Miyazaki, S.; Onodera, K.; Imaizumi, M.; Timmerman, H. *Life Sciences* **1997**, 61 (4), 355-361.
5. (a) Stark, H.; Purand, K.; Huls, A.; Ligneau, X.; Garbarg, M.; Schwartz, J. C.; Schunack, W. *J. Med. Chem.* **1996**, 39, 1220-6. (b) Ligneau, X.; Garbarg, M.; Vizuete, M. L.; Diaz, J.; Purand, K.; Stark, H.; Schunack, W.; Schwartz, J. C. *J. Pharm. Exp. Ther.* **1994**, 271, 452-459.
6. Clitherow, J. W.; Beswick, P.; Irving, W. J.; Scopes, D.; Barnes, J. C.; Clapham, J.; Brown, J.; Evans, D.; Hayes, A.G. *Abstracts of Papers,* 10th Camerino-Noordwijkerhout Symposium On Perspectives In Receptor Research, Noordwijkerhout, The Netherlands; 1995.
7. Tedford, C. E.; Yates, S. L.; Pawlowski, G. P.; Nalwalk, J. W.; Hough, L. B.; Khan, M. A.; Phillips, J. G.; Durant, G. J.; Frederickson, R. C. A. *J. Pharm. Exp. Ther.* **1995**, *275*, 598.

8. Tedford, C. E.; Phillips, J. G.; Khan, M. A.; Frederickson, R. C. A.; Leurs, R.; Timmerman, H.; Yates, S.L. *Abstracts of Papers*, Neuroscience Meeting, Histamine Satellite Symposium, Abstract 21, 1858:729.3, 1995.
9. Schunack, W.; Stark, H. *Eur. J. Drug Metabol. Pharmacokinetics* **1994**, 3, 173-178.
10. Schlicker, E.; Kathman, M.; Bitschnau, H.; Marr, I.; Reidemeister, S.; Stark, H.; Schunack, W. *Naunyn-Schmied. Arch. Pharmacol.* **1996**, 353, 482-488.
11. Arrang, J. M.; Garbarg, M.; Lancelot, J. C.; Lecomte, J. M.; Robba, M. F.; Schwartz, J. C.; U.S. Patent 4 707 487, 1987.
12. Brimblecombe, R. W.; Duncan, W. A. M.; Durant, G. J.; Emmett, J. C.; Ganellin, C. R.; Leslie, G. B.; Parsons, M. E. *Gastroenterology* **1978**, 74, 339-347.
13. Arrang, J. M.; Garbarg, M.; Lancelot, J. C. M.; Lecomte, J. M.; Robba, M. F.; Schwartz, J. C. U. S. Patent 5 290 790, 1994.
14. (a) Durant, G. J.; Khan, A. M.; Tedford, C. E.; U.S. Patent 5 486 526, 1996. (b) Durant, G. J.; Khan, A. M. U.S. Patent 5 639 775, 1997. (c) Durant, G. J.; Khan, A. M. U.S. Patent 5 633 382, 1997.
15. (a) Yanai, K.; Watanabe, T.; Gotoh, T., Sakashita, H.; Murakami, K.; Sugiura, M.; Fukaya, C.; PCT WO 94/17058. (b) Yanai, K.; Ryu, J. H.; Sakai, N.; Takahashi, T.; Iwata, R.; Ido, T.; Murakami, K.; Watanabe, T. *Jpn. J. Pharmacol.* **1994**, 65, 107.
16. Lipp, R.; Schunack, W.; Arrang, J. M. *Abstracts of Papers*, Xth Int. Symp. Med. Chem., Budapest, 239, 1988.
17. Stark, H.; Krause, M.; Arrang, J. M.; Ligneau, X.; Schwartz, J. C.; Schunack, W. *Biorg. Med. Chem. Lett.* **1994**, 4 (24), 2907-2912.
18. Schliker, E.; Kathmann, M.; Reidemeister, S.; Stark, H.; Schunack, W. *Br. J. Pharmacol.* **1994**, 112, 1043-1048.
19. Stark, H.; Purand, K.; Ligneau, X., et al. *J. Med. Chem.* **1996**, 39, 1157-63
20. (a) Jansen, F. P.; Rademaker, B.; Bast, A.; Timmerman, H. *Eur. J. Pharm.* **1992**, 217, 203-205. (b) Jansen, F. P.; Wu, T.S.; Voss, H. P.; Steinbusch, H. W. M.; Vollinga, R. C.; Rademaker, B.; Bast, A.; Timmerman, H. *Br. J. Pharmacol.* **1994**, 113, 355-362.
21. Khan, M. A.; Durant, G. J.; El-Assadi, A. A.; Hoss, W.; Messer Jr., W. S.; Phillips, J. G.; Tedford, C. E.; Frederickson, R. C. A.*Abstracts of Papers*, A. C. S. Meeting, Washington, D. C., Abstract 157, Fall 1994.
22. Young, R. C.; Mitchell, R. C.; Brown, T. H.; Ganellin, C. R.; Griffiths,R.; Jones, M.; Rana, K. K.; Saunders, D.; Smith, I. R.; Sore, N. E.; Wilks, T. J. *J. Med. Chem.* **1988**, 31, 656-671.
23. Ganellin, C. R.; Hosseini, S. K.; Khalaf, Y. S.; Tertiuk, W.; Arrang, J. M.; Garbarg, M.; Ligneau, X.; Schwartz, J. C. *J. Med. Chem.* **1995**, 38, 3342-3350.

24. Bordi, F.; Mor, M.; Morini, G.; Plazzi, P. V.; Silva, C.; Vitali, T.; Caretta, A. *Il Farmaco* **1994**, 49 (3), 153-166.
25. Aslanian, R. G.; Green, M. J.; Shih, N. Y. PCT WO 95/14007.
26. Vollinga, R. C.; Menge, W. M. P. B.; Leurs, R.; Timmerman, H. *J. Med. Chem.* **1995**, 38, 266-271.
27. Leurs, R.; Kathmann, M.; Vollinga, R. C.; Menge, W. M. P. B.; Schlicker, E.; Timmerman, H. *J. Pharm. Exp. Ther.* **1996**, 276, 1-7.
28. Ganellin, C. R.; Andersen, B. B.; Khalaf, Y. S.; Tertiuk, W.; Arrang, J. M.; Garbarg, M.; Ligneau, X.; Rouleau, A.; Schwartz, J. C. *Biorg. Med. Chem. Lett.* **1992**, 2, 10, 1231-1234.
29. Ganellin, C. R.; Fkyerat, A.; Andersen, B. B.; Athmani, S.;Tertiuk, W.; Garbarg, M.; Ligneau, X.; Schwartz, J. C. *J. Med. Chem.* **1996**, 39, 3806-3813.
30. Low, C. M. R.; Vinter, J. G.; Kalindjian, S. B.; Pether, M. J.; Black, J. W. *Abstract* P151 at the joint meeting of the British Society of Pharmacology and the French Society of Pharmacology, Edinburgh, UK, 1997.
31. Mor, M.; Bordi, F.; Silva, C.; Rivara, S.; Crivori, P.; Plazzi, P. V.; Ballabeni, V.; Caretta, A.; Barocelli, E.; Impicciatore, M.; Carrupt, P. A.; Testa, B. *J. Med. Chem.* **1997**, 40, 2571.
32. Stark, H.; Huls, A.; Ligneau, X.; Arrang, J. M.; Schwartz, J. C.; Schunack, W. *Die Pharmazie* **1997**, 52, 7, 495-500.
33. Stark, H.; Ligneau, X.; Lipp, R.; Arrang, J. M.; Schwartz, J. C.; Schunack, W. *Die Pharmazie* **1997**, 52, 5, 1-7
34. Mierzwa, R.; King, A.; Conover, M. A.; Tozzi, S.; Puar, M. S.; Patel, M.; Coval, S. J.; Pomponi, S. A. *J. Nat. Prod.* **1994**, 57, 175.
35. (a) Phillips, J. G.; Tedford, C. E.; Yates, S. L.; Pawlowski, G. P.; Khan, M. A.; Fadnis, L.; Gregory, R.; Nikiforovich, G. *Abstracts of Papers*, Medicinal Chemistry Gordon Conference, Colby Sawyer College, New London, New Hampshire; 1996. (b) Phillips, J. G.; Tedford, C. E.; Yates, S. L.; Pawlowski, G. P.; Khan, M. A.; Fadnis, L.; Gregory, R.; Nikiforovich, G.; Ali, S. Presented at the 11th Camerino-Noordwijkerhout Symposium On Perspectives In Receptor Research, Noordwijkerhout, The Netherlands; 1997. (c) (b) Phillips, J. G.; Yates, S. L.; Pawlowski, G. P.; Khan, M. A.; Fadnis, L.; Gregory, R.; Nikiforovich, G.; Ali, S. M.; Tedford, C. E. Manuscript submitted for publication in *J. Med. Chem.*, **1998**.
36. Lipp, R.; Arrang, J. M.; Garbarg, M.; Luger, P.; Schwartz, J. C.; Schunack, W. *J. Med. Chem.* **1992**, *35*, 4434.
37. Shih, N. Y.; Lupo, Jr., A. T.; Aslanian, R.; Orlando, S.; Piwinski, J. J.; Green, M. J.; Ganguly, A. K.; Clark, M. A.; Tozzi, S.; Kreutner, W.; Hey, J. A. *J. Med. Chem.* **1995**, *38*, 1593.
38. Burger, A.; Bernabe, M.; Collins, P. W. *J. Med. Chem.* 1970, *13*, 33.

39. Khan, M. A.; Yates, S. L.; Tedford, C. E.; Kirschbaum, K.; Phillips, J. G. Accepted for publication in *Biorg. Med. Chem. Lett.*, 1997.
40. Ali, S. M.; Tedford, C. E.; Gregory, R.; Yates, S. L.; Phillips, J. G. submitted for publication in *Biorg. Med. Chem. Lett.*, 1997.
41. Topliss, J. G. *J. Med. Chem.* **1972**, 15 (10), 1006-1011.
42. Ali, S. M.; Tedford, C. E.; Yates, S. L.; Pawlowski, G. P, Phillips, J. G. submitted to *J. Med. Chem.* **1998**.
43. West, R.E.; Zweig, A.; Shih, N. Y.; Siegel, M. I.; Egan, R. W.; Clark, M. A. *Mol. Pharmacol.* **1990**, 38, 610.
44. Brown, J. D.; O'Shaughnessy, C. T.; Kilpatrick, G. J.; Scopes, D. I. C.; Beswick, P.; Clitherow, J. W.; Barnes, J. C. *Eur. J. Pharmacol.* **1996**, 311, 305-310.
45. Rodrigues, A. A.; Jansen, F. P.; Leurs, R.; Timmerman, H.; Prell, G. D. *Br. J. Pharmacol.***1995**, 114 (8), 1523-1524.
46. (a) Ganellin, C. R.; Jayes, D.; Khalaf, Y. S.; Tertiuk, W.; Arrang, J. M.; Defontaine, N., Schwartz, J. C. *Collect. Czech. Chem. Commun.* **1991**, 56, 2448-2455. (b) Bordi, F.; Mor, M.; Plazzi, P. V.; Silva, C. *Il Farmaco* **1992**, 47 (11) 1343-1365.

ACKNOWLEDGEMENTS

The authors are indebted to Professors Stephen Hanessian and David R. Williams for advice and encouragement. We would like to especially acknowledge the opportunity provided by Professor Emeritus Henry Rapoport and Dr. Tom Oesterling to broaden our horizons of imidazole chemistry.

R. Leurs and H. Timmerman (Editors)
The Histamine H_3 Receptor

Molecular modelling studies of histamine H_3 receptor ligands

Iwan J.P. De Esch, Paul H.J. Nederkoorn* and Henk Timmerman

Division Medicinal Chemistry, Leiden/Amsterdam Center for Drug Research (LACDR), Vrije Universiteit, De Boelelaan 1083, 1081 HV Amsterdam, The Netherlands.

1. Abstract

Molecular modelling studies represent a challenging aspect of medicinal chemistry as they facilitate drug design and lead optimisation. Furthermore, these studies may contribute to the unravelling of the molecular mechanisms involved in receptor stimulation. At present, the gene encoding the histamine H_3 receptor has not been cloned and hence, virtually nothing is known about the receptor topography. Therefore, modelling studies regarding the histamine H_3 receptor are limited to receptor mapping, making use of the properties of the ligands. In this chapter, a review will be given of the existing histamine H_3 ligand binding models and complying views on receptor activation mechanisms that evolved from modelling studies.

2. Introduction

Histamine is not only a mediator of several (patho)physiological actions, but also functions as a neurotransmitter, both centrally as peripherally [1, 2]. Feedback mechanisms are crucial to neurotransmission and the presynaptic histamine H_3 receptor not only plays a key role in regulating histamine release but also regulates the release of other neurotransmitters (for further details see chapter 2 and 3). Because inhibitory effects on histamine H_3 receptor-mediated stimuli by G protein toxins (both cholera and pertussin toxin) have been reported, it is most likely that the histamine H_3 receptor also belongs to the superfamily of G protein-coupled receptors [3,4] , *i.e.* coupled to a G protein of the $G_{i/o}$ class [5] . The reader is referred to chapter 6 for more details.

Therapeutic applications (as discussed in chapter 14 and 16) of peripherally active histamine H_3 receptor agonists putatively involve the treatment of asthma and diarrhoea. Therapeutic use of H_3 receptor antagonists in the periphery remains to be established, whereas centrally active H_3 receptor antagonists are predicted to reveal general stimulatory [6] and anti-appetite

* Present address: Shell Exploration & Production International Ventures B.V.
New Business Development, EPB-S, PO Box 663, 2501 CR Den Haag, The Netherlands

[7] effects. It has been speculated that H_3 antagonists may provide the means for treating Alzheimer's disease [8]. A thorough investigation of the possible use of H_3 ligands as therapeutics in brain disorders is hampered by the poor blood-brain-barrier penetration [9] and the interaction with cytochrome P450 isoenzymes of these compounds [10]. Therefore new ligands are needed that can reach the central nervous system more effectively. Rational design of these compounds can be facilitated by molecular modelling. Furthermore, these studies can help to better understand the molecular mechanisms involved in receptor stimulation. Therefore, the molecular structural features that are responsible for affinity and intrinsic activity need to be resolved. In this chapter a review will be given of the literature concerning molecular modelling studies of H_3 ligands that provide such information.

3. Tautomerism of imidazole-containing H_3 ligands

There is general consensus on the fact that the endogenous agonist histamine, 2-(imidazole-4-yl)ethylamine, binds to the receptor in its monocationic state (protonated at the side chain amine group). The monocationic form is predominantly (96.6%) present at pH 7.4 [11]. In this state, the neutral imidazole ring can exist in two different tautomeric forms, denominated by proximal (π) and tele (τ), respectively (Figure 1).

(N^{τ}) (N^{π})

Figure 1. The two tautomeric forms of the monocation of histamine.

For the histamine H_2 receptor, the development of potent histamine analogues that have other functional groups replacing the imidazole heterocycle [11-13] in combination with molecular modelling studies [14,15] have revealed that the N^{π} tautomer of histamine is the biologically active form for this subtype receptor. For the histamine H_3 receptor, information regarding the bio-active tautomer of the ligand is scarce as alteration of the imidazole is not tolerated (for a review see chapter 10 and 11 of this book). Whereas additional substituents on the 4(5)-imidazole ring lead to dramatically reduced H_3 agonistic activity, replacement of the imidazole by other groups results in compounds with no H_3 activity at all. At present, all published potent H_3 receptor antagonists also possess an imidazole moiety and as found for agonists, substitution or alteration of this heterocycle gives rise to severely reduced H_3 activity. We therefore claim that imidazole binding to the histamine H_3 receptor is very strict for both agonists and antagonists and two hydrogen bonds between the imidazole moiety and the receptor can be anticipated. However, no definite conclusions can be drawn concerning the specific tautomer that is recognised by the H_3 receptor.

It has been suggested that the unique tautomeric property of the imidazole ring is essential for efficacy at the histamine H_3 receptor. Investigating simple derivatives of the H_3 ligand proxyfan using molecular modelling techniques, Low *et al.* [16] suggested a correlation between efficacy at H_3 receptors and the tendency to adopt folded conformations (Table 1). The authors postulate that in the folded conformation the presence of intramolecular π-stacking between the imidazole and the aromatic ring in the side chain of the ligands hampers imidazole tautomerisation and thereby may reduce agonistic efficacy.

Table 1
Correlation between the extend of folding and the efficacy (%α) expressed at H_3 receptors as given by Low *et al.* [16].

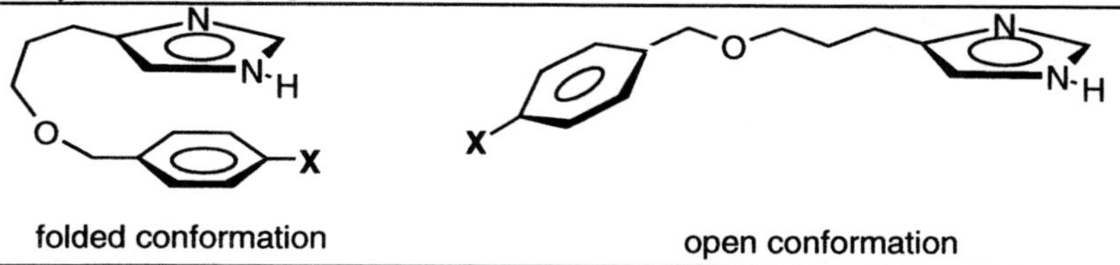

X	% folded conformations	% α[a]
H (proxyfan)	100	33
F	85	15
Cl	56	58
Br	26	84
I	33	82

[a]Relative to R-α-methylhistamine.

4. The bioactive conformation of agonists

Histamine (**1**) is a flexible molecule that can adopt numerous conformations (Figure 2). These can be described in terms of three torsional angles τ_1, τ_2 and τ_3. Quantum mechanical calculations on conformations for histamine [15,17-19] reveal that with respect to τ_2, the histamine monocation is found basically in the trans and gauche conformation. The values for τ_1 and τ_3 are less restricted.

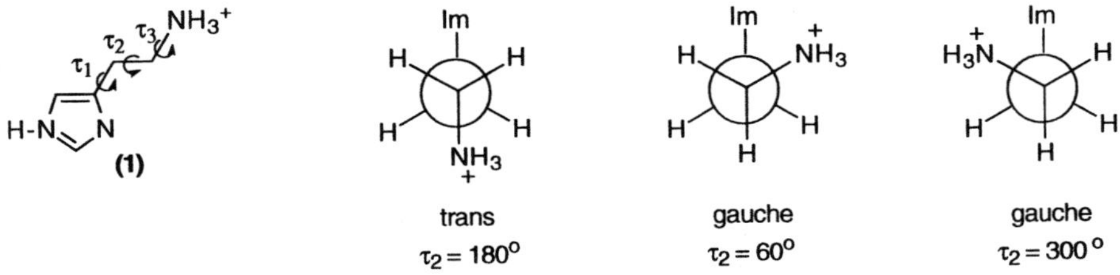

Figure 2. The conformations of histamine **(1)** described by 3 torsion angles (τ_1, τ_2 and τ_3, respectively). Torsion angle τ_2 is found fundamentally in the trans and gauche conformations.

Conformational analysis of the standard H_3-agonist (R)α-methylhistamine **(2)** (table 2) using molecular mechanics reveals a similar result as for histamine [20]. (R)α-methylhistamine **(2)** is predominantly found in three conformations (neglecting the rotational freedom of the τ_1 axis, cf. Figure 2). Only minor energetic differences (<1.5 Kcal) were found between these three rotameric conformations. Based on these conformational studies, Shih and co-workers [20] designed and synthesized a series of conformationally constrained H_3 agonists to investigate the bioactive conformation and to explore the sterical requirements of H_3 receptor agonists (Table 2). Each of the three aforementioned conformations was mimicked by two conformationally restricted analogues of histamine. These results have been depicted in Table 2: compound **I** and **II** aim at mimicking the gauche conformer #1, compound **III** and **IV** should mimic the anti conformer #2 and compound **V** and **VI** mimicking the gauche conformer #3.

Having synthesized these target compounds, Shih *et al.* determined the H_3 activity in terms of H_3 binding (Table 2). The pharmacological data show that compound **III** and **IV** are the most potent H_3 ligands in this series. Therefore, it can be concluded that the bioactive conformation of the studied compound (R)α-methylhistamine **(2)** is the anti-conformer #2. However, it has to be recognized that the activity of these compounds may also be influenced by sterical factors as large substituents on the ethylene chain of histamine are not allowed for agonistic activity on the H_3 receptor (see chapter 10 for details).

Table 2
The design of conformationally restricted H_3 agonists by Shih and co-workers [20]

Conformers of (R)-α-methylhistamine **(2)**	Compounds synthesized[a]

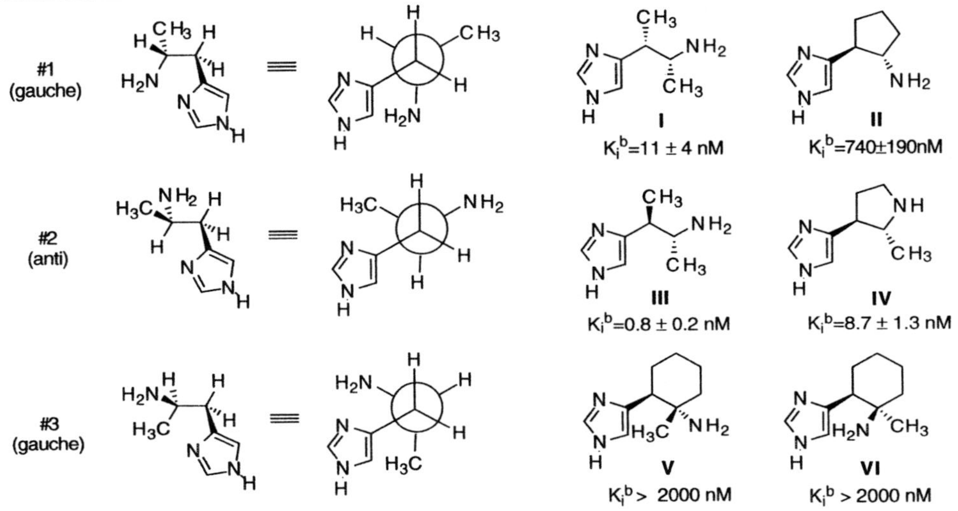

[a] For convenience, only one of the corresponding enantiomers is indicated.
[b] K_i Value for H_3 receptor binding, determined as described by Korte *et al.* [21].

Contrary to the finding of Shih and co-workers, Mazurek *et al.* [22] claimed on basis of *ab initio* molecular orbital calculations that the conformation of (R)α-methylhistamine **(2)** that is recognized by the H_3 receptor has an intramolecular hydrogen bond between the cationic side chain amine and the basic N^{π} atom of the imidazole. This conformation corresponds to one of the gauche-conformers depicted in table 2. To test this hypothesis of Mazurek and co-workers, we used (S)α,(S)β-*trans*-cyclopropylhistamine (VUF 5297; compound **(7)** in Figure 7) which exhibits H_3 agonistic activity [23]. This rigid cyclopropyl-containing histamine analogue **(7)** is unable to form an intramolecular hydrogen bond and reveals that internal hydrogen bond formation is not essential for H_3 activity. This conclusion is in line with the aforementioned model by Low and co-workers as intramolecular hydrogen bonding would hamper a tautomeric shift of the imidazole which is suggested to be essential for histamine H_3 receptor activity.

The development of rigid histamine analogues is important for the determination of the H_3 receptor pharmacophore as the conformations of these compounds only allow restricted spatial orientation of the imidazole ring with respect to the basic nitrogen in the side chain of the ligands. Sippl and co-workers [24] were the first to develop such a pharmacophore model for histamine H_3 agonists. Superimposing all relevant conformations of a series of selected agonists **(1-10)** (Figure 3) and using (R)α,(S)β-dimethylhistamine **(3)** as a template, two pharmacophores evolved in which all imidazole rings and all protonated side-chain nitrogens could be perfectly superimposed. Only one pharmacophore revealed a good overlap of the hyrophobic part of the sidechains (Figure 4) and was therefore selected for further investigation by Sippl *et al.* Histamine reveals a *gauche-trans* conformation in this model (cf. Figure 2, τ_1 and τ_2). The methyl group of Sch 49648 **(5)** occupies the same region of space as the (R)α–methyl group of R(α),S(β)–dimethylhistamine **(3)** while the pyrrolidine ring overlaps with the (S)β–methyl group. The relatively longer side-chains of imetit **(8)** and SKF 91606 **(9)** adopt folded conformations, thereby occupying the same region of space as the (S)β–methyl group of (R)α,(S)β–dimethylhistamine **(3)**. It was noted that of the different stereoisomers of cyclopropylhistamine **(7)**, the (S)α,(S)β-enantiomer fits best in the derived pharmacophore. Recent studies in our laboratories indeed established (S)α,(S)β-cyclopropylhistamine (VUF 5297) as the eutomer of this small and rigid compound [23].

Using the H_3 receptor pharmacophore as a template, Höltje and Sippl [25] built a pseudoreceptor model for the H_3 receptor agonist binding site. The pseudoreceptor concept seeks to construct a model receptor that mimics the essential ligand-macromolecule interactions of the true biological receptor. Starting point for the construction of a psuedoreceptor using the receptor-mapping program YAK [26-28] are the superimposed ligands in their bioactive conformation. In this model, potential binding sites (anchor points) are defined for each of the ligand molecules of the pharmacophore. For this purpose the program YAK uses an extensive database that holds information about ligand-receptor interactions. Suitable binding partners (e.g. amino acids, metal ions and solvent) result from this process and are positioned in three-dimensional space. The collection of binding partners constitutes a pseudoreceptor for the ligands used as the original template.

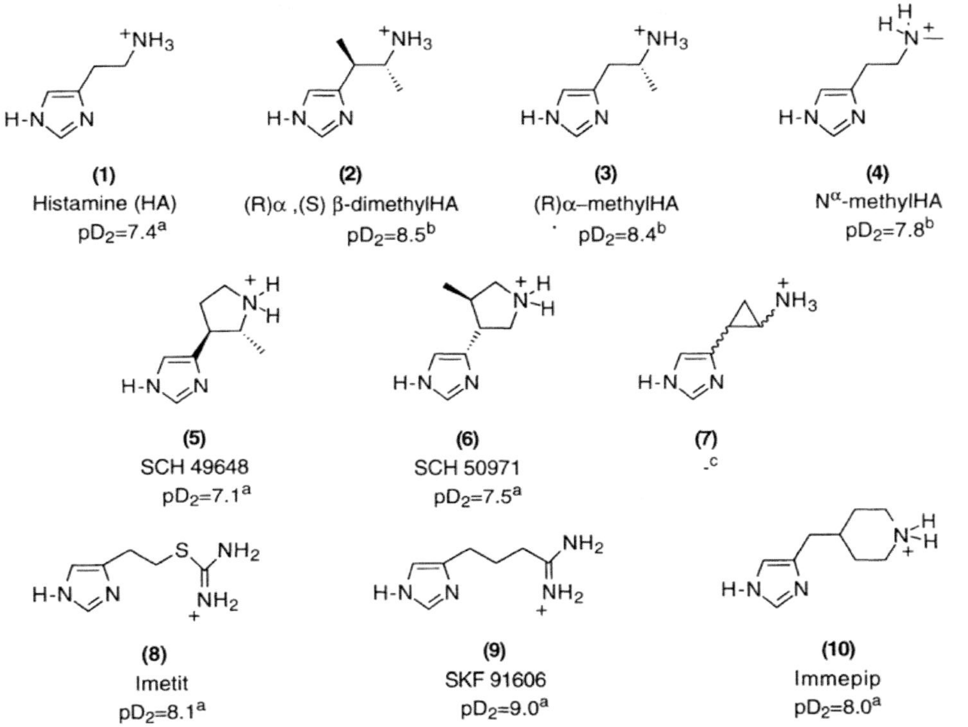

Figure 3. H_3 receptor agonists investigated by Sippl *et al.* [24]. It is noted that the reported biological data have been obtained using different pharmacological systems.

[a] Agonistic activity determined as the inhibition of K^+-stimulated [^{3}H]-histamine release on rat cortex [29].

[b] Agonistic activity determined as the inhibition of electrically evoked, cholinergic contractions of guinea pig intestine preparations [30].

[c] The authors had no information about the H_3 activity of the distinct stereoisomers. See text for details.

Höltje and Sippl selected a representative "training set" of twelve superimposed structures. The aforementioned YAK process resulted in a pseudoreceptor that consists of six amino acid residues (Figure 5). In this pseudoreceptor model, the imidazole ring of the ligands is bound to a tyrosine residue that donates a proton to the imidazole and an asparagine residue that accepts a proton from the imidazole. Hence, two distinct H-bond have been found, which explain the strict binding mode of the imidazole. It has to be noted that the position of these receptor residues relative towards the imidazole is a consequence of

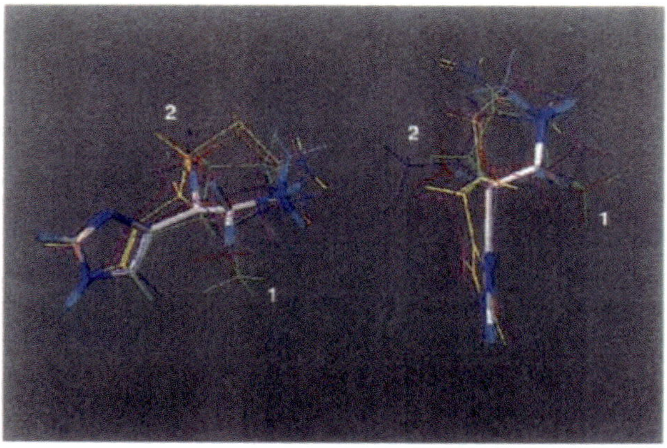

Figure 4. The pharmacophore model for H_3 receptor agonists by Sippl *et al.* [24].

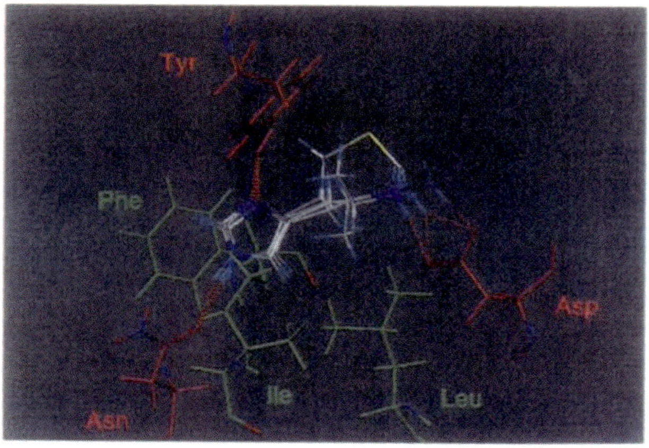

Figure 5. Pseudoreceptor model for histamine H_3 agonists by Höltje and Sippl [25]. Four ligands are shown inside the binding region.
Figure 4 and Figure 5 were kindly provided by Prof. Dr. Höltje.

considering only the τ-tautomer of the ligands while constructing the pharmacophore model. Höltje and Sippl gave no rationale for the selection of the τ tautomer. In addition to the hydrogen-bonding interactions, the imidazole heterocycle has an hydrophobic interaction with a phenylalanine of the pseudoreceptor. Around the hydrophobic part of the side-chains a leucine and a isoleucine fragment are located. The positively charged side chain nitrogens interact with a negatively charged aspartate of the pseudoeceptor.

The YAK program can be used to estimate relative free energies of binding between ligands and pseudoreceptor. Höltje and Sippl calculated for the training set, used to construct the pseudoreceptor, that the correlation coefficient for experimental [31,32] versus calculated free energies of binding equals 0.99 and the RMS deviation is 0.21 Kcal/mol. Subsequently, Höltje and Sippl tested the pseudoreceptor model by predicting biological binding data for a test set of four additional agonists not included in the construction of the pseudoreceptor training set. The test set revealed a RMS deviation of 0.66 Kcal/mol, indicative of the accuracy of the model. It has to be recognised however, that structural diversity between the compounds used to construct the model and the test set is rather limited. We noticed that more severe testing of the predictive power of the pseudoreceptor with structurally more diverse agonists like Immepip **(10)** has not been reported.

5. A qualitative model for histamine H_3 receptor agonists and antagonists

Recently, our group developed a qualitative H_3 ligand binding model for agonists and antagonists [33]. It has already been outlined that the imidazole ring is present in all potent H_3 ligands, agonists and antagonists. Moreover, all H_3 ligands have an identical substitution behaviour concerning the heterocycle (recall our earlier command on the fact that ring alternations are not allowed for H_3 activity). This crucial role of the imidazole ring in both agonists and antagonists strongly suggests that this part of all ligands (both agonists and antagonists) bind to the same receptor site and that this interaction is very strict. Therefore, we suggest two simultaneous H-bonds between the imidazole and the receptor binding site for both agonists and antagonists. Since at present no (experimental) evidence for either the τ nor the π form exists, we aselectively picked the τ form for the sake of choosing one out of these two possible tautomers of the imidazole moiety.

Another substructure that is present in all agonists and many antagonists is a basic nitrogen in the imidazole side-chain. Considering the endogenous agonist histamine and the suggestion that the histamine H_3 receptor is most likely a G-protein coupled receptor (GPCR) [3,4], it can be classified as an aminergic GPCR. Receptors of this type all share a highly conserved aspartic acid residue (Asp) in transmembrane domain three (TM III) [34]. This Asp is involved in binding the positively charged nitrogen atom of the aminergic ligands and can therefore be seen as the main anchoring point for agonist binding [34, 35]. In addition to the predominant role of binding agonists, for some aminergic receptors this highly conserved Asp has been claimed to be involved in binding of antagonists as well [35-37].

Assuming that the imidazole of agonists and antagonists binds in the same manner to the receptor, it was speculated that the Asp that is expected to interact with the amino-group of the agonists might as well be available for binding antagonists having a basic nitrogen atom. In our model, a carboxylate was selected to mimic the interaction of the Asp of the receptor with the basic nitrogen of the ligands. The Cα and Cβ atoms of this carboxylate are fixed with respect to the rigid protein backbone (Figure 6). Within the protein, rotation around the Cα-Cβ bond (and around the Cβ - Cγ bond) is allowed, therefore this degree of freedom was integrated in our model. We used the aforementioned model derived by Sippl *et al.* [24] to position the Asp with respect to histamine in the predicted bioactive conformation. This procedure results in a pharmacophore-pharmacon complex that can be defined by the relative position of the imidazole ring of the ligands with respect to the Cα and Cβ atoms of the carboxylate. Because we suggest that the positions of these atoms is identical for all ligand-carboxylate complexes (due to the aforementioned restrictions in the binding mode), the relative positions of these atoms were fixed for all subsequent calculations. The positions of all other atoms of the complex were optimized by applying a density functional approach. To this end, we used the Amsterdam Density Functional (ADF) program package [38, 39] adapted to parallel computers.

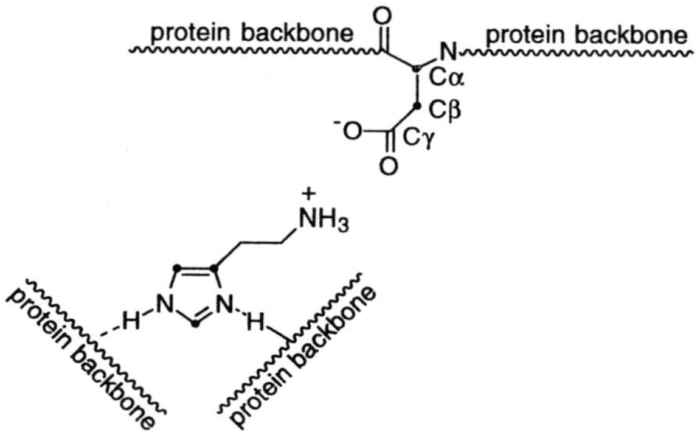

Figure 6. Schematic representation of the pharmacophore-pharmacon complex. (See text for more details).

The interaction modes of a series of characteristic histamine H_3 receptor agonists **(1-10)** (figure 7) and antagonists **(11-18)** (Figure 8) with the flexible receptor residue was investigated. The resulting optimized complexes were superimposed. In this procedure the imidazoles and the Cα and Cβ atoms obviously show a perfect overlap as no variance in their relative position was tolerated during the geometry optimizations.

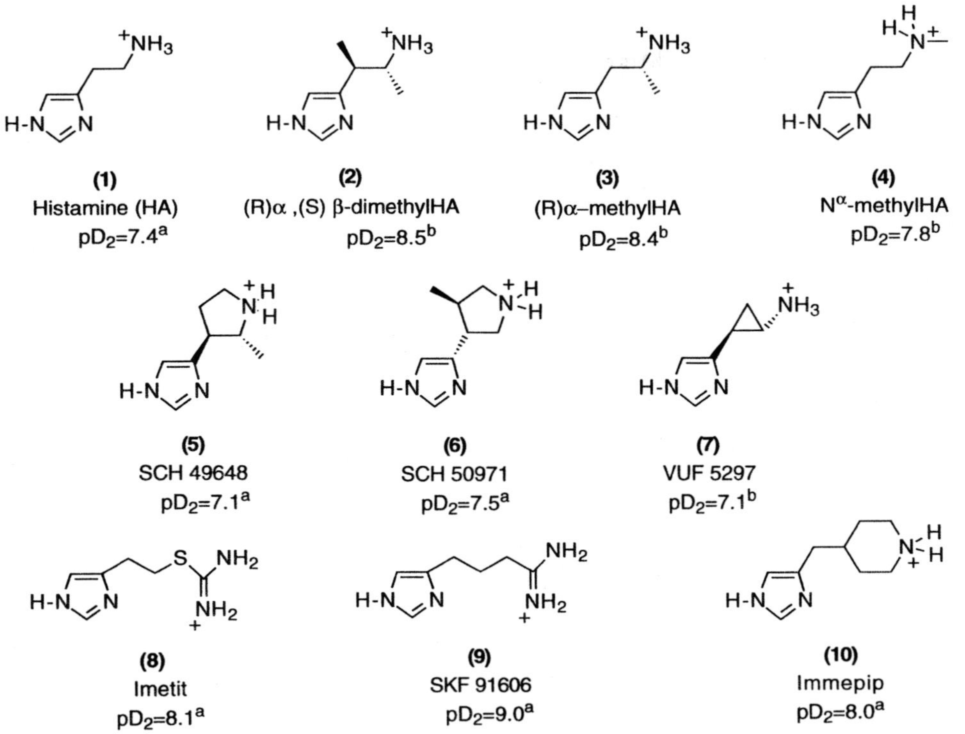

Figure 7. The H_3 agonists (**1-10**) studied by De Esch *et al.* [33]. Biological data have been obtained using different pharmacological systems and direct quantitative comparison is therefore hampered.

[a] Agonistic activity determined as the inhibition of K^+-stimulated [^{3}H]-histamine release on rat cortex [29].

[b] Agonistic activity determined as the inhibition of electrically evoked, cholinergic contractions of guinea pig intestine preparations [30].

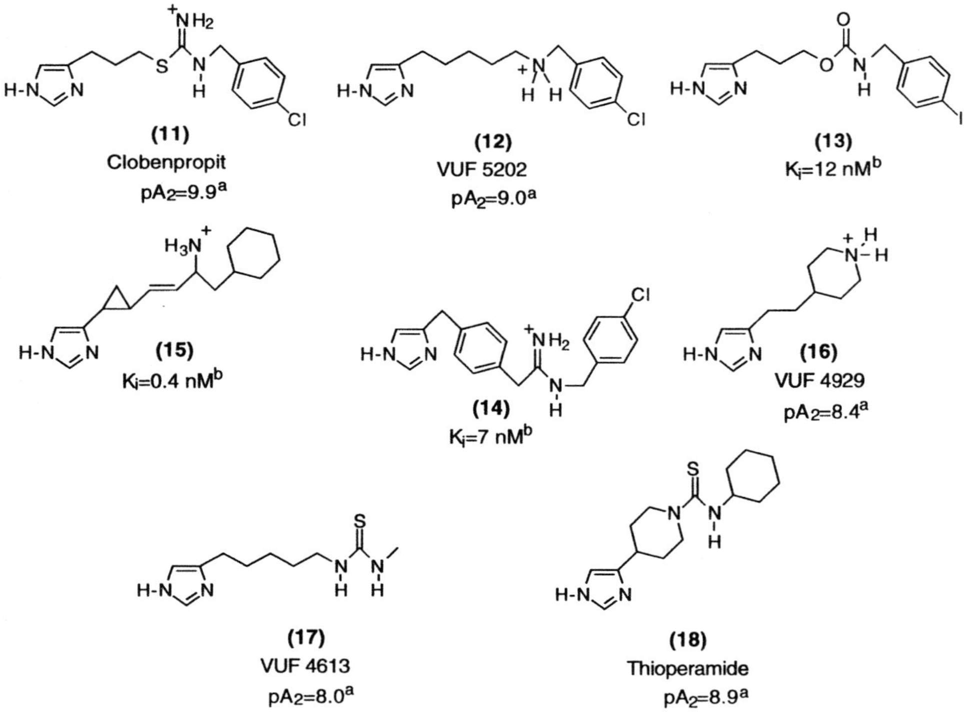

Figure 8. The H_3 antagonists **(11-18)** studied by De Esch *et al.* [33]. Biological data have been obtained using different pharmacological systems and direct quantitative comparison is therefore hampered.

[a] Histamine H_3 receptor activity determined as the effect on K^+-stimulated [^{3}H]-histamine release on rat cortex [29].

[b] Histamine H_3 receptor activity determined as the effect on electrically evoked, cholinergic contractions of guinea pig intestine preparations [30].

Superimposing the agonist-carboxylate complexes (Figure 9) revealed that all agonists interact with the carboxylate in its *all trans* conformation. It is not necessary to superimpose all basic nitrogens in the side-chain of the agonists in the same position in 3D space. A certain degree of positional freedom of this substructure of the ligands is introduced due to flexibility of the interacting carboxylate. An additional advantage of this approach is that the directionality of the hydrogen bond between the basic nitrogen of the ligands and the receptor site point is automatically taken into account and allows for additional spatial flexibility for the nitrogen in 3D space. The conformations of the agonistic structures are similar to those in the aforementioned model by Sippl *et al.* [24] who however did not use any interacting group from the receptor. Comparing Sippl and co-workers' model with our model (cf. Figure 4 and Figure 9), the nitrogen position may be too restricted in the former model.

Thus, we were able to reproduce Sippl and co-workers' model [24] by applying a totally different technique. Furthermore, in our model it was shown that proper hydrogen-bonds between the agonists and a flexible receptor interaction site can be formed. To validate the model, especially with respect to the position of the receptor site-point and the stereoselectivity of the receptor towards the ligands, both enantiomers of the small and rigid compound trans-cyclopropylhistamine were synthesized and the distinct stereoisomers were tested for their H_3 activity [23]. Both enantiomers, that differ significantly in pharmacological activity, give a proper interaction with the carboxylate in the all trans conformation (validating the position of the receptor site-point). Only the configuration of the active stereoisomer (S)α,(S)β-cyclopropylhistamine (VUF 5297) (**7**) is in good sterical agreement with the pharmacophore (validating the stereoselectivity of the receptor). Again, this is in agreement with the findings reported by Sippl *et al.*[24].

Having incorporated a flexible receptor residue in the model, the binding mode of antagonists was studied as well. The selected antagonists all have a protonated or neutral side-chain nitrogen (cf. Figure 8). The lipophilic tail that is attached to the polar group in the side-chain of these antagonists, had to be truncated to a small methyl-group to keep the geometry optimizations using ADF manageable. The methyl groups can of course only give an indication about the position and direction of the lipophilic tail of the antagonists. Antagonists that lack a basic nitrogen in the side-chain (see chapter 11 of this book) probably bind with their imidazole moiety at the imidazole binding-site and reach with their lipophilic tail into a lipophilic pocket that is available for antagonists binding. Because these molecules cannot participate in a monopole-monopole interaction with the aspartic acid residue, these structures were not considered in this study.

Assuming a similar imidazole binding for all H_3 ligands (both agonists and antagonists), it was revealed that the investigated antagonists can all have an interaction with the carboxylate that also binds to the agonists. Superimposing all optimized complexes of agonists and antagonists with the carboxylate revealed that the position of the aspartic acid is clearly different when binding to agonists or antagonists (Figure 10). We therefore propose from this model that the molecular determinant accounting for agonistic *vs.* antagonistic activity is the conformation of the carboxylate. These results suggest an important role of the Asp in receptor stimulation. Similar findings (for the histamine H_1 receptor) have been reported by Ter Laak and co-workers [40].

Figure 9. The ten superimposed agonist-carboxylate complexes. All imidazole rings of the ligands (upper right), all Cα and all Cβ of the Asp (bottom) occupy the same positions in 3D space for all complexes.

Figure 10. The superimposed agonist- and antagonist-carboxylate complexes. The imidazole (upper right) of all ligands occupy the same positions in 3D space and the Cα and Cβ atoms of the carboxylates (bottom) are perfectly superimposed as well. The cluster of carboxylates pointing towards the upper right bind to agonists and the cluster of carboxylates pointing towards the upper left bind to antagonists.

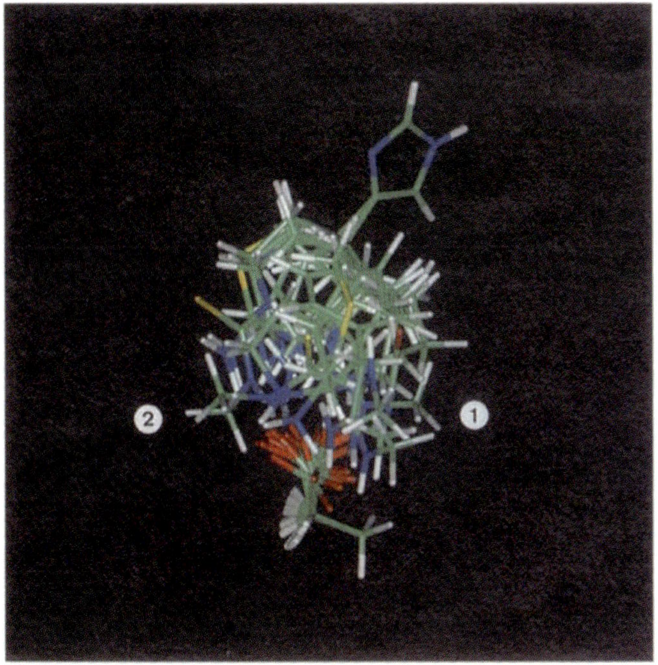

Figure 11. The model developed in our laboratory suggests two distinct lipophilic pockets available for antagonists binding (indicated by 1 and 2). (See text for more details).

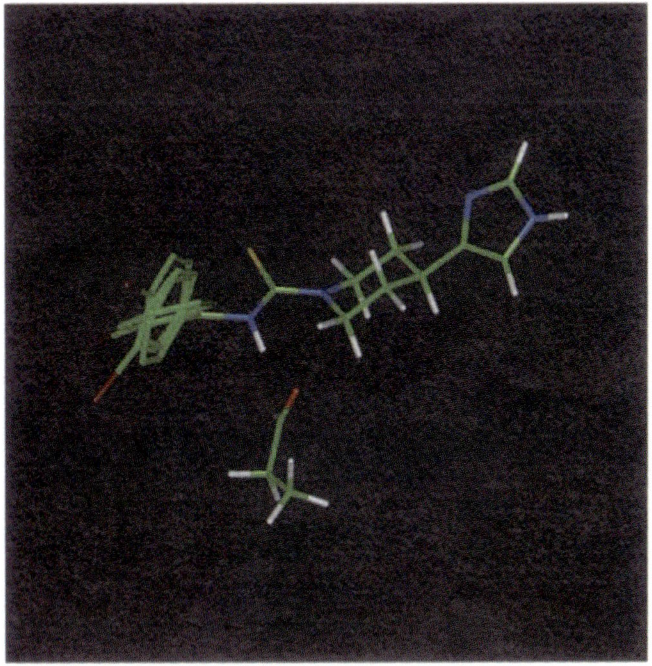

Figure 12. Superposition of the three iodine-substituted benzyl derivatives (cf. Table 3) illustrates the difference in dihedral angle (φ). (See text for more details).

With regard to the antagonistic structures in the model two distinct orientations were found for the methyl groups that represent the lipophilic tails (Figure 11). The methyl groups of VUF 4713 (**17**) and thioperamide (**18**) have a very different position and orientation (these two compounds reach into pocket no. 2, cf. Figure 11) suggesting the existence of two lipophilic pockets which are available for binding antagonists. The presence of two different pockets may explain the differences in SAR observed for the lipophilic moiety of antagonists (for a detailed discussion see literature [41, 42]).

To gain more insight into the binding properties of one of these lipophilic pockets (*i.e.* pocket no. 2, cf. Figure 11), a series of thioperamide analogues was synthesized to determine the SAR of the lipophilic tail of this antagonist [41]. As indicated by the results presented in Table 4, the activity of halogen substituted benzyl analogues depends on the position of substitution and on the specific halogen. Substitution at the ortho position favours the H_3 antagonistic activity, except for fluorine.

Table 3
pA_2 values[a] of several halogen-substituted benzyl analogues of thioperamide as determined by Windhorst and co-workers [41]

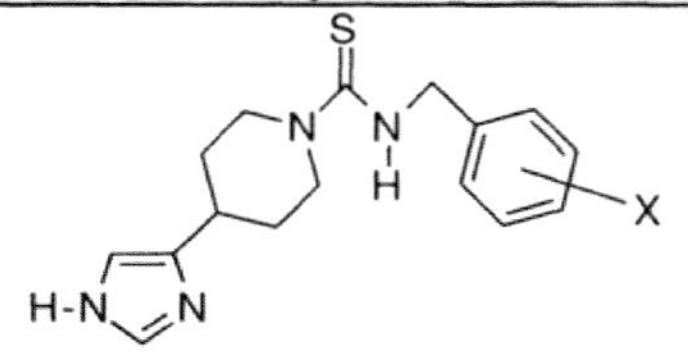

X	ortho	meta	para
H	7.4	7.4	7.4
F	6.0	6.4	6.2
Cl	8.2	7.8	7.2
Br	7.8	7.6	6.8
I	8.2	7.6	6.7

[a] Antagonistic activity determined as the influence on electrically evoked, cholinergic contractions of guinea pig intestine preparations [30].

An additional modelling study was undertaken to quantify these findings [41]. To this end, the already geometry optimized complex of carboxylate and truncated thioperamide (**18**) was used as a template. This fixed template was used to construct the different benzyl analogues shown in Table 4, by attachment of the distinct lipophilic tails. In this additional modelling study, only the geometry of these lipophilic tails were optimized using the aforementioned ADF program [38, 39]. These calculations reveal that the dihedral angle (φ) between the thiourea moiety and the phenyl group differs depending on the position and nature of substitution of the benzyl ring. This effect is illustrated in Figure 12 by superimposing the three iodine analogues (cf. Table 3). An excellent correlation (eq. 1) was found between the antagonistic activity (pA_2) and the dihedral angle φ and the charge δ on the substituted carbon atom of the benzyl group (n=13, r=0.93, F=31.57):

$$pA_2 = -0.02\ \varphi - 0.933\ \delta + 4.72 \qquad \text{(eq. 1)}$$

The results of this QSAR study support the accuracy of the geometry of the ligand-carboxylate complexes in our qualitative H_3 ligand binding model depicted in Figure 9-11. Additional molecular modelling studies will be undertaken to gain more insight into the requirements for binding of the lipophilic part of the other antagonists to the other lipophilic site no. 1 (cf. Figure 11). These molecular modelling studies will also include the binding mode of antagonists that lack a basic nitrogen in the side chain (omitted in the model presented in Figure 9-11). At present, we are developing new antagonists that have two lipophilic tails. These ligands are aimed to have interaction with both putative lipophilic pockets no. 1 and no. 2. This new class of compounds may have improved pharmacological and/or pharmacokinetic profiles.

6. Conclusions

The histamine H_3 receptor has not been cloned yet and hence, virtually nothing is known about the receptor topography. However, ligand based molecular modelling studies have contributed to the understanding of the molecular features involved in ligand-receptor interaction.

All potent H_3 ligands posses an imidazole ring. In it's neutral form, the imidazole can exist in two tautomeric forms (N^{π} and N^{τ}, respectively). At present, it is not clear which tautomeric form is recognised by the receptor.

Several molecular modelling studies and the development of rigid histamine analogues have revealed that the bioactive conformation of the endogenous agonist is a gauche-trans form. Pharmacophore models for H_3 agonists have been developed that give an excellent indication about the sterical requirements of H_3 agonists and can account for the observed stereoselectivity of the receptor. The pharmacophore model for agonists described by Sippl *et al.* and the model developed our group using different theoretical approaches indicates the position of the aspartic acid residue of the receptor that is expected to interact with the cationic amino group of the aminergic ligands. We investigated the binding mode of the flexible aspartic acid residue with both agonists and antagonists. In this model (Figure 9-11), the molecular determinant for agonistic *versus* antagonistic activity seems to be dependent on the conformation of the aforementioned aspartic acid histamine H_3 receptor residue. This study reveals two lipophilic pockets available for antagonist binding.

References

1. Leurs, R.; Timmerman, H. The histamine H_3 receptor: a target for developing new drugs. In *Progress in Drug Research*; Jucker, E. Ed.; Birkhäuser Verlag: Basel, 1992; Vol. 39. pp. 127-165.
2. Schwartz, J.-C.; Arrang, J.-M.; Garbarg, M.; Gulat-Marnay, C.; Pollard, H. Modulation of histamine synthesis and release in brain via presynaptic autoreceptors and heteroreceptors. In *Presynaptic receptors and the question of autoregulation of neurotransmitter release;* Kalsner, S.,Westfall T.C. Eds.; Annals of the New York Academy of Sciences: New York, 1991; Vol. 604; pp. 40-54.
3. Cherifi, Y.; Pigeon, C.; Le Romancer, M.; Bado, A.; Reyl-Desmars, F.; Lewin, M. J. M. *J. Biol. Chem.* **1992,** *267*, 25315.
4. Nozaki, M.; Sperelakis, N. *Am. J. Physiol.* **1989,** *256*, H455.
5. Clark, M. A.; Korte, A.; Egan, R. W. *Agents & Actions* **1993,** *40*, 129.
6. Sakai, N.; Onodera, K.; Maeyama, K.; Yanai, K.; Watanabe, T. *Life Sci.* **1991,** *48*, 2397.
7. Oohara, A.; Yoshimatsu, H.; Kurokawa, M.; Oishi, R.; Saeki, K.; Sakata, T. *J. Neurochem.* **1994,** *63*, 677.
8. Miyazaki, S.; Onodera, K.; Imaizumi, M.; Timmerman, H. *Life Sci.* **1997,** *61*, 355.
9. Garbarg, M.; Trung Tuong, M. D.; Gros, C.; Schwartz, J.-C. *Eur. J. Pharmacol.* **1989,** *164*, 1.
10. Alves-Rodrigues, A.; Leurs, R.; Wu, T. S.; Prell, G. D.; Foged, C.; Timmerman, H. *Br. J. of Pharmacol.* **1996,** *118*, 2045.
11. Ganellin, C. R. Characterization and distribution of histamine receptors. In *The chemical regulation of biological mechanisms;* Creighton, A. M.; Turner, S. Eds.; Royal Society of Chemistry: London, U.K., 1982; pp. 1-15.
12. Eriks, J. C.; Goot, H. v. d.; Sterk, G. J.; Timmerman, H. *J. Med. Chem.* **1992,** *35*, 3239.
13. Eriks, J. C.; Goot, H. v. d.; Timmerman, H. *Mol. Pharmacol.* **1993,** *44*, 886.
14. Nederkoorn, P. H. J.; Vernooijs, P.; Donne-Op den Kelder, G. M.; Baerends, E. J.; Timmerman, H. *J. Mol. Graphics* **1994,** *12*, 242.
15. Nagy, P. I.; Durant, G. J.; Hoss, W. P.; Smith, D. A. *J. Am. Chem. Soc.* **1994,** *116*, 4898.
16. Low, C. M. R.; Vinter, J. G.; Kalindjian, S. B.; Pether, M. J.; Black, J. W. Abstracts P151 at the joint meeting of the British Pharmacological Society and the Société Françoise de Pharmacologie, Edinburgh, UK, **1997**.
17. Topiol, S.; Weinstein, H.; Osman, R. *J. Med. Chem.* **1984,** *27*, 1531.
18. Richards, W. G.; Wallis, G.; Ganellin, C. R. *Eur. J. Med. Chem.* **1979,** *14*, 9.
19. Smeyers, Y. G.; Romero-Sanchez, F. J.; Hernandez-Laguna, A. *J. Mol. Struct. (THEOCHEM)* **1985,** *123*, 431.
20. Shih, N.-Y.; Lupo, A. T. J.; Aslanian, R.; Orlando, S.; Piwinski, J. J.; Green, M. J.; Ganguly, A. K.; Clark, M. A.; Tozzi, S.; Kreutner, W.; Hey, J. A. *J. Med. Chem.* **1995,** *38*, 1593.
21. Korte, A.; Meyers, J.; Shih, N.-Y.; Egan, R. W.; Clark, M. A. *Biochem. Biophys. Res. Commun.* **1990,** *168*, 979.

22. Mazurek, A. P.; Karpinska, G. *Z. Naturforsch.* **1994,** *49 c*, 471.
23. De Esch, I. J. P. *et al. Submitted for publication*
24. Sippl, W.; Stark, H.; Höltje, H.-D. *Quant. Struct.-Act. Relat.* **1995,** *14*, 121.
25. Höltje, H.-D.; Sippl, W. Molecular modelling studies on histamine H_2- and H_3-receptor agonists. In *Proceedings, XIVth International Symposium on Medicinal Chemistry;* Awouters, F. Ed.; Elsevier Science B.V.: Amsterdam, 1997; Vol. 28; pp. 137-148.
26. Vedani, A.; Zbinden, P.; Snyder, J. P. *J. Receptor Res.* **1993,** *13*, 163.
27. Snyder, J. P.; Rao, S. N.; Koehler, K. F.; Vedani, A. Pseudoreceptors. In *3D QSAR in drug design;* Kubinyi, H. Ed.; ESCOM Science Publisher B.V.: Leiden, 1993; pp. 336-354.
28. Vedani, A. *ALTEX* **1994,** *11*, 11.
29. Arrang, J. M.; Garbarg, M.; Schwartz, J. C. *Nature* **1983,** *302*, 832.
30. Vollinga, R. C.; Zuiderveld, O. P.; Scheerens, H.; Bast, A.; Timmerman, T. *Meth. Find. Exp. Clin. Pharmacol.* **1992,** *14*, 747.
31. Ligneau, X.; Garbarg, M.; Vizuete, M. L.; Diaz, J.; Purand, K.; Stark, H.; Schunack, W.; Schwartz, J. C. *J. Pharmacol. Exp. Ther.* **1994,** *271*, 452.
32. Lipp, R.; Stark, H.; Schunack, W. Pharmacochemistry of H_3 Receptors: The Histamine Receptor. In *Receptor Biochemistry and Methodology;* Schwartz, J.-C.; Haas, H. L. Eds.; Wiley-Liss, Inc., 1992; Vol. 16; pp. 57-72.
33. De Esch, I. J. P.; Timmerman, H.; Nederkoorn, P. H. J. *Submitted for publication*
34. Oliveira, L.; Paiva, A. C. M.; Vriend, G. *J.Comp.-Aided Mol.Design* **1993,** *7*, 649.
35. Strader, C. D.; Sigal, I. S.; Candelore, M. R.; Rands, E.; Hill, W. S.; Dixon, R. A. F. *J. Biol. Chem.* **1988,** *263*, 10267.
36. Ter Laak, A. M.; Venhorst, J.; Donné-op den Kelder, G. M.; Timmerman, H. *J. Med. Chem.* **1995,** *38*, 3351.
37. Leurs, R.; Smit, M. J.; Menge, W. M. B. P.; Timmerman, H. *Biochem. Biophys. Res. Com.* **1994,** *201*, 295.
38. Baerends, E. J.; Ellis, D. E.; Ros, P. *Chem. Phys.* **1973,** *2*, 41.
39. te Velde, G. *Numerical integration and other methodological aspects of bandstructure calculations*, Vrije Universiteit 1990.
40. Ter Laak, A. M.; Timmerman, H.; Leurs, R.; Nederkoorn, P. H. J.; Smit, P. H. J.; Donné-Op den Kelder, G. M. *J. Comp. -Aided Mol. Design* **1995,** *9*, 319.
41. Windhorst, A. *et al. Submitted for publication.*
42. Vollinga, R. C.; Menge, W. M. P. B.; Leurs, R.; Timmerman, H. *J. Med. Chem.* **1995,** *38*, 2244.

R. Leurs and H. Timmerman (Editors)
The Histamine H_3 Receptor

Brain histamine in pathophysiological conditions and brain diseases

P. Panula[a], T. Sallmen[a], O. Anichtchik[a], K. Kuokkanen[a], M. Lintunen[a], J.O. Rinne[b], M. Mättö[b], J. Kaslin[a], K.S. Eriksson[a] and K. Karlstedt[a]

[a]Department of Biology, Åbo Akademi University, Tykistökatu 6, 20520 Turku
[b]Department of Neurology, University of Turku, 20520 Turku, Finland

1. INTRODUCTION

Histamine is widely distributed in the human brain[24]. The histaminergic neurons lie in the posterior hypothalamus, where about 64,000 cells form a dispersed group referred to as nucleus tuberomammillaris[1] (Fig. 1). Histamine-containing neurons have not been found in other areas of the human brain[40]. During fetal development another transient histamine system is present in at least rat brain, where raphe neurons contain histamine[59]. The general organization of the histaminergic system is similar in all mature vertebrates: cell bodies located in the tuberomammillary nucleus[43] provide almost all parts of the CNS with varicose fibers containing histamine. In human brain, histamine-containing projections have so far been shown to extend to various parts of the cerebral cortex[40] and cerebellar cortex[44].
Histamine may regulate higher brain functions by at least two mechanisms. It enhances hippocampal N-methyl-D-aspartate (NMDA)-mediated synaptic currents in cultured hippocampal neurons[4,61], an effect mediated through the polyamine-binding site on the NMDA receptor complex[61]. The significance of this mechanism *in vivo* is not yet fully understood. Histamine also switches thalamic neuronal activity from rhythmic burst firing to single-spike activity through histamine H_1 and H_2 receptors, thus promoting accurate transmission of thalamocortical relay neurons and processing of sensory inputs and cognition[27]. Histamine may thus mediate its effects through specific histamine H_1, H_2 and H_3 receptors (for a recent review, see [22,52]) and through modulatory actions on other receptors.

2. BLOOD BRAIN BARRIER AND BRAIN TRAUMA

Histamine displays potent effects on the brain vessels *in vivo* and *in vitro*. It increases pinocytosis and synthesis of prostaglandins in brain capillaries[18], and promotes trypan blue-albumin diffusion through endothelial cells[49]. High levels of histamine have been detected in microvessel-enriched fractions isolated from

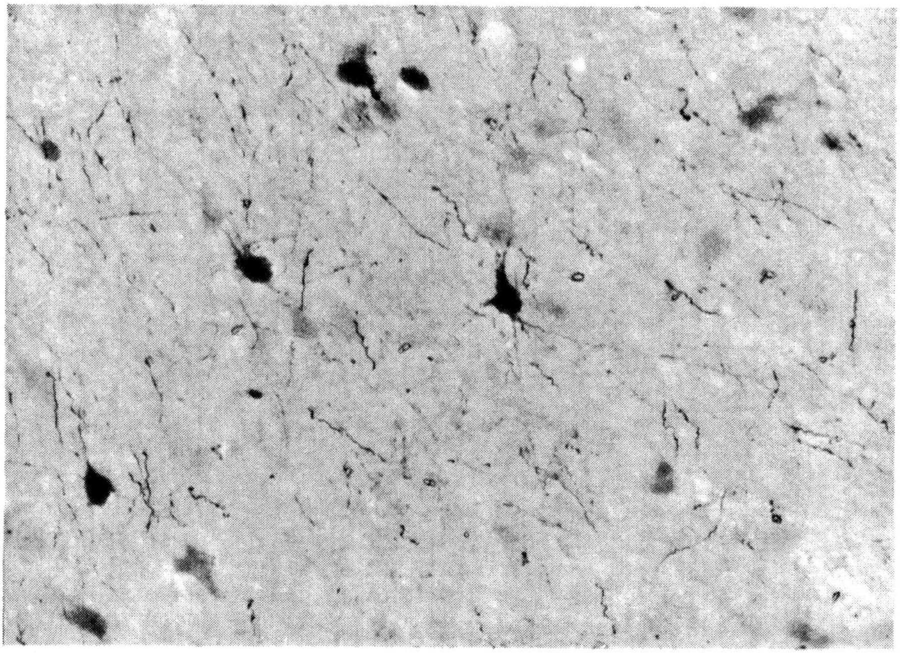

Figure 1. Histamine-containing tuberomammillary neurons and nerve fibers in normal human posterior hypothalamus.

guinea pig or bovine brain [48,17]. However, direct evidence of histamine synthesis in brain capillary endothelial cells is still lacking. In an immortalized brain endothelial cell line, RBE4[12], no L-histidine decarboxylase (HDC) mRNA was detected by PCR or *in situ* hybridization, and histamine was undetectable by HPLC and immunocytochemistry[20]. However, both H_1 and H_2 receptor mRNA:s were present, and the expression appeared to be downregulated by dexamethasone. Given that steroids have beneficial effects on cerebral ischemia and edema, and histamine receptors may mediate increased permeability, this downregulation of histamine receptors may contribute to the effects of steroids on vascular permeability.

3. SEIZURES

Some H_1 receptor antagonists induce seizures in humans[62,53,30] and increase epileptic discharges in patients suffering from epilepsy[51,63]. H_1 ligand uptake is also increased in PET-images of epileptic foci from patients with seizures[16]. In the rat brain, both autoradiographic binding studies[39] and *in situ* hybridization studies with specific oligonucleotides for H_1 receptor mRNA[23] suggest that H_1 receptors are abundant in areas involved in seizure activity (Fig. 2). Moreover,

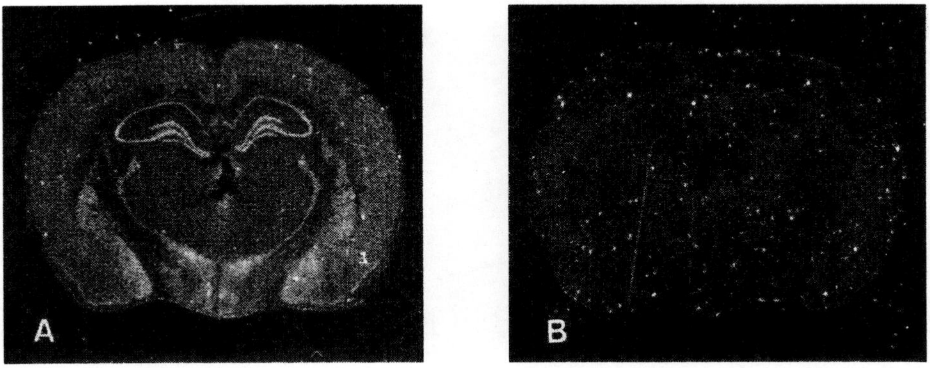

Figure 2. A) Expression of H_1 receptor mRNA in normal adult rat brain as revealed by *in situ* hybridization with an oligonucleotide probe. High expression is evident in the dentate gyrus and hippocampus, moderate expression in the reticular thalamic nucleus, striatum, zona incerta and cerebral cortex.
B) No signal can be seen in a consecutive section hybridized with a control probe.

activation of the central histaminergic system by treatments that increase brain histamine levels reduces convulsions in epileptic animal models, and suppression of the histaminergic system increases seizure duration and/or sensitivity[58,57,50,63]. The existing evidence thus suggests that histamine contributes to the modulation of seizure activity through H_1 receptors.

4. DEGENERATIVE DISEASES

4.1. Alzheimer's disease

Previous studies on histamine in Alzheimer's disease report conflicting results. In one study, significant increases in histamine concentrations were found in almost all brain areas except for the corpus callosum and globus pallidus[6], whereas another study reported decreases in the frontal, temporal and occipital cortices and in the caudate nucleus[26]. In a recent study, histamine concentrations were significantly lower in the hypothalamus, hippocampus and temporal cortex of Alzheimer brains than in control brains[42]. The differences in the prefrontal cortex, occipital cortex, putamen, pars compacta and pars reticulata of the substantia nigra were not significant, although a tendency to reduced levels was seen. No difference in the concentrations was seen in the caudate nucleus.

In the same study, the distribution of histamine-containing nerve fibers was examined in normal and Alzheimer brains. In normal brains all areas of the temporal lobe contained histamine-immunoreactive nerve fibers (Fig. 3). There was no obvious difference in the morphology or distribution of nerve fibers between normal brains and those of patients suffering from Alzheimer's disease Figs. 4 and 5). Long varicose fibers immunoreactive for histamine entered the

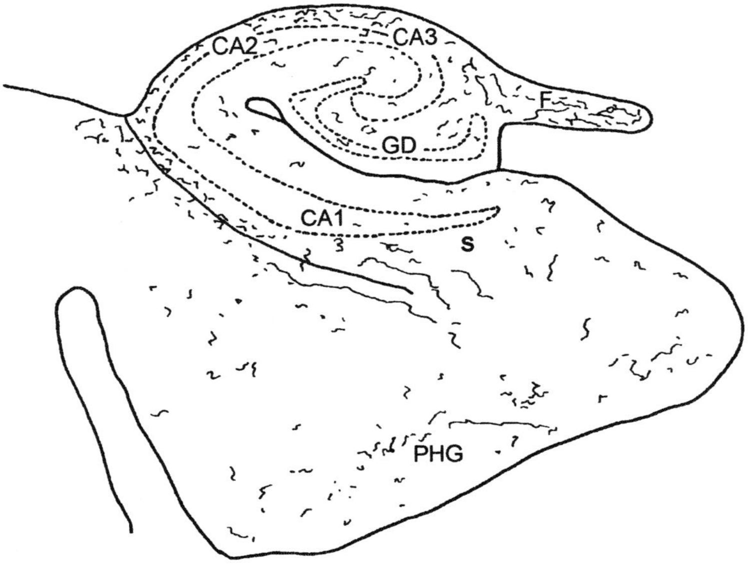

Figure 3. Distribution of histamine-immunoreactive nerve fibers in normal human hippocampal structures. F, fimbriae; CA1-CA3, cornu ammonis, areas 1-3; GD, gyrus dentatus; S, subilucum; PHG, parahippocampal gyrus. From Panula *et al.*[42]

hippocampus both via the fimbriae and the perforant pathway. The density of these fibers in the fimbriae and subiculum was higher than in other areas of the hippocampus. Scattered fibers were seen in hippocampal fields CA1-CA3 and in the dentate gyrus. Moderately dense fibers innervated all layers of the parahippocampal gyrus and entorhinal cortex. No histamine-immunoreactive mast cells or capillary endothelial cells were seen in these brain areas.

Other areas of the human brain have not been studied in detail. In Alzheimer's disease, neurofibrillary tangles are found to colocalize with histamine in the tuberomammillary areas in the posterior hypothalamus[1], and significantly reduced neuron numbers of the tuberomammillary nucleus in Alzheimer's brains[33] have been reported.

Based on current knowledge, nerve fibers are the primary storage site of histamine in the human hippocampus and associated temporal lobe structures. It appears that the hypothalamic histaminergic neurons project to the hippocampus both through the subiculum and fimbriae, which is in agreement with previous findings in the rat and with abundant fiber bundles originating from the hypothalamic histamine-containing neurons in the human brain[1].

Lack of histamine-immunoreactive mast cells suggests that the observed changes occur in the neuronal pool. Obviously, the postmortem time and storage

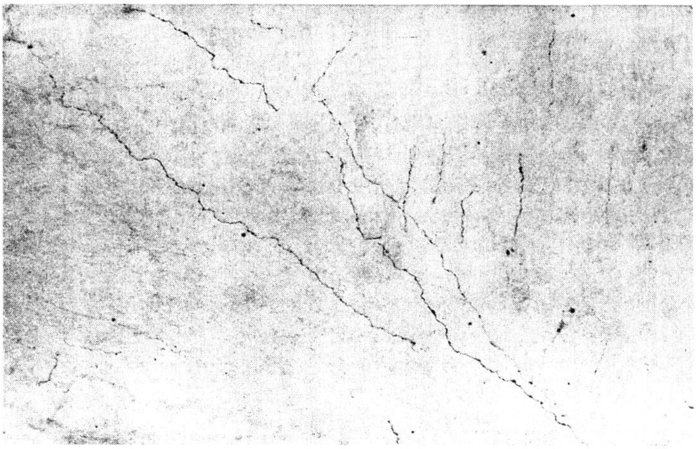

Figure 4. Histamine-immunoreactive varicose nerve fibers in the subiculum of an Alzheimer brain. Modified from Panula *et al.*[42].

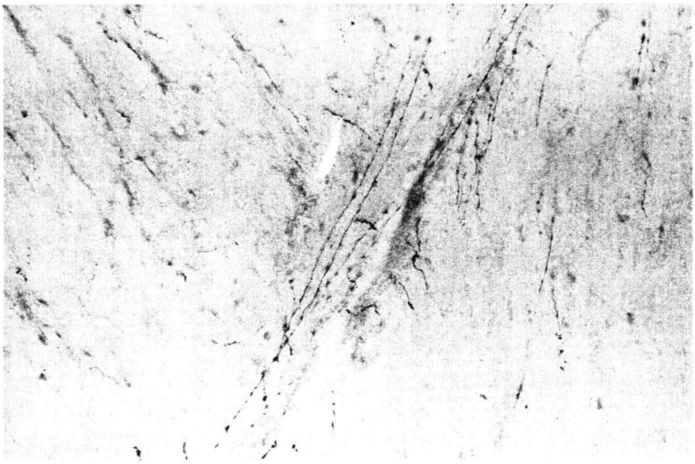

Figure 5. Histamine-immunoreactive varicose fibers in the alveus and subiculum of a normal human brain. Modified from Panula *et al.*[42].

conditions are essential factors that contribute to previous conflicting results, as a significant increase in brain histamine concentration occurs with increasing postmortem time.

The cholinergic system is commonly considered to be the one most severely affected in Alzheimer's disease, but loss of dopamine, serotonin and noradrenaline have also been reported in a number of studies (for review, see e.g.[14]). It is interesting to note that tetrahydroaminoacridine (THA), a non-specific cholinesterase inhibitor that affects cholinergic functions[35] and is found beneficial in Alzheimer's disease[56], also increases the action potential duration of histaminergic neurons[47] and inhibits histamine N-methyltransferase[8], and may thus affect histaminergic transmission in the brain. Decreased histaminergic input may also affect cholinergic activation of cortical and hippocampal neurons, as histamine excites cholinergic nucleus basalis neurons[21] and stimulation of the tuberomammillary histaminergic neurons increases hippocampal acetylcholine release in rats, an effect inhibited by an H_1 receptor antagonist, pyrilamine[29]. The presence of a widespread histaminergic neuronal system in the temporal lobe, as shown here and in other parts of the cerebral cortex[40], may also have implications in other disorders involving cortical functions.

Taken together, significant reductions, up to 55% in the hippocampus, are found in brain neuronal histamine content in Alzheimer's disease. Lack of brain histamine may contribute to the cognitive decline in Alzheimer's disease. However, activation of different histamine receptors may exert different modulatory effects on other systems. For example, extended use of H_2 blocking agents has been reported to delay the onset of Alzheimer's disease among siblings at high risk[5].

4.2. Parkinson's disease

Histamine is a known cataleptogen (for a review, see[37]), and the mechanism appears to be related to H_1 receptor rather than H_2 receptor function[36]. 1-Methyl-4-phenyl-1,2,3,6-tetrahydropyridine (MPTP) induces degeneration of the dopaminergic neurons of the substantia nigra[10,15]. The oxidative conversion of MPTP to MPP+, which is a substrate for dopamine transporters, requires monoamine oxidase[25]. Inhibitors of MAO B, but not of MAO A, prevent neurotoxicity of MPTP[55], which renders MAO B a potentially important enzyme in the MPTP-induced destruction of dopaminergic neurons. Interestingly, the sites of MPTP oxidation in Wistar rats were limited to a few systems, namely the histaminergic tuberomammillary neurons, serotonin neurons of the raphe nuclei, and the noradrenergic medullary neurons[34]. The tuberomammillary histamine neurons are also known to contain MAO B in several species, and these neurons innervate the substantia nigra[41,2,3]. These findings suggest that the MAO B activity responsible for the oxidation of MPTP to MPP+ may reside in the nigral projection fibers and/or cell bodies of the tuberomammillary histamine neurons. HDC activity in postmortem parkinsonian brains has been reported to be normal[13], and histamine levels in the neocortex, hippocampus and hypothalamus of MPTP-treated mice were unaffected[8], which suggests that the histamine neurons do not undergo degeneration after MPTP treatment. However, they may still be responsible for the toxic effect, provided that the levels of MPP+ in histaminergic neurons do not reach toxic levels. If astrocytes are the predominant site of MPP+ formation as has been suggested, it remains to be explained how MPP+, a polar compound, gets from astrocytes to the extracellular space. In histaminergic neurons, it could be packaged in secretory vesicles. A fairly high density of histaminergic fibers is characteristic of the substantia nigra of many mammals, including humans. The histamine levels in postmortem parkinsonian brains are considerably higher than those of control brains, whereas no differences can be seen in multiple system atrophy[31]. Although this may in part be due to medication, the possibility for specific pathology of the histaminergic system and associated changes in histamine levels and metabolism merit further investigation.

5. SCHIZOPHRENIA

Schizophrenia is currently viewed as a complex, at least in part neurodevelomental disorder of unknown etiology. A number of structural changes in schizophrenic brains have been described, and changes in dopaminergic functions are associated with many other neurotransmitter and receptor changes[54,60,7]. Reduced B_{max} of histamine H_1 receptor binding in postmortem schizophrenic frontal cortex[32], and a 2.6-fold increase in the cerebrospinal fluid concentration of *tele*-methylhistamine, the only known proximal histamine metabolite in the brain[46], suggest that the histamine turnover may be increased and followed by downregulation of H_1 binding. Famotidine, a histamine H_2 receptor antagonist, has reduced negative symptoms in some patients[19], a finding supported by an open-labeled trial[11]. An allelic variant, H2R649G, is 1.8 times more frequent in schizophrenics than in normal controls[38]. Histaminergic nerve fibers are present in all layers of the human neocortex[40], but autoradigraphic or *in situ* hybridization studies on regional differences in histamine receptors in normal or diseased brains have not yet been published.

6. CONCLUSIONS

Distinct changes in the histaminergic system of the brain have been reported in pathophysiological conditions and human brain diseases, including major degenerative disorders. However, in none of these disorders is there as yet evidence of primary involvement of specific pathology of the tuberomammillary histamine neurons. Recent cloning of the two histamine receptors enables studies on receptor expression and mutations. Studies on these will cast further light on the significance of the changes in the histamine system of the brain.

ACKNOWLEDGEMENTS

The authors' original research has been supported by the Medical Research Council of the Academy of Finland and Sigrid Juselius Foundation, CIMO and the Signal Transduction Program of Åbo Akademi University.

REFERENCES

1. Airaksinen M.S., Paetau A., Paljärvi L., Reinikainen K., Riekkinen P.J., Suomalainen R. and Panula P. (1991), Neuroscience **44**: 465-481.

2. Airaksinen M.S. and Panula P. (1988), J. Comp. Neurol. **273**: 163-186.

3. Airaksinen M.S., Flugge G., Fuchs E. and Panula P. (1989), J. Comp. Neurol. **286**: 289-310.

4. Bekkers J.M. (1992), Science **261**: 104-106.

5. Breitner J.C., Welsh K.A., Helms M.J., Gaskell P.C., Gau B.A., Roses A.D., Perick-Vance M.A. and Saunders A.M. (1995), Neurobiol. Aging **16**: 523-530.

6. Cacabelos R., Yamatodani A., Niigawa H., Hariguchi S., Tada K., Nishimura T., Wada H., Brendeis L. and Pearson J. (1989), Meth. Find. Exp. Clin. Pharmacol. **11**: 353-360.

7. Carpenter W.T. and Buchanan R.W. (1994), New Engl. J. Med. **330**: 681-690.

8. Cumming P., Jakubovic A. and Vincent S.R. (1989), Eur. J. Pharmacol. **166**: 299-301.

9. Cumming P., Reiner P.B. and Vincent S.R. (1990), Biochem. Pharmacol. **40**: 1345-1350.

10. Davis G.C., Williams A.C., Markey S.P., Ebert M.H., Caine E.D., Reichert C.M. and Kopin I.S. (1979), Psychiatry Res. **1**: 249-254.

11. Deutsch S.I., Rosse R.B., Kendrick K.A., Fay-McCarthy M., Collins J.P. Jr. and Wyatt R.J. (1993), Clin. Neuropharmacol. **16**: 518-524.

12. Durieu-Trautman O., Fédérici C., Creminon C., Foignant-Chaverot N., Roux F., Claire M. and Strosberg A.D. (1993), J. Cellular Physiol. **155**: 104-111.

13. Garbarg M., Javoy-Agid F., Schwartz J.C. and Agid Y. (1983), Lancet **1**: 74-75.

14.Hardy J., Adolfsson R., Alafuzoff I., Bucht G., Marcusson J., Nyberg P., Perdahl E., Wester P. and Winblad B. (1985), Neurochem. Int. **7**: 545-563.

15.Heikkila R.E., Hess A. and Duvoisin R.C. (1983), Science **224**: 1451-1453.

16.Iinuma K., Yokoyama H., Otsuki T., Yanai K., Watanabe T., Ido T. and Itoh M. (1993), Lancet **341**: 238.

17.Jarrott B., Hjelle J.T. and Spector S. (1979), Brain Res. **168**: 323-330.

18.Joo F., Dux E., Karnushina I.L., Halász N., Gecse Á., Ottlecz A. and Mezei Zs. (1981), Agents Actions **11**: 129-134.

19.Kaminsky R., Moriarty T.M., Bocine J., Wolf D.E. and Davidson M. (1990), Lancet **335**: 1351-1352.

20.Karlstedt K., Sallmen T., Eriksson K.S., Lintunen M., Wasowicz K., Fukui H., Joó F. and Panula P. (1996), Soc. Neurosci. Abstr. Vol. **22**, Part 1, p53.

21.Khateb A., Fort P., Pegna A., Jones B.E. and Muhlethaler M. (1995), Neuroscience **69**: 495-506.

22.Leurs R., Smit M. and Timmerman H. (1995), Pharmac. Ther. **66**: 413-463.

23.Lintunen M., Karlstedt K., Fukui H. and Panula P. (1996), Eur J. Neurosci. Suppl. 9, p75.

24.Lipinski J.F., Schaumburg H.H. and Baldessarini R.J. (1973), Brain Research **52**: 403-408.

25.Markey S.P., Johannessen J.N., Chiueh C.C., Burns R.S. and Herkenheim MA (1984), Nature **311**: 464-467.

26.Mazurkiewicz-Kwilecki I.M. and Nsonwah S. (1989), Can. J. Physiol. Pharmacol. **67**: 75-78.

27.McCormick D.A. and Williamson A. (1991), J. Neurosci. **11**: 3188-3199.

28.Mirra S.S., Heyman A., McKel D., Sumi S.M., Crain B.J., Brownlee L.M., Vogel F.S., Hughes J.P., van Helle G., Berg L. and participating CERAD neuropathologists (1991), Neurology **41**: 479-486.

29.Mochizuki T., Yamatodani A., Okakura K., Horii A., Yamamoto Y. and Yamatodani A. (1994), J. Neurochem. **62**: 2275-2282.

30.Mueller S. M. (1983), New Engl. J. Med. **308**: 653.

31.Mättö M., Panula P., Relja M., Eriksson K.S., Kalimo H., Röyttä M., and Rinne J. (1997), Movement Disord. **12**: 829.

32.Nakai T., Kitamura N., Hashimoto T., Kajimoto Y., Nishino N., Mita T., Tanaka C. (1991), Biol. Psychiat. **30**: 349-356.

33.Nakamura S., Takemura M., Ohnishi K., Suenaga T., Nishimura M., Akiguchi I., Kimura J. and Kimura T (1993), Neurosci. Lett. **151**: 196-199.

34.Nakamura S. and Vincent S.R. (1986), Neurosci. Lett. **65**: 321-325.

35.Nordberg A., Nilsson-Håkanson L., Adem A., Lai Z. and Winblad B. (1989), Progr. Clin. Biol. Res. **317**: 1169-1178.

36.Onodera K. (1991), Meth. Find. Exp. Clin. Pharmac. **13**: 397-403.

37.Onodera K., Yamatodani A., Watanabe T. and Wada H. (1994), Progr. Neurobiol. **42**: 685-702 .

38.Orange P.R., Heath P.R., Wright S.R., Ramchand C.N., Kolkeiwicz L. and Pearson R.C. (1996), Mol. Psychiat. **1**: 466-469.

39.Palacios J.M., Warmsley J.K. and Kuhar M.J. (1981), Neuroscience **6**: 15-37.

40.Panula P., Airaksinen M.S., Pirvola U. and Kotilainen E. (1990), Neuroscience **34**: 1127-1132.

41.Panula P., Pirvola U., Auvinen S. and Airaksinen M.S. (1989), Neuroscience **28**: 585-610.

42.Panula P., Rinne J., Kuokkanen K., Eriksson K.S., Sallmen T., Kalimo H. and Relja M. (1998), Neuroscience In Press.

43.Panula P., Yang H.-Y.T. and Costa E. (1984), Proc. Natl. Acad. USA **81**: 2572-2576.

44.Panula P., Takagi H., Inagaki N., Yamatodani A., Tohyama M., Wada H. and Kotilainen E. (1993), Neurosci. Lett. **160**: 53-56.

45.Panula P., Häppölä O., Airaksinen M.S., Auvinen S. and Virkamäki A. (1988), J. Histochem. Cytochem. **36**: 259-269.

46.Prell G.D., Green J.P., Kaufmann C.A., Khandelwal J.K., Morrishow A.M., Kirch D.G., Linnoila M. and Wyatt R.J. (1995), Schizophr. Res. **14**: 93-104.

47.Reiner P.B. and McGeer E.G. (1988), Eur. J. Pharmacol. **155**: 265-270.

48.Robinson-White A. and Beaven M.A. (1982), J. Pharmacol. Exp. Ther. **223**: 440-445.

49.Rotrosen D. and Gallin J.I. (1986), J. Cell Biol. **103**: 2379-2387.

50.Scherkl R., Hashem A. and Frey H.H. (1991), Epilepsy Res. **10**: 111-118.

51.Shimoda Y., Koizumi A. and Tanaka K. (1960), Yonago Acta Med. **4**: 99-102.

52.Schwartz J.C., Arrang J.M., Garbarg M., Pollard H. and Ruat M. (1991), Physiol. Rev. **71**: 1-51.

53.Schwartz J.C. and Patterson J.H. (1978), Am J. Dis. Child. **132**: 37-39.

54.Seeman P (1993), Arch. Neurol. **50**: 1093-1095.

55.Sonsalla P.K. and Nicklas W.J. (1992), In: Handbook of parkinson's disease, WC Koller Ed., Marcel Dekker, New York, 1992, pp. 319-340.

56.Summers W.K., Majowski V., March G.M., Tackiki K. and Kling A. (1986), N. Engl. J. Med. **315**: 1241-1245.

57.Tuomisto L. and Tacke U. (1986), Neuropharmacology **25**: 955-958.

58.Tuomisto L., Tacke U. and Willamn A. (1987), Agents Actions **14**, 558-560.

59.Vanhala A., Yamatodani A. and Panula P. (1994), J. Comp Neurol. **347**: 101-114.

60.Williamson P.C. (1993), Arch. Neurol. **50**: 1096-1097.

61.Vorobjev V.S., Sharonova I.N., Walsh I.B. and Haas H. (1993), Neuron **11**: 837-844.

62.Wyngaarden J.B. and Seevers M.H. (1951), J. Am. Med. Assoc. **145**: 277-283.

63.Yokoyama H., Onodera K., Maeyama K., Yani H., Inuma K., Tuomisto L. and Watanabe T. (1992), Naunyn-Schmiedbergs Arch. Pharmacol. **346**: 40-45.

R. Leurs and H. Timmerman (Editors)
The Histamine H_3 Receptor
1998 Elsevier Science B.V.

Histamine H_3 Antagonists as Potential Therapeutics in the CNS

Kenji Onodera[a] and Takehiko Watanabe[b]

[a] Department of Pharmacology, Tohoku University School of Dentistry, Seiryo-machi 4-1, Aoba-ku, Sendai 980-77, Japan

[b] Department of Pharmacology I, Tohoku University School of Medicine, Seiryo-machi 2-1, Aoba-ku, Sendai 980-77, Japan

1. INTRODUCTION

In 1983, Schwartz and his coworkers demonstrated the presence of a novel presynaptic autoreceptor designated as the H_3 receptor [1,2]. A few years later, the existence of the histamine H_3-autoreceptor was further confirmed by the introduction of the first selective ligands such as (R)-α-methylhistamine as an agonist and thioperamide as an antagonist, which allowed pharmacological studies on function of histamine H_3-receptors [3]. Nowadays, more specific H_3-antagonists have become available such as clobenpropit, GT-2016, AQ0145, and FUB 181 [4-7], which will be described in other chapters in details. On the other hand, recent neuropharmacological studies of the central histaminergic system have shown a variety of physiological roles of histamine in learning and memory, convulsion, thermoregulation, circadian rhythm, locomotion, neuroendocrine regulation, and others (8,9). In addition, the clinical importance of histaminergic drugs in treatments of Parkinson's disease, Alzheimer disease, epilepsy, motion sickness, etc. has been indicated (10,11). In this chapter, we described studies on certain physiological roles of brain histamine elucidated by the application of histamine H_3-receptor antagonists and discussed their possibilities as therapeutics to CNS diseases, particularly to disorders of learning and memory and convulsions.

2. LEARNING AND MEMORY

Recent studies showed that the central histaminergic system plays an important role in learning and memory in rodents. In brief, activation of the histaminergic system by intracerebroventricular (i.c.v.) administration of histamine or intraperitoneal (i.p.) administration of L-histidine, a precursor of histamine, leads to improve learning and memory in rodents [12-14]. These effects were antagonized by histamine H_1-receptor antagonists. Moreover, inhibition of the central histaminergic system by blocking the H_1 receptors or histamine synthesis resulted in disturbance of learning and memory [13]. However, little is known about the effects of histamine H_3 antagonists on various physiological events, especially on learning and memory, except that Meguro *et al.* showed that thioperamide improved learning/memory in passive avoidance test of P/8 senescence accelerated (SAM) mice [15]. In this section, we showed the effects of thioperamide and clobenpropit on a scopolamine-induced learning deficit using an step through-passive avoidance test in mice.

2.1. Effects of thioperamide and clobenpropit on the scopolamine-induced learning deficit in the step-through passive avoidance test in mice.

Treatment with scopolamine (1 mg/kg) significantly shortened step-through latency in the retention trial compared with the vehicle-treated control group in all cases. Thioperamide, clobenpropit or in combination with zolantidine did not affect the step-through latency at the dose tested in the retention trial compared with the vehicle-treated control group [16,17]. Thioperamide (20 mg/kg) alone [16] or clobenpropit (10 and 20 mg/kg) alone showed a tendency to reverse the scopolamine-induced shortening of step-through latency in the retention trial (Figure 1) [17]. Notably, in combination with zolantidine (20 mg/kg, i.p.), thioperamide (20 mg/kg) or

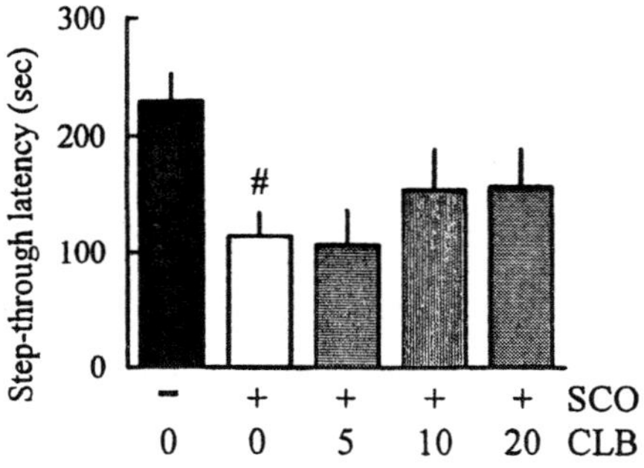

Figure 1 Effect of clobenpropit on scopolamine-induced shortening of the step-through latency in the passive avoidance test. Male ICR mice (Clea Japan, Inc., Tokyo, Japan), aged 6 weeks and weighing 30-35 g were housed under standard conditions (23±1℃, light-dark cycle with the light on from 7:00 to 19:00) with free access to water and food in their home cage. between 13:00 and 17:00. The step-through passive avoidance test was performed between 13:00 and 17:00 as described previously (16). Briefly, an acquisition trial was performed as follows: the mice were placed in a light compartment facing away from a dark compartment. When the mice entered the dark compartment, an electrical foot shock (constant voltage: 75 V) was delivered to the grid. Twenty-four hours later, a retention trial was performed in the same manner as an acquisition trial, and the latency for entering the dark compartment (step-through latency) was recorded. If the mice did not enter the dark compartment within 300 sec in the retention trial, the test was stopped and the step-through latency was recorded as 300 sec. Clobenpropit (CLB; 5, 10, and 20 mg/kg) and scopolamine (SCO, 1 mg/kg) were administered i.p. 60 and 15 min, respectively, before the acquisition trial. Physiological saline was injected into the reference group. Each column and bar represent the step-through latency in the retention trial as the mean ± S.E. of 10 mice. Significant difference: #P < 0.05 vs. vehicle-treated control group.

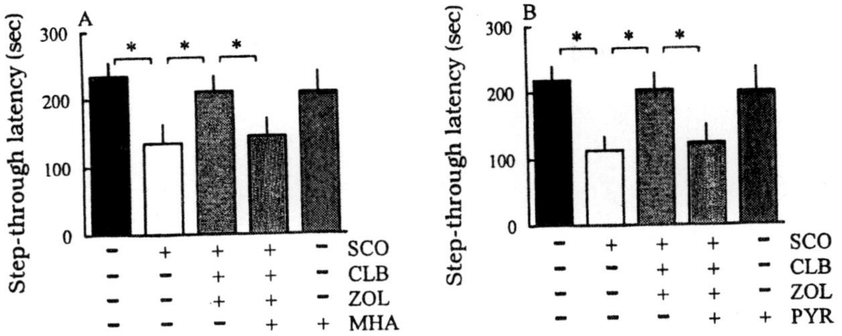

Figure 2 Effect of clobenpropit plus zolantidine on scopolamine-induced shortening of the step-through latency and the antagonism of (R)- α-methylhistamine (A) and pyrilamine (B) in the passive avoidance test. (R)- α-Methylhistamine (MHA, 20 mg/kg) or pyrilamine (PYR, 20 mg/kg), zolantidine (ZOL, 20 mg/kg), clobenpropit (CLB, 10 mg/kg), and scopolamine (SCO, 1 mg/kg) were administered i.p. 80, 70, 60, and 15 min, respectively, before the acquisition trial. Physiological saline was injected into the reference groups. Each column and bar represent the step-through latency in the retention trial as the mean ± S.E. of 15 mice. Significant difference: $^*P < 0.05$.

clobenpropit (10 mg/kg) significantly improved the scopolamine-induced shortening of step-through latency in the retention trial, and these ameliorating effects were antagonized by (R)-α-methylhistamine (20 mg/kg) and pyrilamine (20 mg/kg) (Figure 2) [16,17]. (R)-α-Methylhistamine or pyrilamine alone did not affect the step-through latency at the dose tested in the retention trial compared with the vehicle-treated control group. Zolantidine (20 mg/kg) alone affected neither the step-through latency in the retention16,17].

Thus, thioperamide or clobenpropit in combination with zolantidine significantly improved the learning deficit produced by scopolamine. This is consistent with our previous finding that thioperamide plus zolantidine ameliorated a scopolamine-induced learning deficit using the elevated plus-maze test in mice [18]. These data show that the potency of clobenpropit against a scopolamine-induced learning deficit is roughly 2-fold higher than that of thioperamide. Since the ameliorating effect of H_3 antagonists in combination with zolantidine was antagonized by (R)- α-methylhistamine, the ameliorating effect of H_3 antagonists and zolantidine is probably due to the increased release of endogenous histamine *via* autoreceptors on histaminergic neurons (Figure 3) [19]. This is supported by our data in which clobenpropit dose-dependently decreased histamine levels and increased histidine decarboxylase activity in the mouse brain (Table 1) [1,2,20] and the blockade of histamine H_3 receptors leads to enhance neuronal histamine release, resulting in lower histamine levels in tissue homogenates [1, 2,20].

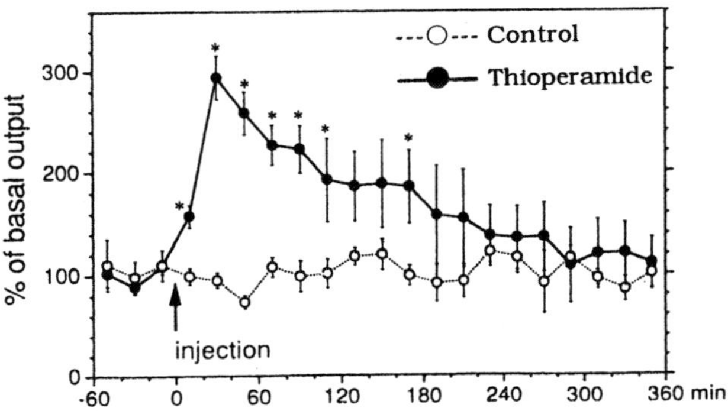

Figure 3 Effect of thioperamide on histamine release measured by *in vivo* microdialysis in the anterior hypothalamus. *P< 0.05 *vs.* control group. The basal output is defined as an average of the first 3 fractions before injection of saline or thioperamide, and subsequent fractions are expressed as percentages of it (mean±S.E.). Data from Mochizuki et al. (1991) [19].

Table 1
Effects of clobenpropit on histamine level and histidine decarboxylase (HDC) activity of mouse brain

Dose (mg/kg, i.p.)	Histamine levels (pmol/g)	HDC activity (pmol/min/mg protein)
0	312.1±20.5	0.190±0.008
0.3	273.5±22.3	0.193±0.008
1.0	262.0±15.8 [a]	0.230±0.04 [a]
3.0	231.7±18.5 [b]	0.304±0.04 [a]

The mice (n=6) were killed 60 min after clobenpropit. [a] $P < 0.05$ and [b] $P < 0.01$ vs. the saline-treated group.

Moreover, the ameliorating effect was antagonized by pretreatment with pyrilamine, a histamine H_1-receptor antagonist, in this study. This result suggested that the ameliorating effect of H_3 antagonists plus zolantidine is also mediated through postsynaptic histamine H_1 receptors. However, it is notable that clobenpropit or thioperamide alone could not significantly improve the scopolamine-induced learning deficit. The H_3 antagonists block histamine H_3 receptors, and then increase neuronal histamine, which in turn stimulates both postsynaptic histamine H_1 and H_2 receptors. 2-Methylhistamine, a histamine H_1-receptor agonist, facilitated memory, and 4-methylhistamine, a histamine H_2-receptor agonist, worsened memory (21). Therefore, it is conceivable that stimulation of histamine H_1 receptors may improve the learning deficit, but that of histamine H_2 receptors may have the opposite effect. Although we have the data that even thioperamide alone improved learning deficit in SAM-mice in the passive avoidance test [15], we have no data to explain why thioperamide alone showed opposite effects in the same method, except the differences in strain and manipulation of learning deficit.

In summary, histamine H_3-antagonists such as thioperamide and clobenpropit in combination with zolantidine, a histamine H_2-receptor antagonist, ameliorated the scopolamine-induced effect. This ameliorating effect was mediated through histamine H_3 receptors and/or histamine H_1 receptors.

2.2 Involvement of histamine H_3 receptors as heteroreceptors in learning and memory

Histamine H_3-receptors have been reported to regulate not only the release and turnover of histamine *via* autoreceptors on histaminergic nerve endings [1-3], but also the releases of noradrenaline, dopamine, serotonin, and acetylcholine *via* heteroreceptors on non-histaminergic axon terminals [22-26]. Thioperamide increased the release of these neurotransmitters, while histamine and (R)- α -methylhistamine decreased them *via* histamine H_3 heteroreceptors *in vitro* [22-26].

2.2.1 The relationship between histaminergic and cholinergic systems: Effect of H_3 antagonists on brain acetylcholine or choline levels in mice

The central cholinergic system is also known to play an important role in learning and memory [27, 28], and thus it is highly possible that there is a close relationship between the cholinergic and histaminergic neuron systems. For example, the memory facilitating effect of 2-methylhistamine was attenuated by a muscarinic antagonist, pirenzepine [21]. Conversely, activation of the central histaminergic system could antagonize the scopolamine-induced learning deficit, as shown here [14,16-18]. *In vitro* and *in vivo* studies showed that thioperamide-induced increase of acetylcholine release was mediated by histamine H_3 heteroreceptors [25,26,29]. Therefore, we examined the possibility that H_3 antagonists, in addition to acting on the histaminergic system, acted *via* the cholinergic system in producing its ameliorating effect.

Thioperamide (20 mg/kg) in combination with zolantidine (20 mg/kg) significantly increased contents of choline in the brain. However, in case of clobenpropit, we could not detect any changes in acetylcholine or choline levels. Therefore, the changes of choline induced by thioperamide may not be due to histamine H_3 heteroreceptors or probably heteroreceptors played a minor role in the modulation of acetylcholine release.

These data suggest that the contribution of the cholinergic system to the ameliorating effect induced by histamine H_3-receptor antagonists is not mediated by histamine H_3 heteroreceptors, but by postsynaptic histamine H_1 receptors. This is supported by previous reports that histamine excites cholinergic neurons through histamine H_1 receptors [30, 31].

2.2.2 The relationship between histaminergic and monoaminergic systems: Effect of H_3 antagonists on brain levels of monoamines and their metabolites in mice

Thioperamide is suggested to enhance ^3H-overflow after incubation of cortical slice with ^3H-noradrenaline or ^3H-serotonin, and H_3 antagonist inhibited the releases *in situ* (22,24,26). Compton *et al.* reported that bilateral lesions of the locus coeruleus, a noradrenergic nucleus, impaired learning and memory (32). Thus, we considered the possibility that H_3 antagonists, in addition to the histaminergic system, acted *via* the monoaminergic system in producing its ameliorating effect.

Thioperamide has no influence on the monoaminergic system in mice (33). Oishi *et al.*, (34) reported that thioperamide did not induce any significant changes in the contents of monoamines and their metabolites, but actually enhanced the histamine turnover rate in mice and rats. Clobenpropit alone or in combination with zolantidine did not affect the ratio of (HVA+DOPAC)/dopamine or 5-HIAA/5-HT at doses tested in any regions examined. Clobenpropit (10 and 20 mg/kg) alone or in combination with zolantidine significantly increased the MHPG/noradrenaline ratio in the midbrain and/or pons and medulla oblongata, indicating an increase in noradrenaline turnover in these regions [17].

In relation to the monoaminergic systems we observed that clobenpropit increased turnover rate of noradrenaline only in some brain regions [17], although histamine H_3 heteroreceptors modulate the releases of noradrenaline, dopamine, and serotonin [23-26]. Thus, it appears that the contribution of histamine H_3 heteroreceptors on the modulation of monoaminergic neurotransmitters may be minor, just being similar to the cholinergic system.

3. CONVULSIONS

Several lines of evidence have indicated that the central histaminergic system plays an important role in inhibition of convulsions [10,20,35-40]. Tuomisto and Tacke [35] showed that metoprine, an inhibitor of histamine *N*-methyltranseferase that increases in histamine levels after its systemic administration, inhibited hindlimb extension after maximal electroshock in rats. In mice, activation of the histaminergic system by i.p. administration of L-histidine showed the anticonvulsive effects on the duration of the clonic and convulsive coma phases in mice [36]. These effects were found to depend on the brain levels of histamine (Figure 4) and to be mediated through central histamine H_1-receptors [36,37]. Moreover, inhibition of histamine synthesis leads to enhance the clonic and convulsive coma phases in younger mice [36]. Concerning histamine H_3-ligands and seizures, Scherkl et al. reported that neither an agonist nor an antagonist influenced the seizure threshold for electrically-induced convulsions [41].

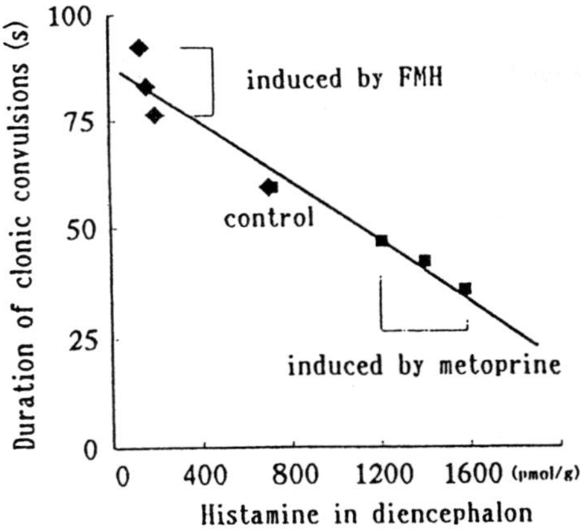

Figure 4 Correlation between the duration of chronic convulsion and the histamine level in the diencephalon in mice.

3.1. Effects of thioperamide and clobenpropit on electrically-induced convulsions in mice.

Thioperamide (3.75-15 mg/kg) or clobenpropit (0.3-3 mg/kg) showed a dose-dependent inhibition of convulsions (Figure 5) [20,40]. The time course of suppressive effects of thioperamide (7.5 mg/kg) on tonic, clonic and convulsive coma phases in mice. The maximum effects of thioperamide (7.5 mg/kg) or clobenpropit (1 mg/kg) on convulsions were observed one hour after the administration (Figure 6): Clobenpropit was approximately 10 times more potent than thioperamide against electrically-induced convulsions. Anticonvulsive effects by thioperamide or clobenpropit were antagonized by pyrilamine and (R)-α-methylhistamine.

In summary, anticonvulsive effects by thioperamide or clobenpropit were antagonized by pyrilamine and (R)-α-methylhistamine, indicating that histamine is released from histaminergic nerve terminals through histamine H_3 receptors and interacts with histamine H_1 receptors on postsynaptic neurons [20,40]. These findings support the hypothesis that the central histaminergic neuron system is involved in the inhibition of seizures. In other words, histamine is an endogenous anticonvulsant, which is the basis of possibilty of H3 antagonists as anticonvulsants.

3.2. Effects of AQ 145 on electrically-induced convulsions in mice

A new H_3 antagonist AQ 145 (N-1-adamantyl-N'N"[1,5-(3-4,(5)-1H imidazolyl)-petanediyl]formamidine dihydrochloride) was synthesized by Green Cross Pharmac. Co, Osaka, Japan. AQ 145 has 5-fold higher affinity to H_3 receptors than thioperamide in *in vitro* binding assay. The effect of this compound on electrically-induced convulsions in mice was examined as similarly as described above. The durations of tonic, clonic and convulsive come phases were significantly decreased by i.p. administration of 30 mg/kg dose [6].

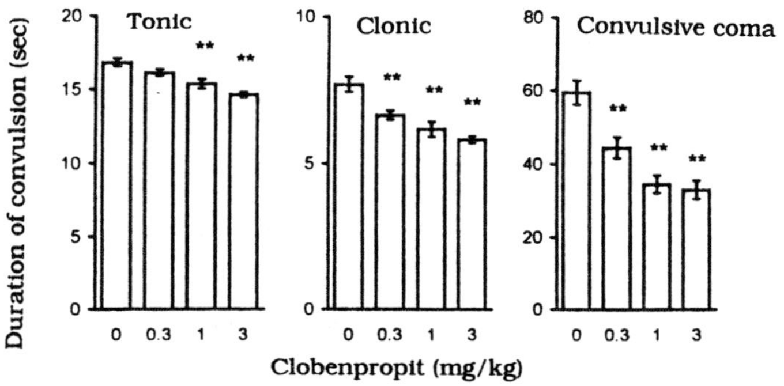

Figure 5 Effects of clobenpropit on the duration of tonic, clonic and convulsive coma phases of electrically induced convulsions in mice (n=8). Six-week-old male ddY mice (Funabashi Farm Co., Funabashi, Japan) weighing 23-26 g were used. The animals were housed under standard conditions (22±2℃, light-dark cycle with the light on from 7:00 to 19:00) with free access to water and food in their home cages. Experiments were performed between 13:00 and 16:00. Convulsions were induced as described earlier [20,40]. Briefly, electroconvulsive shock was induced by applying an electric current (110-Hz square waves of 30 mA for 0.1 s) through ear-clip electrolodes attached with electroencephalograph paste. Seizure susceptibility was evaluated as the duration of the various phases of convulsions. The tonic phase was regarded as the period between the onset of hindlimbs extension and the start of myoclonic jerks, the clonic phase as that during myoclonic jerks, and the convulsive coma phase as that between the end of myoclonic jerks and recovery of the righting reflex. The mice were subjected to electroshock 60 min after the i.p. administration of clobenpropit. *P < 0.05 and ** P < 0.01 vs. the saline-treated group.

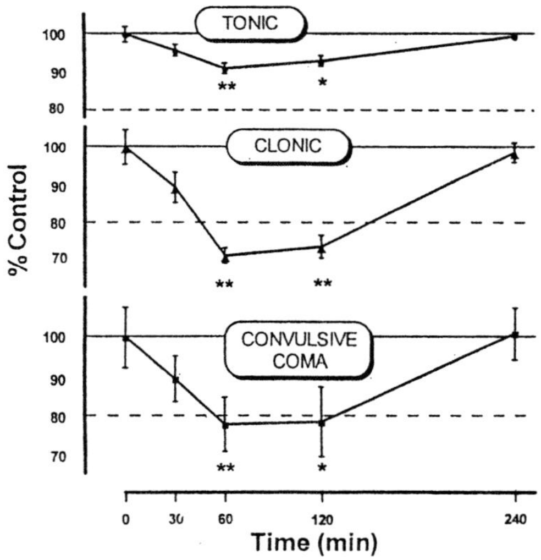

Figure 6 Time course of effect of thioperamide on electrically-induced convulsions in mice. *P < 0.05 and ** P < 0.01 vs. control group.

4. MISCELLANEOUS AND PERSPECTIVES

In this chapter, we can clearly demonstrate that the histamine H_3 antagonists are effective in experimental models of dementia and epilepsy. The potency of clobenpropit against a scopolamine-induced learning deficit is roughly 2-fold higher than that of thioperamide, while clobenpropit was approximately 10 times more potent than thioperamide against electrically-induced convulsions. The pharmacological actions of histamine H_3 antagonists are almost the same as those by i.c.v. administration of histamine or its agonists in animals [44,45]. Sakurai *et al.* reported that thioperamide does not penetrate the brain easily by their pharmacokinetic study [42]. Recently, Mochizuki *et al.* studied the penetration of histamine H_3-ligands by *ex vivo* experiments, showing that clobenpropit is worse in penetration through the blood-brain barrier than thioperamide does [43]. Nevertheless, the results of the pharmacokinetic studies does not always agree with the results from pharmacological studies. In fact, we obtained positive effects of histamine H_3-antagonists such as thioperamide and clobenpropit on electrically-induced convulsions in mice. The i.c.v. administration of histamine directly stimulates the postsynaptic histamine receptors [44], whereas histamine H_3 antagonists increase the release of neuronal histamine, and the released histamine stimulates the postsynaptic histamine receptors (Table 2) [1,2,45]. The i.c.v. administration of histamine produces behavioral arousal in conscious state and decreases the duration of narcosis of pentobarbital-treated animals [10]. However, in humans, it is impossible to administer histamine through the brain. Therefore, histamine H_3 antagonists provide a good tool to activate the central histaminergic neuron system. Previously, we suggested the clinical importance of central histaminergic system in epilepsy, Alzheimer's disease, Parkinson's disease, motion sickness, schizophrenia, nalcolepsy and depression [10]. Since classical antiepileptics such as phenytoin and phenobarbital have serious side effects such as drowsiness, sedation and depression [46], histamine H_3 antagonists will be a candidate of therapy of epilepsy without these side effects. That is because increased release of endogenous histamine induced by thioperamide produced wakefulness, and the effect was diminished by pretreatment with an histamine H_1-antagonist [47,48]. Moreover, this wakeful action may also be useful to treat the patients of nalcolepsy.

In addition, Kathmann et al. [49] and Rodriges et al. [50] pointed out the important role of histamine H_3-receptors in the atypical profile of clozapine, which has an efficacy to the neurolpetic-nonresponsive schizophrenia patients. Clozapine acts as an antagonist with apparent K_B value of 79.5 nM in a study on H_3-mediated inhibition of (^{3}H)-5-hydroxytryptamine release from rat brain cortex slices [50]. Ryu *et at.* suggested that histamine H_3 receptors were closely related to the dopaminergic neuron system, especially dopamine D_1 receptors, demonstrating the heterogeneous distribution of histamine H3- and dopamine D_1- and D_2-receptors in the extrapyramidal system of rat [51, 52]. The increases of the histamine H_3-, dopamine D_1- and D_2-receptors binding sites were reported in treatment with quinolinic acid and 6-hydroxydopamine. The upregulation of histamine H_3-receptors occurred in the substantia nigra and striatum after 6-hydroxydopamine injection into the rat brain [51]. The treatment of quinolinic acid resulted in similar increases in histamine H_3- and dopamine D_1-receptor binding sites in the striatum and ipsilateral substantia nigra. Dopamine D_2-receptor binding sites were relatively well conserved, whereas H_3-receptors increased considerably [52]. These data suggest that histamine H_3- and dopamine D_1-receptor binding sites are localized on the striatonigral projection neurons which are together sensitive to quinolinic acid, and that the distributional compartment of dopamine D_2-receptor binding sites is quite different from those of histamine H_3- and dopamine D_1-receptors [51-53]. Further approaches in clinics will be needed to clarify the effect of histamine H_3-antagonists on disorders related to central D_1-receptors.

In conclusion, from the findings of various basic researches, histamine H_3 antagonists may be useful for the therapies of Altzheimer's disease, narcolepsy, schizophrenia, and depression in addition to dementia and epilepsy.

Table 2
Functional changes by stimulation of postsynaptic histamine receptors with histamine release induced by thioperamide or clobenpropit

	Involved postsynaptic receptors			References
	H_1-receptors		H_2-receptors	
Ameriolating effects on scopolamine-induced learning deficits	**Potentiation**	>>	Suppression	14,16,17
Electrically-induced convulsions	**Suppression**		No change	20,40
Feeding behavior	**Suppression**		No change	54
Arousal pattern in EEG	**Increase**		No change	47
Morphine-induced antinociception	**Suppression**	>>	Potentiation	55

Bold characters show the effects of histamine H_3 antagonists finally.

REFERENCES

1. J.C. Schwartz, J.M. Arrang, M. Garbarg, and H. Pollard, Histaminergic Neurons: Morphology and Function. T . Watanabe and H . Wada eds. CRC Press: Boca Raton. (1991)85.
2. J.M. Arrang, M. Garbarg and J.C. Schwartz,Nature 302 (1983)832

3. J.M. Arrang, M. Garbarg and J.C. Schwartz, Neuroscience 15 (1985) 553.
4. H. Van Der Goot, M.J.P. Schepers, G.J. Sterk, and H. Timmerman, Eur. J.Med. Chem. 27 (1992) 511.
5. C.E. Tedford, S.L. Yates, G.P. Paulowski, J.W. Nalwalk, L.B. Hough, M.A. Khan, G.J. Durant and R.E.A. Frederickson, J. Pharmacol. Exp. Ther. 275 (1995) 598.
6. K. Murakami, H. Yokoyama, K. Onodera, K. Iinuma, and T. Watanabe, Meth. Find. Exp. Clin. Pharmacol.17(Supl.C) (1995) 70.
7. W. Schunack, H. Stark, K. Purand, A. Hüls, X. Ligneau, J.M.Arrang, J.C. Schwartz, Abstract of XXVI European Histamine Research Society (Seville) (1998).
8. J.C. Schwartz, J.M. Arrang, M. Garbarg and E. Traiffort. Psychopharmacology; the fourth Generation of Progress. Bloom, F.E. and Kupfer, D.J. MA, (eds). Raven Press, New York, (1995) 397.
9. H. Wada, N. Inagaki, A. Yamatodani, and T. Watanabe, Trends Neurosci. 14 (1991) 415
10. K, Onodera, A, Yamatodani, T . Watanabe and H. Wada, Prog. Neurobiol. 42 (1994) 685
11. R.Leurs and H. Timmerman, Prog. Drug Res. 39 (1992) 127.
12. M.A.M.R. De Almeida, and I. Izquierdo, Arch.Int.Pharmacodyn.Ther. 291 (1988) 202
13. K. Tasaka, New Advances in Histamine Research. K. Tasaka (ed), Springer-Verlag, Tokyo (1994) 27.
14. S. Miyazaki, M. Imaizumi and K. Onodera, Life Sci. 57 (1995) 1563.
15. K. Meguro, K. Yanai, H. Yokoyama, A. Sakurai, K. Maeyama, T. Watanabe and T. Matsuzawa, 8 (1992) 299.
16. S. Miyazaki, M. Imaizumi and K. Onodera, Meth. Find. Exp. Clin. Pharmacol. 17 (1995) 653.
17. S. Miyazaki, K. Onodera, K. M. Imaizumi and H. Timmerman ,Life Sci. 61 (1997) 355.
18. S. Miyazaki, M. Imaizumi and K. Onodera, Life Sci. 57 (1995)2137.
19. T. Mochizuki, A. Yamatodani, K. Okakura, M. Takemura, N. Inagaki and H. Wada, Naunyn-Schmied.Arch.Pharmacol. 343 (1991) 190.
20. H. Yokoyama, K. Onodera, K. Maeyama, E. Sakurai, K. Iinuma, R. Leurs, H. Timmerman and T. Watanabe, Eur. J. Pharmacol.

260(1994) 23

21. S.K. Bhattacharya, Eur. J. Pharmacol. 183 (1990) 925.
22. E. Schlicker, A. Behling, G Lummen, and M. Göthert, Naunyn-Schmied. Arch.Pharmacol. 345 (1992) 489.
23. E. Schlicker, K. Fink, M. Detzner, and M. Göthert, J.Neural Transm. 93 (1993) 1
24. E. Schlicker, R. Betz and M. Göthert, Naunyn-Schmied. Arch. Pharmacol. 337 (1988) 588
25. J. Clapham and G.J. Kilpatrick, Br.J.Pharmacol. 107 (1992) 919
26. A. Philippu, Annals of Psychiatry: Basic and Clinical Neuroscience 3 (1992) 79.
27. E. Giacobini, Prog. Brain Res. 98 (1993) 447.
28. A.G. Karczmar, Prog. Neuro-Psychopharmacol. Biol. Psychiat. 19 (1995) 187.
29. T. Mochizuki, K. Okakura-Mochizuki, A. Horii, Y. Yamamoto and A. Yamatodani, J Neurochem. 62 (1994) 2275,
30. A. Khateb, P. Fort, A. Pegna, B.E. Jones and M. Mühlethaler, Neuroscience 69, (1995) 495.
31. N. Gorelova and P.B. Reiner, J. Neurophysiol. 75, (1996) 707.
32. D.M. Compton, K.M. Dietrich, J.S. Smith and K. Davis, NeuroReport 7 (1995) 177.
33. M. Imaizumi and K. Onodera, Life Sci. 53 (1993) 1675.
34. R. Oishi, M. Nishibori, Y. Itoh, S. Shishido and K. Saeki, Eur. J. Pharmacol. 184 (1990)135.
35. L. Tuomisto, and U. Tacke, Neuropharmacology, 25 (1986)955.
36. H. Yokoyama, K. Onodera, K. Maeyama, K. Yanai, K. Iinuma, L, Tuomisto, and T. Watanabe, Naunyn-Schmied. Arch. Pharmacol. 346 (1992)40.
37. H. Yokoyama, K. Onodera, K. Iinuma, and T. Watanabe, Psychopharmacology. 112 (1993)199.
38. H. Yokoyama, K. Onodera, K. Iinuma, and T. Watanabe, Pharmacol. Biochem. Behav. 47 (1994)503.
39. H. Yokoyama, K. Iinuma K. Yanai, T. Watanabe, E. Sakurai, K. Onodera, Meth. Find. Exp. Clin. Pharmacol. 15 (1993)183.
40. H. Yokoyama, K. Onodera, K. Iinuma, and T. Watanabe, Eur. J. Pharmacol. 234 (1993)129.

41. R. Scherkl, A. Hashem and H.H. Frey, Epilepsy Res. 10 (1991)111.
42. E. Sakurai, E. Gunji, Y. Iizuka, N. Hikichi, K. Maeyama and T. Watanabe, J. Pharmacol. Pharmac. 46(1994)209.
43. T. Mochizuki, F.P. Janssen, R. Leurs, A.D. Windhorst, A. Yamatodani, K. Maeyama and H. Timmerman. Brain Res. 743(1997)178.
44. K. Onodera and Y. Ogura Advances in the Biosciences. Vol. 33. B. Uvnäs and K. Tasaka (eds.), Pergamon Press, Oxford, 127(1982).39
45, K. Onodera and T. Watanabe. Jpn. J. Psychopharmacol., 15 (1995)87.
46. T.W. Roll and L.S. Schleifer, The Pharmacological Basis of Therapeutics (6th Ed.),. A. F. Gilman, L. S. Goodman and A.Gilman (Eds.). Macmillan, New York. (1975) 448.
47. J.S. Lin, K. Sakai and M. Jouvet, Neuropharmacology, 27 (1988) 111.
48. J.S. Lin, K. Sakai, G. Vanni-Mercier, J.M. Arrang, M. Garbarg, J.C. Schwartz, and M. Jouvet, Brain Res. 523(1990)325.
49. M. Kathman, E. Schliker and M. Göthert. Psychopharmacology, 116 (1994) 464.
50. A.A. Rodrigues, F.P. Jansen, R. Leurs, H. Timmerman and G.D. Prell. Br J Pharmacol 114 (1995) 1523.
51. J.H. Ryu, K. Yanai and T. Watanabe. Neurosci. Lett. 178:19, (1994).
52. J.H, Ryu, K. Yanai, R, Iwata, T, Ido and T. Watanabe, NeuroReport, 5 (1994) 621.
53. J.H. Ryu, K. Yanai, X.L. Zhao and T. Watanabe. Br J Pharmacol, 118 (1996) 585.
54. T. Sakata, Histaminergic Neurons: Morphology and Function. Eds. T. Watanabe and H. Wada (eds). CRC Press: Boca Raton. (1991) 297.
55. T. Suzuki, K. Takamori, Y. Takahashi, M. Narita, M. Misawa and K. Onodera Life Sci. 54 (1994) 203.

R. Leurs and H. Timmerman (Editors)
The Histamine H_3 Receptor

Clinical application of HA H_3 receptor antagonists in learning and memory disorders

C. E. Tedford

Gliatech Inc., 23420 Commerce Park Road, Cleveland, Ohio 44122

1. INTRODUCTION

Histamine (HA) is known to influence many homeostatic processes including arousal (sleep/wakefulness), eating and drinking, attentional or learning and memory processes, neuroendocrine, locomotor as well as other CNS behaviors. [1, 2] To date, several lines of evidence strongly suggest a role of neuronal HA in cognitive processes[3-9] and more recently the use of H_3 antagonists in learning and memory disorders has been suggested.[8-12] This chapter will attempt to summarize previous findings and describe some recent results in context with HA's influence on whole brain activity with emphasis on cognitive processes. More specifically, the potential use of HA H_3 receptor antagonists in cognitive disorders will be discussed.

In addition, several reviews on the neuronal mechanisms of learning and memory exist and are beyond the scope of this chapter.[13-16] For example, Sarter reviewed the complex and critical components of early stage information processing and its relationship to known clinic disorders involving higher cognitive function.[13] As described, the initial selective processing of information in which relevant information is captured and non-relevant information is discarded is essential for appropriate cognitive function. Several CNS disorders can be attributed to impairments in information processing and the role of HA and the potential use of the H_3 antagonist will be discussed in those terms.

2. HISTAMINE'S MODULATORY ROLE IN WHOLE BRAIN ACTIVITY

Immunohistochemical, *in situ* hybridization, and retrograde tracer studies have clearly established the localization of histaminergic cell bodies in the tuberomammillary nuclei of the posterior hypothalamus with widespread projections to most regions of the diencephalon and telencephalon.[17-24] The widespread distribution of these histaminergic projections resembles that of other biogenic amines. However, in contrast to the well-defined monoamine terminals, an overlapping topography is seen at the level of the terminal fields from the projections originating from the E1-E5 histaminergic nuclei. This has been suggested to indicate that the HA neuronal system may provide a single unified modulatory effect on whole brain activity.[25-28] The influence of HA may also be neuromodulatory in nature as suggested by the paucity of terminal dendrite-dendrite synapses.

Indirectly, evidence of HA's role in cognitive processes is also seen by the temporal nature of histaminergic neuronal activity. The activity of histaminergic neurons is maximal during periods of wakefulness and reduced during sleep, suggesting a circadian rhythm for the histaminergic neurons.[5, 29-30] The activity of the histaminergic neurons and HA release is greater in rats during the night or dark period corresponding to their most active or aroused state.[31] In contrast, in the diurnal Rhesus monkey, HA transmission paralleled waking, being higher during the daytime.[5, 32] In humans, CNS penetrating H_1 receptor blockers will lead to marked sedation or impairment in cognitive performance during the day.[33-35] These findings suggest that the HA system is particularly involved in maintaining wakefulness and may promote enhanced attention or vigilance during the wake state.[36-38] Conversely, impairment in histaminergic tone during the wake state leads to performance impairment and decreased vigilance.

3. HISTAMINE H_3 RECEPTOR-INVOLVEMENT IN COGNITIVE PROCESSES

3.1 H_3 Receptors and CNS Localization

As reviewed in this book, early studies demonstrated the existence of a third subtype of HA receptor, the H_3 receptor.[2, 39-41] Receptor distribution studies further demonstrated that the H_3 receptor is found in the highest levels in the brain and in very low levels in the periphery. The regional distribution of H_3 receptors in the brain parallels the areas known to receive histaminergic innervation.[21, 42] The cerebral cortex being one of the areas in which high levels of H_3 receptors were found. A rostrocaudal gradient of H_3 cortical receptors was seen, with the frontal cortex receiving the highest density, and within the cortex, deep layers (IV-VI) having the greatest density.[42] Limbic regions of the basal forebrain including the caudate nucleus, globus pallidus, olfactory tubercle and nucleus accumbens also contained high levels of H_3 receptors as well as the reticular part of the substantia nigra.[42] The hypothalamus, where the densest network of histaminergic fibers are found, contained a moderate density of H_3 receptors.[42]

The HA H_3 receptor is further localized on the histaminergic nerve terminals in the brains of rats.[40-41, 43-46] Of the three subclasses of HA receptor that have been identified in the brain, the H_1 and H_2 receptors are thought to be exclusively post-synaptic.[46-49] The H_3 receptor found on the histaminergic nerve terminal was defined as an autoreceptor and is thus a HA receptor subtype that is uniquely positioned to regulate the amount of HA synthesized and released from the HA neurons.[45, 50] Moreover, the presence of H_3 receptors on several non-histaminergic nerve terminals (heteroreceptors) has now been further established. Modulation of neurotransmitter release has been shown by the H_3 receptor for several systems (i.e., norepinephrine, dopamine, serotonin, acetylcholine, GABA, etc.) as reviewed earlier in this book[51-55].

Recent studies investigating the ontogenic development of the H_1 and H_3 receptors have also established marked mismatches and independent developmental patterns as well as possible modulation of the various attentional and motor monoamine systems by the H_3 receptors.[56] The high levels of H_3 receptors found in well conserved limbic regions suggest involvement in arousal, emotion, motor and cognitive functions. H_3 receptors found in the cortex and hippocampus, also imply a role in higher learning function. Together, the neuroanatomical localization of the H_3 receptor and its modulatory influence on neurotransmitter release strongly suggests an involvement in higher learning processes.

3.2 H_3 Receptors and Cortical Activation

Numerous lines of evidence support a role of neuronal HA in arousal/vigilance or sleep/wake mechanisms.[27, 37, 57-60] Administration of intracere-broventricular (icv) HA induces the appearance of a cortical arousal EEG pattern in rabbits[61] and increases spontaneous locomotor activity, grooming and exploratory behavior in both saline and pentobarbital-treated rats.[57] The cortical EEG dysynchronization pattern seen after administration of HA is similar to the vigilant pattern acetylcholine produces after local administration. An interaction between cholinergic and histaminergic neurons to facilitate cortical activation is supported by the demonstration that HA excites nucleus basalis cholinergic neurons and increases tonic firing of the cortical projecting cholinergic neurons.[62] Histamine has also been shown to modulate the threshold for induction of LTP in hippocampal slices by enhancing NMDA-gated currents.[7]

The most recent advances in understanding the role of HA in arousal mechanisms have been gained through the use of selective HA H_3 receptor ligands.[44, 63-64] (R)-α-methylhistamine (RAMHA) was the first potent selective agonist for the HA H_3 receptor. Oral administration of RAMHA caused a significant increase in deep slow wave sleep in the cat[65] consistent with a reduction in HA release and diminution in histaminergic tone. Conversely, thioperamide, a prototypical selective H_3 antagonist enhances wakefulness in a dose-dependent fashion in the cat[65] and in the rat.[60] The arousal properties of thioperamide are prevented by pretreatment with mepyramine, a H_1 receptor antagonist, suggesting that the increase in HA release caused by thioperamide was acting on post-synaptic H_1 receptors.[65] These findings suggest that H_3 receptor blockade might provide enhanced neuronal firing and provide improvements in vigilance or cognitive processing.

3.3 H_3 Receptors Antagonists and Animal Models of Learning and Memory

Several studies initially indicated that HA was involved in higher learning and memory functions. Histamine given immediately post-training has been shown to increase recall in a step-down inhibitory avoidance task.[3] Likewise, pre-testing administration of HA enhances retention performance of rats in an active avoidance task while H_1 antagonists impair retention.[66-68] Both HA and acetylcholine can also prevent the impairing effects of H_1 antagonists on memory retrieval.[66-68] These early findings clearly implicate neuronal HA in higher learning processes. In 1996, Prast[9] further showed that histidine and HA (icv) facilitated social memory in rats. However, bilateral lesions of the tubermammilary nucleus appear to produce improvements in learning suggesting that removal of the inhibitory HA tone is facilitatory for cognitive processes.[69] These results are intriguing and somewhat counter to the previous results on HA direct effects on EEG activation and improvements in learning and animal models. However, the extent and selectivity of HA neuronal cell loss should be established for the TM lesions. Acute and chronic lesion effects on HA receptor subtypes should also be assessed in light of the distinct and opposing modulatory effects HA receptor subtypes have on non-HA neurotransmitter systems.

The role in learning and memory processes of neuronal versus mast cell-derived HA has also been investigated. Histamine, administered icv immediately post-training have been shown to increase memory in a step-down inhibitory avoidance task.[3] However, 48/80, a mast cell HA releaser, was ineffective in producing cognition enhancement when administered icv.[4] These findings indicate that neuronally derived HA is responsible for the improvements in learning and memory tasks.

Recently, the role of the H_3 receptor in cognition has been reported. Thioperamide, the prototypical H_3 antagonist and now other improved H_3 antagonists has been shown to enhance recall in a variety of studies. Glaxo has reported that two selective H_3 antagonists, clobenpropit and thioperamide, produced significant improvements in a delayed non-matching to position task in rats, a model of short term memory.[70] Thioperamide also attenuated the learning deficit induced by scopolamine in an elevated plus-maze in mice alone, and in combination with the H_2 antagonist, zolantidine.[8] These effects were blocked by pretreatment with H_1 antagonists. More recently, thioperamide was shown to improve learning and memory in a senescence-accelerated genetic mouse strain in a step-through passive avoidance response (PAR).[11] Thioperamide increased histidine decarboxylase (HDC) activity and improved response latency. In those studies, the senescence-accelerated control mice had reduced forebrain levels of HDC activity suggesting that improvement in HA neuronal activity could be useful in age-related memory decline.

Several other studies have demonstrated an involvement of HA-acetylcholine (ACh) interactions in areas of higher learning which further support the role of HA in cognitive processing. The cholinergic hypothesis of learning and memory has been proposed for several years to support the loss of memory seen in normal aging as well as diseases such as Alzheimer's disease (AD).[71] This model has been supported by a wealth of studies describing the importance of basal forebrain cholinergic afferents to hippocampal and cortical areas in cognitive processes as well as the effects of pharmacological manipulations with muscarinic agonists, antagonists and AChE inhibitors on learning and memory tasks. This hypothesis appears to be finally somewhat validated with the recent clinical approval and demonstrated improvements in AD patients with the AChE inhibitors, Cognex® and Aricept®. Stimulation of H_3 receptors with RAMHA and imetit has been shown to decrease ACh from the frontal cortex and impair cognitive performance in both object recognition and PAR.[12] In contrast, RAMHA was shown to improve recall in a water maze, suggesting opposite effects in a hippocampal-driven spatial learning paradigm.[72] These findings suggest differential influences by HA H_3 receptor activation or blockade in either short term memory paradigms such as the object recognition and PAR versus spatial learning tasks such as the Morris water maze. The involvement of the frontal cortex in short term/working memory as well as the developmental role of the cholinergic amygdaloid system in "short term memory" of the passive avoidance learning in the rat has been described and both areas contain high to moderate densities of H_3 receptors and would support their role in short term memory tasks.[12, 73-75]

Together, the recent neuroanatomical, biochemical and behavioral data emerging support the role of neuronal HA and H_3 receptor antagonists in cognition modulation. Studies at Gliatech have focused on the development on non-thiourea H_3 receptor antagonists.[10] GT-2016 was demonstrated to be a selective H_3 receptor antagonist which penetrated the CNS very effectively and increased cortical HA release (Fig.1).

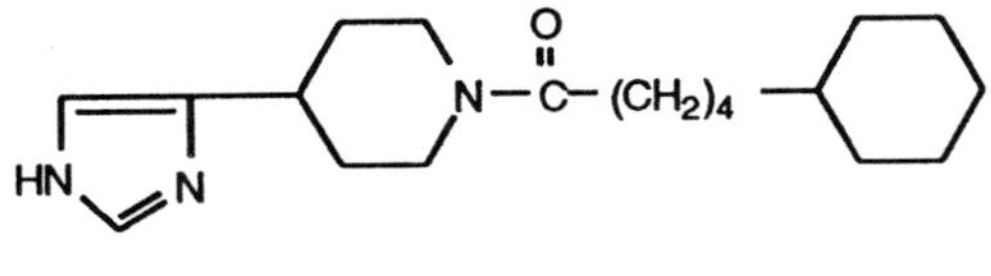

Fig. 1 Structure of GT-2016.

We were further interested in the potential EEG cortical activation and cognitive properties of GT-2016. EEG studies confirmed the unique wake-promoting or vigilant properties of GT-2016.[76] Subsequently, we conducted studies in collaboration with Dr. James McGaugh at the Univ. of California, Irvine to establish the memory-enhancing properties of GT-2016 in mice. Four double-blind studies were conducted in normal and amnesiac mice, using both an inhibitory avoidance response and a Y-Maze reversal paradigm for assessment of cognitive performance. Cognitive deficits were induced by pretraining administration of either scopolamine or diazepam. GT-2016 was administered post-training in all studies.

Post-training administration of GT-2016 (10 and 30 mg/kg, i.p.) increased recall in normal mice in the PAR (Fig. 2).

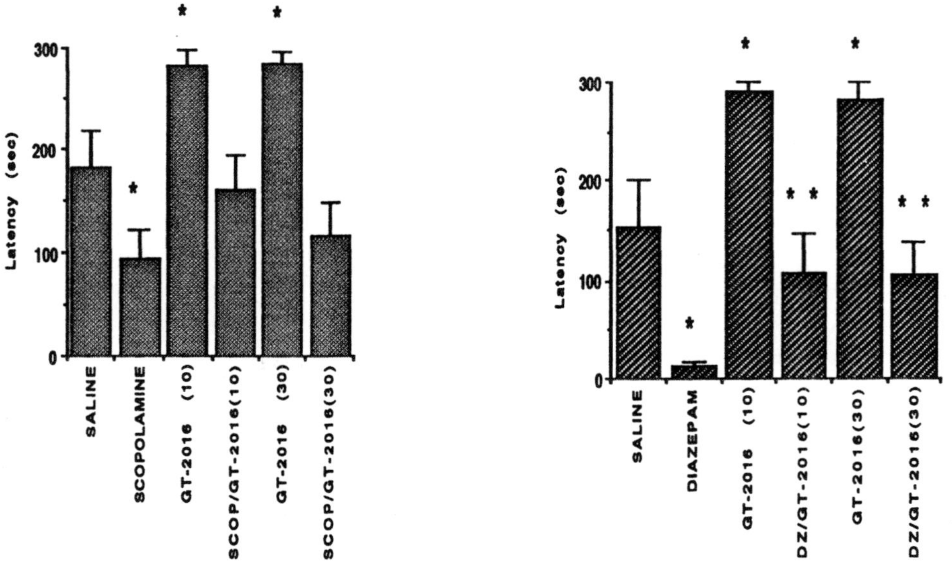

Fig. 2 Effect of GT-2016 on Scopolamine- or Diazepam-Induced Amnesia in the PAR in Mice. Effect of GT-2016 on 48 hr retention in the inhibitory avoidance task. A total of six groups of mice were tested per experiment (n = 12-18 mice per group). Left: All groups received either scopolamine (1.0 mg/kg, ip) or saline thirty min prior to training. Right: All groups received either diazepam (2.0 mg/kg, ip) or saline thirty min prior to training. Additionally, GT-2016 (10 and 30 mg/kg, ip) or vehicle was administered immediately following a single-trial training session. Retention latency was measured in a single trial 48 hrs following training. * indicates significant difference from saline-treated control group, $p < 0.05$. ** indicates significant difference from diazepam-treated control group, $p < 0.05$.

GT-2016 also attenuated the PAR deficit induced by pretraining administration of either diazepam or scopolamine. Likewise, memory-enhancing properties were indicated in the Y-Maze reversal paradigm with GT-2016, (data not shown). These findings demonstrate the memory-enhancing properties of GT-2016 in normal adult mice as well as attenuation of cognitive deficits in established drug-induced amnesiac models.

To further establish the effects of the H_3 antagonists on various cognitive processes, we established a juvenile pup model at Gliatech to evaluate the attentional aspects of the GT-2016 on cognitive processing in immature animals. Breese[77] in an excellent early study, examined the biochemical time course development of the monoamine systems in rats from 6 days prenatal to 80 days postnatal. Their studies indicated that the norepinephrine (NE) and dopamine (DA) systems completed maturation by about day 60. The greatest rise per unit time occurred between days 7 and 18, from this time forward the rise was relatively constant. The % of adult NE and DA content raised from approximately 20 to 60% during days 7-18. This represents a substantial maturation of the monoamine systems in a brief period, and coincides with the development of normal exploratory behavior and habituation profiles as well as cognitive function.

Attention deficit disorders with hyperactivity can be produced in various juvenile rat models.[78-80] One model utilizes selective lesioning of the dopaminergic neurons 5 days postnatally. This results in increases in motor activity and impairments in habituation in t-maze and shuttle box paradigms by 3 weeks postnatal.[78] The hyperactive profile is specific to the dopamine system as selective lesioning of the norepinephrine system leads to habituation of the hyperactivity.[81] Cognitive deficits in the passive avoidance response can also be seen in 6-OH-dopa treated juvenile rats and suggests the involvement of monoamines in the modulation of learning and memory during development.[79] Moreover, the amygdaloid complex contains a moderate density of H_3 receptors and the developmental role of the cholinergic amygdaloid system in passive avoidance learning in the rat has been described to occur during this same time period.[73-75]

We initially saw deficits in a 48 hr recall of a single trial PAR test in the rat pups up to day 17 postnatally (Fig. 3). Rats (17-35 days) displayed maximal retention in a one-trial PAR. These findings are in agreement with others who have described the cognitive impairment of developing rat pups.[74] Subsequently, the rate of learning was examined in a single-day 10 trial PAR to evaluate cognitive capabilities (Fig. 4).

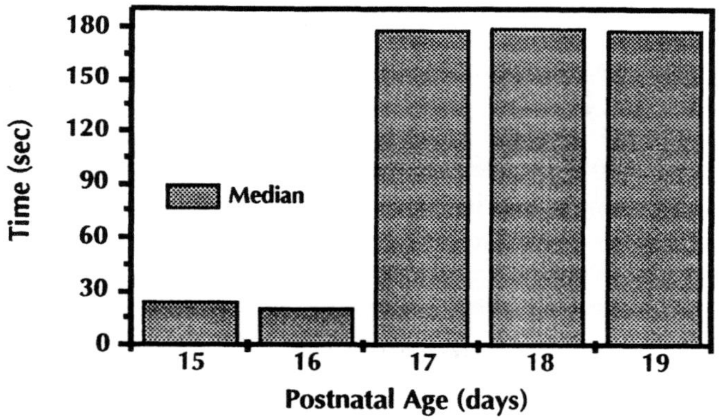

Fig. 3 Development of Cognitive Function in a Single-Trial PAR in Juvenile Rats. Rats of various ages were tested for retention of a single trial PAR, 48 hr after acquisition. Median latencies (sec) are shown 48 hr following the training trial. A 180 sec maximum cutoff was utilized in these experiments (n = 8-12/group).

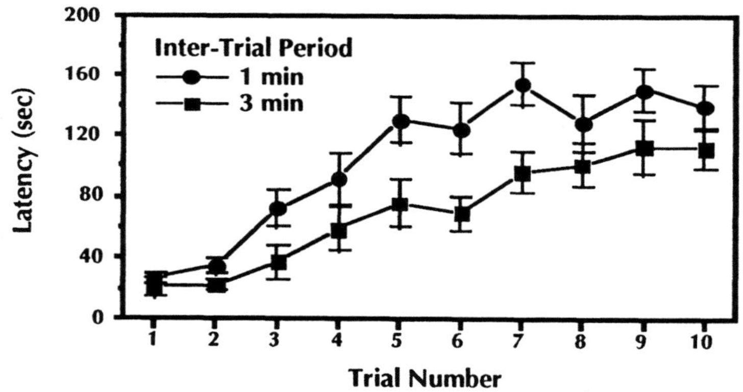

Fig. 4 Acquisition of the Multiple-Trial PAR in Day 15-16 Rat Pups. Juvenile rat pups from two litters were tested for acquisition of a multi-trial PAR. Littermates were equally divided into two groups in which intertrial time was either 1 or 3 min. Animals were returned to their home cage with their littermates for the intertrial time period (n = 4-6/group).

Acquisition of the multiple-trial PAR was complete by trial 10, however, sustained learning deficits were clearly apparent in the 15-16 day old rat pups.

Minimal recall of the first trial exposure was seen even with a 1 min retest period (second trial). Intertrial time was then varied in littermates and results indicated that the acquisition rate in the juvenile pups was further reduced at longer intertrial time periods (Fig. 4). Our results suggest that juvenile pups can learn, but extinction is rapid for cues important in learning and recalling new tasks.

Finally, litters (n ≥ 12 pups/treatment group) were equally divided into vehicle- or GT-2016-treated pups and tested for acquisition in the multi-trial PAR. GT-2016 was administered 30 min prior to training (1 min inter-trial period). GT-2016 was tested at doses of 5, 7.5, 10, 20 and 30 mg/kg, ip, (n=12-16/group), (Fig. 5). These doses would produce significant cortical H_3 receptor occupancy (~15-90%) and thereby enhance HA release.[10]

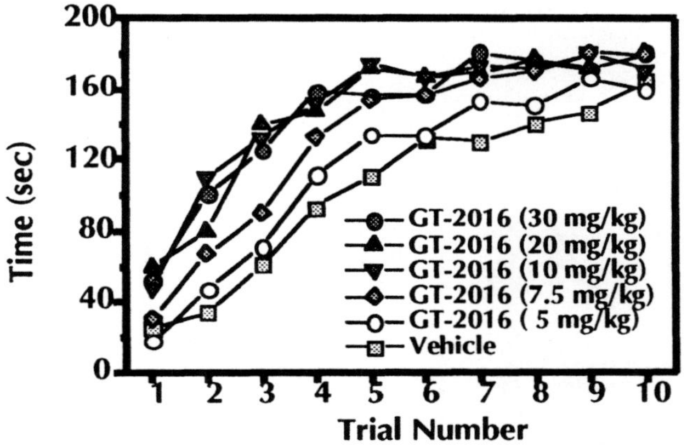

Fig. 5 Effect of GT-2016 on Acquisition of the PAR in juvenile rat pups. Juvenile rat pups (day 15-16) were tested for acquisition of a multi-trial PAR. Littermates were equally divided into vehicle or drug treatment groups. GT-2016 salt was given ip at a dose of 5-30 mg/kg (base), 30 mins prior to training. Animals were returned to their home cage with their littermates for the intertrial time period (n = 12-18/group). * indicates statistically significant differences between drug-treatment group and vehicle-treatment group at the specific trial #. Non-parametric statistical analysis (Kruskal-Wallis test) was conducted on median latencies (sec). Mean + SEM entry latencies (sec) are presented.

No significant enhancement in acquisition rates were seen at the 5 mg/kg dose, slight improvements were seen at 7.5 mg/kg dose and statistically significant increases in the acquisition rate were seen by trial two after the 10, 20 and 30 mg/kg dose of GT-2016. No significant differences in entry times were seen at trial 1 indicating no effect of GT-2016 on motor performance. Additionally, GT-2016-treated pups tended to maintain maximal latencies after initial acquisition of the task (trials 5-10).

These data demonstrate that H_3 antagonists, like GT-2016 can improve acquisition of a novel task in the immature juvenile rats. Further, these findings are the first to correlate H_3 receptor occupancy with a behavioral outcome.

The utility of this model was further evaluated by establishing the procognitive effects of established attention deficit hyperactivity disorder (ADHD) agents in the juvenile rat pups.[82-83] Methylphenidate (Ritalin®) was tested in the juvenile pup model to assess its effects on acquisition in the PAR. Methylphenidate (3 mg/kg, ip) was chosen as an intermediate dose, which in adult rats provided clear evidence of psychostimulant activity. Pretreatment (20 min) of the pups with methylphenidate produced a significant improvement in the acquisition of the PAR (Fig. 6).

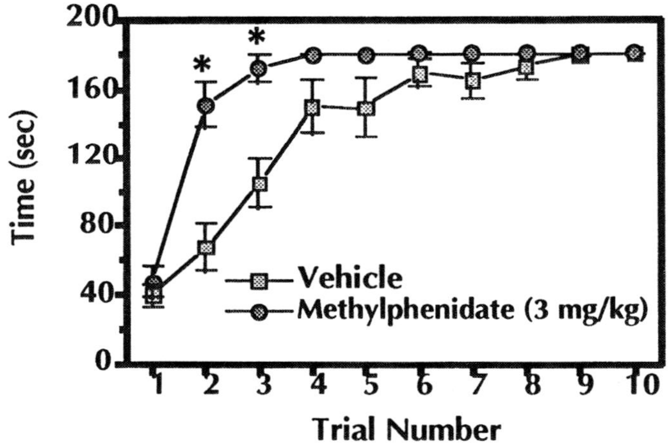

Fig. 6 Effect of methylphenidate on Acquisition of the PAR in juvenile rat pups. Juvenile rat pups (day 15-16) were tested for acquisition of a multi-trial PAR. Littermates were equally divided into vehicle or drug treatment groups. Methylphenidate salt was given ip at a dose of 3 mg/kg (base), 30 mins prior to training. Animals were returned to their home cage with their littermates for the intertrial time period. * indicates statistically significant differences between drug-treatment group and vehicle-treatment group at the specific trial #. Non-parametric statistical analysis (Kruskal-Wallis test) was conducted on median latencies (sec). Mean ± SEM entry latencies (sec) are presented (n = 12-18/group).

No significant differences were seen in the entry time for trial 1, suggesting, no impairment in motor or behavioral properties. Latencies to enter the dark room were significantly increased after the initial exposure to the footshock, similar to adult levels. Furthermore, no extinction was seen throughout the ten-trial testing period.

These findings demonstrate that methylphenidate improves acquisition in the learning impaired juvenile rat pups analogous to the H_3 antagonists. Recently, Smith and Coffin[84] described a variety of CNS agents on a single trial PAR in juvenile rats. Only established cognition enhancing agents showed any improvement (i.e. E2020, tacrine, arecoline, etc.) but other CNS agents (pargyline, diazepam, piracetam, NMDA, etc.) were devoid of procognitive activity, indicating the selective influence of these agents on learning and memory processes in the developing animals. Neither methylphenidate nor any H_3 antagonists were evaluated.

4. H_3 Receptor Antagonists and Clinical Utility:

The data presented from several independent laboratories suggest that blockade of the H_3 receptor would lead to increased neuronal firing and EEG activation or arousal of higher learning centers. Moreover, several studies in various animal models have indicated improvements in acquisition as well as recall following use of a selective H_3 antagonist. Modulatory effects on ACh and monoamine systems intimately involved in higher learning centers are also seen following blockade of the H_3 receptor. Together, these findings strongly support the involvement of HA in learning and memory processes.

Several models of human information-processing systems have been proposed and these models all recognize that several important variables in cognitive processing are involved.[13-16] One important variable is the attentional processes that allow the initial selective sorting of critical information. Critical information will achieve a higher level of processing and can then enter into additional stages of cognitive processing. The body of literature decribing the CNS modulatory effects of the histaminergic system support a role for neuronal HA in these early attentional processes and H_3 antagonists could be envisioned to be useful in clinical disorders that exhibit dysfunction in attentional processes.

CNS diseases have been identified with imbalances or the inability to *attend* to the initial stimuli and correctly process critical stimuli.[16] For example, schizophrenia has been proposed to be reflective of a hyperattentive state in which the occurrence of psychotic symptoms may be due to the inability to limit or selectively process critical stimuli versus non-critical stimuli. Conversely, age-related memory decline and progressive dementia as seen in diseases such as Alzheimer's would clearly be associated with a hypoattentive state.[16, 85] Whether through the natural aging process or in disease states where progressive loss in neuronal function occurs inevitably severe cognitive disorders would result. Additionally, diseases such as ADHD where potential asymmetry in higher learning centers disrupt the normal selective processing of stimuli and lead to reduced attentional regulation.[86-88]

4.1 Alzheimer's Disease and Age-related memory disorders

Alzheimer's disease is a neurodegenerative brain disorder which typically leads to progressive memory loss, dementia and, ultimately, death.[89] AD was first described in 1907 by Dr. Alois Alzheimer, a German psychiatrist, who discovered large numbers of unusual microscopic deposits in the brain of a demented patient upon autopsy. These deposits, now called senile neuritic plaques, contain highly insoluble beta-amyloid protein aggregates that form in particular regions of the brains of AD patients, including those involved with memory and cognition. Affected brain regions demonstrate significant loss of specific populations of neurons.[89] This neurotoxicity, characterized by the loss of neurons and the impaired function of surviving neurons, is a likely key contributor to the dementia that characterizes Alzheimer's disease. As described, the cholinergic hypothesis of memory decline has been proposed as a primary explanation for the loss in cognitive function as the disease progresses. However, while the cholinergic system is clearly compromised in AD, loss of other neuronal components clearly contribute to the decline in cognitive processes. Postmortem studies with brains from Alzheimer patients have indicated a significant decrease in HA levels, which indicates that alterations in the central histaminergic system may be one of the contributing factors to the impairments in cognition.[90]

In terms of early stage losses in cognitive processes in AD, it has been clearly established that attentional abilities in AD patients are compromised.[85, 91-94] This loss in the AD patient's ability to properly process stimuli may result in the initial "forgetfulness" signs of AD and gradually worsen as the disease progresses until significant losses in several information processes are apparent. The palliative approaches of restoring or improving cognitive function using AChE inhibitors and selective M_1 muscarinic agonists are the main focus of the newer AD drugs. However, agents like the H_3 antagonist may provide additional benefit in restoring cognitive function in AD patients by virtue of the modulatory effect HA has on whole brain activity. In essence, they may be considered as agents that would increase neuronal firing and improve the gain/noise ratio allowing appropriate selection of critical stimuli and further cognitive processing. In addition, disruptions in sleep/wake cycles and attentional aspects are particularly evident in the elderly and can negatively impact on daily function. This may reflect the normal neuronal loss associated with aging and similar utility could be seen for the use of H_3 antagonists in cognitive impairments associated with aging. The use of an H_3 antagonist might provide an alternative approach to a wide range of age-related impairments.

4.2 Attention Deficit Hyperactivity Disorders

ADHD is a complex developmental disorder with underlying emotional, attentional and learning disabilities. ADHD occurs in 3-6% of the school age population.[95-96] A common myth is that all ADHD children outgrow the disorder. It is estimated however, that over 50% of those children diagnosed with ADHD will continue to experience attentional problems as an adult.[97-98]

The disorder is characterized by a delay in the age-appropriate control of behavior and the characteristic traits include deficits in sustained attention/vigilance, impulse control, rule-governed behavior and the regulation of activity in accordance with situational demands. ADHD is believed to be the result of neurotransmitter abnormalities, particularly the monoamines.[96, 99-101]

The primary drug therapies are psychostimulants which are indicated for both emotional based sleep disorders (i.e., narcolepsy) as well as ADHD. The drugs of choice are Ritalin® (methylphenidate), dextroamphetamine or Cylert® (pemoline), all CNS stimulants that effect the monoamine systems. The current therapies provide symptomatic relief but the current medications are not without side effects, including abuse potential, cardiovascular effects, insomnia, appetite suppression, head and stomach aches, crying and nervous mannerisms.

The resemblance of ADHD to patients with lesions of the prefrontal cortex (PFC) has also been noted.[86, 88] Poor attention regulation, disorganized behavior and impulsivity can be demonstrated in animals and humans with PFC lesions. More recently, PET studies in ADHD children have indicated an asymmetry in the prefrontal cortex and caudate regions of the brain consistent with the attentional deficits and hyperactivity experienced by the patients.[87-88, 102] The right PFC has been shown to be smaller in patients with ADHD and ADHD patients also show impairments in test of prefrontal lobe function but not parietal attentional abilities. Imaging studies have further suggested that decreased activity in the PFC and striatum can be restored by methylphenidate.[83] Thus, the ADHD patient may experience an imbalance in the initial processing of stimuli and treatment with agents that increase DA availability may normalize this imbalance. As described, the presence of high levels of H_3 receptors in these regions suggest involvement in both attentional and motor systems. The ability of the HA system to unify neurotransmitter release and improve vigilant or attentional processes would further suggest the utility of H_3 therapeutics in ADHD and other attentional disorders. Together, with the recent evidence of arousal and cognitive enhancing properties by several laboratories after administration of H_3 antagonists, the use of H_3 antagonists in the treatment of ADHD seems plausible.

5. Concluding remarks

In conclusion, the role of HA in modulating states of CNS arousal and vigilance is quite apparent. However, the complexity of human information-processing systems precludes the expectation that a single neurotransmitter system is responsible for the many aspects involved in cognitive processes. Therefore, the development of useful agents for the treatment of various cognitive disorders will require the continual assessment of disease-specific symptomology and possible intervention at a multitude of levels and stages. The discovery of the H_3 receptor with its unique brain localization and function provides an opportunity to develop a new class of therapies. The need for novel agents for the treatment of debilitating sleep disorders, ADHD and cognitive disorders including Alzheimer's disease is clear. The profile of the H_3 antagonists offered to date support the potential use of these agents in treating these diseases.

REFERENCES

1. Timmerman H (1990) J. Med. Chem. 33: 4-11.
2. Schwartz JC Arrang JM, Garbarg M, Pollard H and Ruat M (1991) Physiol. Rev. 71: 1-51.
3. de Almeida MA and Izquierdo I (1986) Arch. Int. Pharmacodyn. Ther. 283: 193-198.
4. de Almeida MA and Izquierdo I (1988) Arch. Int. Pharmacodyn. Ther. 291: 202-207.
5. Onodera K, Yamatodani A, Watanabe T and Wada H (1994) Progress in Neurobiology 42: 685-702.
6. Alvarez EO and Banzan AM (1995) J Neural Trans. 101: 201-211.
7. Brown RE, Fedorov NB, Haas HL and Reymann KG (1995) Neuropharmacol. 34: 181-190.
8. Miyazaki S, Imaizumi M and Onodera K (1995) Life Sci 57: 2137-2144.
9. Prast H, Argyriou A and Philippu A (1996) Brain Research 734: 316-318.
10. Tedford CE, Yates SL, Pawlowski GP, Nalwalk JW, Hough LB, Khan MA, Phillips JG, Durant GJ and Frederickson RCA (1995). J. Pharmacol. Exp. Ther. 275: 598-604.
11. Meguro K-I, Yanai K, Sakai N, Sakurai E, Maeyama K, Saski H and Watanabe T, (1995) Pharmacol. Biochem. and Behavior, 50: 321-325.
12. Blandina P, Giorgetti M, Bartolini MC, Timmerman H, Leurs R, Pepeu G and Giovannini MG (1996) Br. J. Pharmacol. 119: 1656-1664.
13. Nisson MJ, and Bullemer P (1987) Cogn Psychol. 19: 1-32.
14. Cowen N (1988) Psychol Bull. 104: 163-191.

15. Mesulam MM, (1990) Ann Neurol. 28: 597-613.
16. Sarter M, (1994) Psychopharmacol. 114: 539-550.
17. Garbarg M, Barbin G and Schwartz JC (1974) Science 186: 833-835.
18. Garbarg M, Barbin G, Liorens C, Palacios JM, Pollard H and Schwartz JC (1980) In: Neurotransmitters, receptors and Drug Action. (ed., Essman, W.B.) 179-202; Spectrum.
19. Panula P, Yang HYT and Costa E (1984) PNAS 81: 2572-76.
20. Watanabe T, Taguchi Y, Shiosaka S, Tanaka J, Kubota H, Terano Y, Tohyama J and Wada H (1984) Brain Res. 295: 13-25.
21. Pollard H, Pachot I and Schwartz JC (1985) Neurosci. Lett. 54: 53-58.
22. Bayliss DA, Wang YM, Zahnow CA, Joseph DR and Milhorn DE (1990) Mole. Cell. Neurosci. 1: 3-9.
23. Castren E and Panula P (1990) Neurosci. Lett. 120: 113-116.
24. Erickson H, Blomqvist A and Kohler C (1991) J. Comp. Neurol. 311: 45-64.
25. Kohler C, Swanson LW, Haglund L and Wu Y (1985) Neurosci. 16: 85-110.
26. Inagaki N, Yamaatodini A, Ando-Yamatodani M, Tohyama M, Watanabe T, and Wada H, (1988) J.Comp. Neurol. 273: 283-300.
27. Wada H, Inagaki N, Yamatodani A and Watanabe T (1991) TINS 14: 415-418.
28. Wada H, Inagaki N, Itowi N, and Yamatodani A (1991) Brain Research Bulletin 27: 367-370.
29. Orr EL and Quay WB (1975) Endocrinology 96: 941-945.
30. Vanni-Mercier G, Sakai K and Jouvet M (1984) C. r. Acad. Sci. 298: 195-200.
31. Mochizuki T, Yamatodani A, Okakura K, Takemura M, Inagaki N and Wada H (1991) Naunyn Schmiedebergs Arch. Pharmacol. 343: 190-195.
32. Onoe H, Yamatodani A, Watanabe Y, Ono K, Mochizuki T, Wada H and Hayaishi O (1992) Abstr. Eur. Soc. for Sleep Res. (Helsinki).
33. Nicholson AN, Pascoe PA and Stone BM (1985) Neuropharmacol. 24: 245-250.
34. Nicholson AN and Stone BM (1986) Eur. J. Clin. Pharmac. 30: 27-32.
35. Nicholson AN, Pascoe PA, Turner C, Ganellin CR, Greengrass PM, Casy AF and Mercer AD (1991) Br. J. Pharmac. 104: 270-276.
36. Monti JM, Pellejero T, Jantos H and Pazos S (1985) In: Sleep: Neurotransmitters and Neuromodulators. Wauguier A, Gaillard JM, Monti JM, Radulovacki M(eds), Raven Press, New York, 197-209.
37. Monti JM (1993) Life Sci. 53: 1331-1338.
38. Pascoe, PA (1994) Pharmac. Ther. 61: 227-236.
39. Schwartz J, Arrang J and Garbarg M (1986) TIPS 8: 24-28.
40. Schwartz JC, Arrang JM, Garbarg M and Pollard H (1990) Agents and Action 30: 13-23.

41. Bertacccini G, Coruzzi G, Adami M, Pozzoli C, and Gambarelli E (1991) Ital. J. Gastroenterol. 23: 378-385.
42. Pollard H, Moreau J, Arrang JM, and Schwartz JC (1993) Neurosci. 52: 169-189.
43. Arrang JM, Garbarg M and Schwartz JC (1983) Nature 302: 832-837.
44. Arrang JM, Garbarg M, Lancelot JC, Lecomte J-M, Pollard H, Robba M, Schunack W and Schwartz JC (1987) Nature 327: 117-123.
45. Arrang JM, Garbarg M and Schwartz JC (1987) Neurosci. 23: 149-157.
46. Schwartz J, Garbarg M and Pollard H (1986) In: Handbook of Physiology, The Nervous System. Vol. IV. (eds. Bloom, F.E. et al.) American Physiological Society, Bethesda, 257-316.
47. Ash, A. and Schild, H.(1966) Br.J. Pharmac. Chemo. 27: 427-439.
48. Black JW, Duncan WAM, Durant GJ, Ganellin CR and Parsons ME (1972) Nature 236: 385-390.
49. Prell GD and Green JP (1986) Rev. Neurosci. 9: 209-254.
50. Arrang JM, Devaux B, Chodkiewicz JP and Schwartz JC (1988) J. Neurochem. 51: 105-108.
51. Fink K, Schlicker E, Neise A and Göthert M (1990) N-S. Arch. Pharmacol. 342: 513-519.
52. Fink K, Schlicker E and Gothert M (1994) N-S Arch of Pharmacol. 349: 113-7.
53. Schlicker E, Behling A, Lummen G and Gothert M (1992) N-S Archives of Pharmacology 345(4): 489-493.
54. Schlicker E, Behling A, Lummen G, Malinowska B and Gothert M (1992) N-S Archives of Pharacology, 345: 639-646.
55. Schlicker E, Fink K, Detzner M and Gothert M (1993) J. Neural Trans. General Section, 93: 1-10.
56. Ryu JH, Yanai K, Sakurai E, Kim C-Y and Watanabe T (1995) Developmental Brain Research 87: 101-110.
57. Kalivas P (1982) J.Pharmacol. Exp. Ther. 222: 37-42.
58. Lin JS, Sakai K and Jouvet M (1988) Neuropharmacology 27: 111-122.
59. Sakai K, Mansari ME, Lin JS, Zhang JG, and Vanni-Mercier G (1990) In: The Diencephlon and Sleep. (eds. Mancia M and Marini G) Raven Press, Ltd. New York. 171-198.
60. Monti JM, Jantos H, Boussard M, Altier H, Orellana C and Olivera S (1991) Eur. J. Pharmacol. 205: 283-287.
61. Monnier M, Sauer R and Hatt AM (1970) Int. Rev. Neurobiol. 12: 265-305.
62. Khateb A, Fort P, Pegna A, Jones BE, and Muhlethaler M (1995) Neurosci 69: 495-506.
63. Arrang JM, Garbarg M, Schwartz J-C, Lipp R, Stark H, Schunack W and Lecomte JM (1991) In: New Perspectives in Histamine Research. 55-67.
64. Hill SJ (1992) Biochem Soc. Trans. 20: 122-125.

65. Lin JS, Sakai K, Vanni-Mercier G, Arrang JM, Garbarg M, Schwartz JC and Jouvet M (1990) Brain Research 523: 325-30.
66. Kamei C, Chung YH and Tasaka K (1990) Psychopharmacol., 102: 312-318.
67. Kamei C and Tasaka K (1991) Japan J. Pharmacol., 57: 473-482.
68. Kamei C, Okumura Y and Tasaka K (1993) Psychopharmacol., 111: 376-382.
69. Klapdor K, Hasenohrl RU and Huston JP (1994) Behav. Brain Res. 61: 113-116.
70. Barnes JC, Brown JD, Clarke NP, Clapham J, Evans DJ, and O'Shaughnessy CT (1993) Eur. J. Pharmacol. 250: 147-152.
71. Fibiger HC, (1991) Trends Neurosci. 14: 220-223.
72. Smith CPS, Hunter AJ and Bennet GW (1994) Psychopharmacology 114: 651-656.
73. Dumery V, and Blozovski D (1987) Exp. Brain Res. 67: 61-69.
74. Dumery V, Derer P and Blozovski D (1988) Dev. Psychobiol. 21: 553-565.
75. Blozovski D and Dumery V (1984) Behav Brain Res. 13: 97-106.
76. Tedford CE and Edgar D (1997) In: Neurobiology of Histamine. Humana Press, Totowa, NJ eds. Prell G, in press.
77. Breese GR and Traylor TD (1972) Br. J. Pharmacol. 44: 210-222.
78. Shaywitz BA, Teicher MH, Cohen DJ, Anderson GM, Young JG and Levitt P (1984) Psychopharmacol. 82: 73-77.
79. Cornwell-Jones CA, Decker MW, Chang JW, Cole B, Goltz KM, Tran, T and McGaugh JL (1989) Brain Res. 493: 258-268.
80. Sagvolden T, Metzger MA, Schiorbeck HK, Rugland AL, Spinnangr I and Sagvolden G (1992) Behav. Neural. Biol. 58: 103-112.
81. Raskin LA, Shaywitz BA, Anderson GM, Cohen DJ, Teicher MH and Linakis J (1983) Pharmacol. Biochem. Behav. 19: 743-749.
82. Callaway E (1984) Biol. Psychiatry 19: 649-662.
83. Matochik JA, Liebenauer LL, King CA, Szymanski HV, Cohen RM and Zametkin AJ (1994) Am J Psychiatry 151: 658-664.
84. Smith RD and Coffin VL (1995) Soc. for Neurosci. 21: 158.
85. Filoteo JV, Delis DC, Massman PJ, Demadura T, Butters N, Salmon DP, (1992) J. Clin. Exp. Neuropsychol. 14: 871-883.
86. Barkley RA, Grodzinsky G, and DuPaul GJ (1992) J. Abnorm Child Psychol. 20: 163-188.
87. Hynd GW, Hern KL, Novey ES, Eliopulos D, Marshall R, Gonzalez JJ and Voeller KK, (1993) J Child Neurol. 8: 339-347.
88. Arnsten FT, Steere JC, and Hunt RD (1996) Arch. Gen. Psychiatry 53: 448-455.
89. Selkoe, DJ (1994) In: Basic Neurochemistry: Molecular, Cellular and Medical Aspects. 5th Ed., Raven Press, LTD. NY, 919-933.

90. Mazurkiewicz-Kwilecki IM and Nsonah S (1988) Can. J. Physiol. Pharmacol., 67: 75-78.
91. Freed DM, Corkin S, Growden JH and Nissen MJ (1988) Neuropsychology 26: 895-902.
92. Grady CL, Grimes AM, Patronas NP, Sunderland T, Foster NL and Rapaport SI (1989) Arch. Neurol. 46: 317-320.
93. Della Salla S, Laiacona M, Spinnler H and Ubezio C (1992) Psychol. Med. 22: 895-901.
94. Foldi NS, Jutagir R, Davidoff D and Gould T (1992) J. Gerontol. 47: 146-153.
95. Meller WLK (1987) Prim. Care 14: 745-759.
96. Oades, RD (1987) Prog. Neurobiol. 29: 365-391.
97. Bellak L and Black RB (1992) Clinical Therapeutics 14: 138-147
98. Levin GM (1995) American Pharmacy NS35: 8-20.
99. Swanson JM, Cantwell D, Lerner M, McBurnett K and Hanna G (1991) J. Learning Disabilities 24: 219-230.
100. Wilens TE and Biederman J (1992) Pediatric Psychopharmacol. 15: 191-223.
101. Shenker A (1992) In: Advances in Pediatrics. Vol. 39; Mosby-Year Book, Inc., 337-383.
102. Lou HC, Henriksen L and Bruhn P (1990) Lancet 335: 8-11.

ACKNOWLEDGEMENTS

We would like to thank Drs. James McGaugh and Ines Collison for their work on GT-2016 and Dr. McGaugh for his early enthusiasm and support for the project.

AUTHOR INDEX

SUBJECT INDEX